Probability Theory and Random Processes

Probability Theory and Random Processes

Kattaswamy Mergu

Lecturer
Department of Electrical and Computer Engineering
Wolaita Sodo University
Wolaita Sodo
Ethiopia

CBS Publishers & Distributors Pvt Ltd

New Delhi • Bengaluru • Chennai • Kochi • Kolkata • Mumbai
Hyderabad • Nagpur • Patna • Pune • Vijayawada

Probability Theory and Random Processes

ISBN: 978-93-86217-46-2

First Edition: 2017

Published by Satish Kumar Jain and produced by Varun Jain for
CBS Publishers & Distributors Pvt Ltd
4819/XI Prahlad Street, 24 Ansari Road, Daryaganj, New Delhi 110 002, India.
Ph: 23289259, 23266861, 23266867 Website: www.cbspd.com
Fax: 011-23243014 e-mail: delhi@cbspd.com; cbspubs@airtelmail.in.
Corporate Office: 204 FIE, Industrial Area, Patparganj, Delhi 110 092
Ph: 4934 4934 Fax: 4934 4935 e-mail: publishing@cbspd.com; publicity@cbspd.com

Branches

- **Bengaluru:** Seema House 2975, 17th Cross, K.R. Road, Banasankari 2nd Stage, Bengaluru 560 070, Karnataka
  Ph: +91-80-26771678/79 Fax: +91-80-26771680 e-mail: bangalore@cbspd.com
- **Chennai:** 7, Subbaraya Street, Shenoy Nagar, Chennai 600 030, Tamil Nadu
  Ph: +91-44-26680620, 26681266 Fax: +91-44-42032115 e-mail: chennai@cbspd.com
- **Kochi:** Ashana House, 39/1904, AM Thomas Road, Valanjambalam, Ernakulam 682 016, Kochi, Kerala
  Ph: +91-484-4059061-62-64-65 Fax: +91-484-4059065 e-mail: kochi@cbspd.com
- **Kolkata:** 6/B, Ground Floor, Rameswar Shaw Road, Kolkata-700 014, West Bengal
  Ph: +91-33-22891126, 22891127, 22891128 e-mail: kolkata@cbspd.com
- **Mumbai:** 83-C, Dr E Moses Road, Worli, Mumbai-400018, Maharashtra
  Ph: +91-22-24902340/41 Fax: +91-22-24902342 e-mail: mumbai@cbspd.com

Representatives

- **Hyderabad** 0-9885175004
- **Nagpur** 0-9021734563
- **Patna** 0-9334159340
- **Pune** 0-9623451994
- **Vijayawada** 0-9000660880

Printed at:

Rashtriya Printers, Delhi-95

to

my mother
Late Smt Laxmi Mergu
and
my family members

Preface

Probability Theory and Random Processes is an introductory textbook to provide detailed concepts of probability, random variables, and random processes for BE and BTech (electronics engineering, and electronics and communication engineering) students, who have a strong mathematical background. The concepts, which are covered in this textbook, are very useful for the advanced courses like digital communications and advanced digital signal processing.

One of the main objectives of this book is to provide the students a clear and logical presentation of the basic concepts in a simple and lucid language.

The book has several remarkable features. It provides an organized and broad coverage of theory as well as illustrations of the applications. Several appendices provide background material on related topics of mathematics, and signals and systems. Many examples are included to illuminate subtle points that are often overlooked or not fully explained in basic textbooks. Each chapter also contains an exercise for practice.

The book has eight chapters, starting with basic mathematical concepts needed for understanding probability theory, followed by fundamentals of random variables and random processes and their applications.

Probability concepts through set theory are introduced in Chapter 1. This chapter also provides in detail the various probability concepts like sample space, events, conditional probability, total probability and Baye's theorem. A clear understanding of this chapter helps the students in understanding the concepts of random variables and random processes.

Random variables are defined in Chapter 2, which includes a description of many well-known and some not so well-known continuous and discrete parametric families of distribution and density functions. Important properties and applications of distribution and density functions are also discussed in this chapter.

Chapter 3 starts with the definition of expectation of a random variable and extends to functions of random variables. Moments, characteristic functions, conditional expectation and their important properties are discussed with proofs. Several techniques for deriving the distributions of transformations of single random variable are also investigated.

Chapter 4 extends the single random variable theory to several random variables, while Chapter 5 defines operations on multiple random variables.

By extending the characterization of multiple random variables to be indexed by time, we introduce random processes in Chapter 6. Various properties of a random process are covered, such as independence, stationarity and ergodicity. Specific well-known random processes such as Poisson and Gaussian processes are developed and investigated.

The power spectral density of a random process is defined in Chapter 7, which is used to describe the behavior of signals when processed (filtered) by a system and also provides various properties of PSD.

Chapter 8 focuses on linear time invariant systems, though some nonlinear processing is also mentioned. This chapter also provides the basic concept of noise and its measurements.

Finally, the book closes with a number of appendices that contain material helpful to the students in solving problems.

I am thankful to CBS Publishers and Distributors. I would like to put on record the sincere efforts of Mr YN Arjuna and his team comprising Mrs Ritu Chawla, Mrs Poonam Kapoor Bhatia, Mr Manish and Mr Parmod Kumar for bringing out the book in the present form.

I welcome comments, suggestions and corrections from the readers.

Kattaswamy Mergu

Contents

1

Introduction to Probability Theory

1.0 INTRODUCTION

In our daily life, we come across many uncertainties, like weather forecasting, winning a race and lifetime of a plant in a laboratory. The outcomes of the above examples are not known in advance. Probabilities are associated with experiments where the outcomes are not known in advance or cannot be predicted. The value of probability is a number between 0 and 1, both the values being inclusive.

Probability is the branch of mathematics devoted to the study of such events. People have always been interested in games of chance and gambling. The existence of games such as dice is evident since 3000 BC. But such games were not treated mathematically till fifteenth century. During this period, the calculation and theory of probability originated in Italy. Later in the seventeenth century, French contributed to this literature of study. The foundation of modern probability theory is credited to the Russian mathematician, Kolmogorov. In 1993, he proposed the axioms on which the present subject of probability is based.

1.1 INTRODUCTION TO SET THEORY

It is very convenient to introduce probability through set theory. Most of the people who study probability have already encountered set theory and are familiar with terms such as *set, element, union, intersection,* and *complement*. In this paragraph, we introduce the basic definition and properties of sets, which are important in the study of probability.

1.1.1 Set

A well-defined collection of objects is called a *set*. These objects are called elements or members of the set. Generally, uppercase letters are used to represent a set and lowercase letters are used to represent the elements of a set.

Example 1.1.1: $A = \{a, b, c, d\}$ is a set and a, b, c and d are its elements.

Note that the elements of a set are enumerated within a pair of curly brackets as shown in this example.

Some commonly used sets and their standard notations are as follows:

1. Set of natural numbers: N
2. Set of integers: Z

3. Set of rational numbers: Q
4. Set of real numbers: R
5. Set of complex numbers: C
6. Open interval on real number line:

$$(a, b) = \{x : a < x < b, a, b \in R\}$$

7. Closed interval on real number line:

$$[a, b] = \{x : a \leq x \leq b, a, b \in R\}$$

8. Semi open intervals:

$$(a, b] = \{x : a < x \leq b, a, b \in R\}$$

$$[a, b) = \{x : a \leq x < b, a, b \in R\}$$

1.1.2 Representation of a Set

A set can be represented by two methods. They are:

i. Roster method
ii. Set builder form

i. Roster Method

Roster method is also known as tabular method. In this method, a set is represented by listing all the elements of the set, the elements being separated by commas and are enclosed within flower brackets { }. Suppose, a set containing elements 1, 2, 3, 4, 5 is represented as $\{1, 2, 3, 4, 5\}$.

ii. Set Builder Method

Set builder method is also known as property method. Instead of listing all the elements, we can represent a set in the set builder notation specifying some common property satisfied by the elements. Thus, the set $B = \{x \mid 0 \leq x \leq 7$ and x is an odd integer$\}$ or $B = \{x: 0 \leq x \leq 7$ and x is an odd integer$\}$ represents the set $\{1, 3, 5, 7\}$. We read '$x|$' or 'x:' as 'all x such that'. Particularly, if a set is having infinite number of elements, listing all the elements is cumbersome; for such a case this representation is useful.

1.1.3 Finite and Infinite Sets

If a set contains finite number of elements then it is said to be a *finite set*. Otherwise, it is an *infinite set*.

$A = \{a, e, i, o, u\}$ and set of binary digits are examples of finite set. Examples of infinite set are set of rational numbers between the integers 1 and 10, and set of all points on a line segment of length 2 m.

1.1.4 Countable and Uncountable Sets

If all elements of a set can be put in one-to-one correspondence with the natural numbers, then the set is said to be *countable*. If a set is not countable, it is called *uncountable*.

1.1.5 Null Set or Empty Set or Void Set

If a set does not contain at least one element then it is said to be a null set or empty set. In other words, a set with no elements is an empty set and is denoted by ϕ or { }.

Note: $\{\phi\}$, $\{0\}$ are not null sets.

Example 1.1.2: $E = \{x : x^2 = 9 \text{ and } 3x = 6\}$ is an empty set because there is no number that can satisfy both $x^2 = 9$ and $3x = 6$.

$\therefore \quad E = \phi.$

1.1.6 Singleton Set (Singlet) and Cardinality

A set consisting of a single element is called a singleton set or singlet.

The number of elements in a finite set A is called the cardinality of A and is denoted by $n(A)$. The cardinality of the singleton set is 1. For example, consider a set $A = \{4\}$. It consist only one element, i.e. 4. Hence this set is called singleton and the cardinality of A is $n(A) = 1$.

1.1.7 Equivalent Sets

Two finite sets A and B are said to be equivalent if cardinality of both the sets are equal, i.e.

$$n(A) = n(B)$$

1.1.8 Equal Sets

Two sets A and B are said to be equal if and only if they contain the same elements, i.e. if every element of A is in B and every element of B is in A. We denote the equality by $A = B$.

Example 1.1.3: $A = \{1, 3, 5, 7, 9\}$ and $B = \{9, 1, 7, 3, 5\}$

$$A = B$$

Note: Elements in both the sets need not be in same order.

1.1.9 Universal Set

In any application of the theory of sets, the members of all sets under consideration usually belong to some fixed large set called the universal set. Usually we denote the universal set by the letter U. Suppose, if we consider dice experiment, the universal set consists of all possible outcomes, $\{1, 2, 3, 4, 5, 6\}$.

1.1.10 Subsets

A set A is called a *subset* of B (or B is called the *superset* of A), denoted by $A \subseteq B$, if all the elements of A are also elements of B. Thus $A \subseteq B$ if and only if $x \in A \Rightarrow x \in B$.

If A is a *subset* of B and there is at least one element in B which is not an element of A, then A is called a *proper subset* of B. We write $A \subset B$.

Example 1.1.4: Let $A = \{1, 2, 3, 5\}$, $B = \{1, 2, 3, 4, 5, 6, 7\}$ and $C = \{2, 4, 6, 8\}$. Then $A \subset B$, $C \not\subset B$, $B \not\subset C$, etc.

A set A is a subset of itself $A \subseteq A$.

$A \subseteq B$ and $B \subseteq C$ implies that $A \subseteq C$.

The null set ϕ is a subset of every set.

If the set A is *finite* w. h n number of elements, then A has 2^n subsets. For example, the set of binary digits $B = \{0, 1\}$ has $2^2 = 4$ subsets.

These are ϕ, $\{0\}$, $\{1\}$ and B.

If an element x is a member of the set A, we write $x \in A$. If x is not a member of A, we write $x \notin A$. In example 1.1.1, $a \in A$ and $e \notin A$.

1.1.12 Power Set

The set of all subsets of any set S is called the power set of S. We denote the power set of S by $P(S)$.

i. Let $S = \{1, 2\}$, then $P(S) = \{\{1\}, \{2\}, \{1, 2\}, \phi\}$

ii. Let $R = \{a, b, c\}$ then $\{\{a\}, \{b\}, \{c\}, \{a, b\}, \{b, c\}, \{c, a\}, \{a, b, c\}, \phi\}$

If a set S is finite, say S has n elements, then the power set of S can be shown to have 2^n elements.

1.2 SOME THEOREMS ON SETS

Theorem 1: Null set is the subset of every set.

Proof: Let A be any set. In order to prove that $\phi \subseteq A$ we must show that there is no element of ϕ which is not present in A. And since ϕ contains no element at all, no such element can be found out. Hence $\phi \subseteq A$.

Theorem 2: For any set A we have $\phi \subseteq A \subseteq U$.

Proof: Every set A is a subset of universal set U since by definition all elements of A belong to U. Also the null set $\phi \subseteq A$.

Theorem 3: If A is a subset of ϕ then $A = \phi$.

Proof: The null set ϕ is a subset of every set, in particular $\phi \subseteq A$. By hypothesis, $A \subseteq \phi$. The two conditions imply $A = \phi$.

Theorem 4: The total number of all possible subsets of a given set containing n elements is 2^n.

Proof: The number of subsets each containing r elements is

$$^nC_r = \frac{n!}{r!(n-r)!}, r = (0, 1, 2, 3,..., n)$$

Hence, the total number of subsets is

$$= {}^nC_0 + {}^nC_1 + {}^nC_2 + {}^nC_3 + \cdots + {}^nC_n = 2^n$$

1.3 SET OPERATIONS

We can combine events by set operations to get other events. Following set operations are useful:

Union: The union of two sets A and B is defined as the set of elements that are either in A or in B or in both and it is denoted by $A \cup B$. In set builder notation, it can be represented as

$$A \cup B = \{x : x \in A \text{ or } x \in B\} \quad ...(1.1)$$

Intersection: The set of common elements in any two sets A and B is called intersection of the two sets. It is denoted by $A \cap B$ and can be written as

$$A \cap B = \{x : x \in A \text{ and } x \in B\} \quad ...(1.2)$$

Difference: The *difference* of two sets A and B, denoted by A/B is the set of those elements of A which do not belong to B. Thus,

$$A/B = \{x : x \in A \text{ and } x \notin B\} \qquad ...(1.3)$$

Similarly,
$$B/A = \{x : x \in B \text{ and } x \notin A\} \qquad ...(1.4)$$

Complement: The complement of a set A, denoted by A^c, is defined as the set of all elements which are not in A.

Let U be a universal set and A be any subset of U, then the elements of U which are not in A, i.e. $U - A$ is the complement of A w.r.t. U and is written as $A' = U - A = A^c$.

$$A^c = \{x : x \notin A\} \qquad ...(1.5)$$

Clearly, $A^c = U/A$

Disjoint: Two sets A and B are called *disjoint* if $A \cap B = \phi$.

Example 1.3.1: Let $U = \{1, 2, 3, 4, 5, 6, 7, 8, 9\}$, $A = \{1, 2, 3, 5\}$, $B = \{2, 3, 4, 5\}$ and $C = \{6, 8\}$. Find:

i. $A \cup B$ ii. $A \cap B$ iii. $B \cap C$ iv. A/B v. A^c

Solution: i. $A \cup B = \{1, 2, 3, 4, 5\}$ ii. $A \cap B = \{2, 3, 5\}$ iii. $B \cap C = \phi$
iv. $A/B = \{1\}$ v. $A^c = \{4, 6, 7, 8, 9\}$.

Algebraic Properties of Set Operations

Idempotent laws

If A is any set, then
i. $A \cup A = A$
ii. $A \cap A = A$

Identity laws

i. $A \cup \phi = A$
ii. $A \cap U = A$

Commutative laws

If A, B are two sets, then
i. $A \cup B = B \cup A$
ii. $A \cap B = B \cap A$

Associative laws

If A, B, C are three sets, then
i. $A \cup (B \cup C) = (A \cup B) \cup C$
ii. $A \cap (B \cap C) = (A \cap B) \cap C$

Distributive laws

If A, B, C are three sets, then
i. $A \cup (B \cap C) = (A \cup B) \cap (A \cup C)$
ii. $A \cap (B \cup C) = (A \cap B) \cup (A \cap C)$

De Morgan's Laws

i. $(A \cup B)' = A' \cap B'$

Proof: Let $x \in (A \cup B)'$

$$\begin{aligned} \Rightarrow \quad & x \notin A \cup B \\ \Rightarrow \quad & x \notin A \text{ or } x \notin B \\ \Rightarrow \quad & x \in A' \text{ and } x \in B' \\ \Rightarrow \quad & x \in A' \cap B' \\ \therefore \quad & (A \cup B)' = A' \cap B' \end{aligned}$$

ii. $(A \cap B)' = A' \cup B'$

Proof: Let $x \in (A \cap B)'$

$$\begin{aligned} \Rightarrow \quad & x \notin A \cap B \\ \Rightarrow \quad & x \notin A \text{ and } x \notin B \\ \Rightarrow \quad & x \in A' \text{ or } x \in B' \\ \Rightarrow \quad & x \in A' \cup B' \\ \therefore \quad & (A \cap B)' = A' \cup B' \end{aligned}$$

1.4 VENN DIAGRAMS

A pictorial representation of sets by set of points in the plane is called Venn diagram. Universal set U is represented pictorially by interior of a rectangle and the other sets are represented by closed figures, viz. circles or ellipses or small rectangles or some curved figures lying within the rectangle.

The procedure to draw Venn diagrams is as follows:

Step 1: Draw the rectangle that mentions the universal set.

Step 2: Draw the circles inside the rectangle based on the conditions for indicating the sets.

Step 3: Point out the value of each circle and find out the results according to the given condition.

1.4.1 Representation of Venn Diagrams

Universal Set

Set theory is the subdivision of mathematics that learns about the sets, which are the group of objects. The universal set should be the set of all elements that under deliberation. These are represented by the capital alphabet U, sometimes E.

Fig.1.1: Venn diagrams: (a) Universal set (b) A is a subset of B ($A \subseteq B$)

Disjoint Sets

Let A and B be any two arbitrary sets, such that some elements are in A but not in B, some are in B but not in A, some are in both A and B, and some are in neither A nor B. We represent A and B in the pictorial form as in shown in the Venn diagram (Fig. 1.2). If the intersection of sets A and B is an empty set, then the sets A and B are called disjoint sets (Fig. 1.2a).

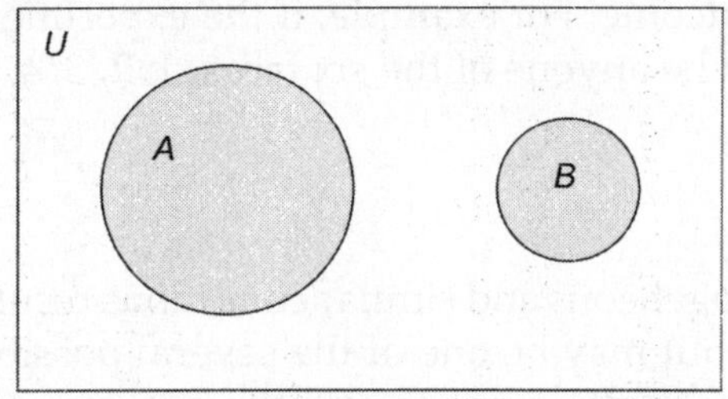

(a)

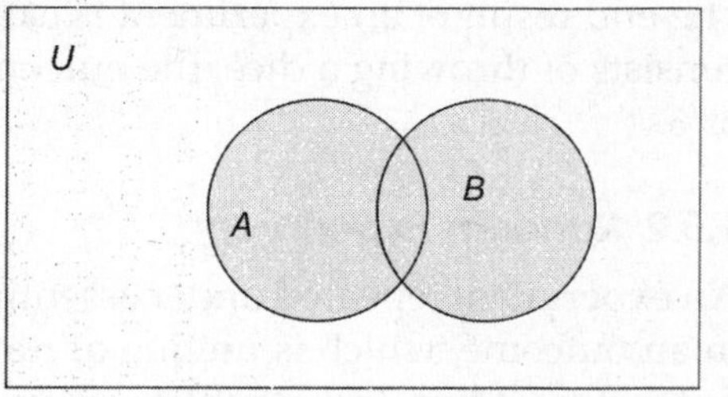

(b)

Fig. 1.2: (a) A and B are disjoint sets (b) Some elements are common in both A and B

Example 1.4 1: Given universal set, $U = \{1, 2, 3, 4, 5, 6, 7, 8, 9, 10\}$, list the elements of the following sets:

a. $A = \{x : x \text{ is greater than } 7\}$
b. $B = \{x : x \text{ is a multiple of } 2\}$

Solution: Given $U = \{1, 2, 3, 4, 5, 6, 7, 8, 9, 10\}$

The elements of sets A and B should be selected from the universal set U.

a. $A = \{x : x \text{ is greater than } 7\}$
   i.e. $A = \{8, 9, 10\}$.
b. $B = \{x : x \text{ is a multiple of } 2\}$
   i.e. $B = \{2, 4, 6, 8, 10\}$.

Eample 1.4.2: Draw the Venn diagram for the condition of set V = {Tuesday, Wednesday, Sunday}.

Solution: Given set V = {Tuesday, Wednesday, Sunday}

For this problem, the universal set is

U = {Sunday, Monday, Tuesday, Wednesday, Thursday, Friday, Saturday}

The Venn diagram is as follows:

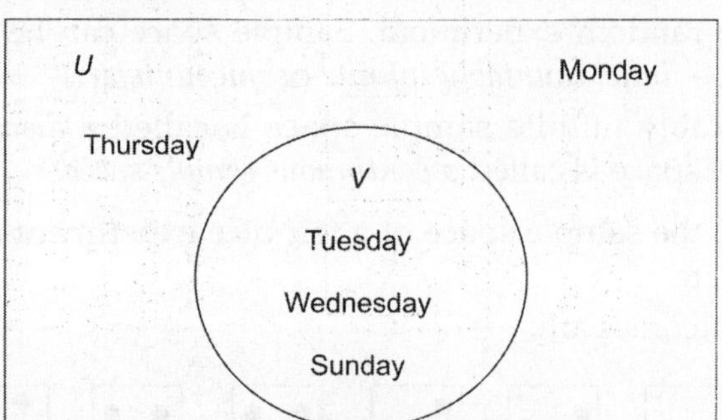

1.5 BASICS OF PROBABILITY

We introduce some of the basic concepts of probability theory by defining some terminology relating to *random experiments* (i.e. experiments whose outcomes are not predictable).

1.5.1 Outcome

The end result of an experiment is called an outcome. For example, if the experiment consists of throwing a dice, the outcome would be anyone of the six faces, 1, 2, 3, 4, 5, or 6.

1.5.2 Random Experiment

An experiment repeated under essentially homogeneous and similar conditions results in an outcome, which is unique or not unique but may be one of the several possible outcomes. When the result is unique, then the experiment is called a *deterministic experiment*. If the outcome is one of the several possible outcomes, then such an experiment is called a *random experiment* or *nondeterministic experiment*.

Consider that a voltage of V volts is applied to a simple resistive circuit. The mathematical model which describes the flow of current in the circuit is $I = \frac{V}{R}$ (Ohm's law) where R is the resistance in the circuit. This model predicts the value of I, when V and R are given.

If we repeat the above experiment a number of times, each time using the same circuit, i.e. keeping V and R as constant, we would presumably expect to observe same value of I. Thus, the above model is a deterministic model and the experiment is referred to as deterministic experiment.

Now, consider an experiment such as tossing a coin. The result of this experiment will be either tail or head. So, in some experiments, we are not able to control the value of certain variables so that the results will vary from one performance of the experiment to the next even though most of the conditions are same. These experiments are called nondeterministic or random experiments.

1.5.3 Sample Space (*S*)

The set of all possible outcomes of a random experiment is called the sample space associated with the random experiment. Sample space can be represented by S. A sample space may be *finite, countably infinite* or *uncountable.*

A finite or countably infinite sample space is called a *discrete sample space.* An uncountable sample space is called a *continuous sample space.*

Example 1.5.1: Find the sample space of a fair dice experiment and determine if it is continuous or discrete.

The possible 6 outcomes are:

The associated sample space is $S = \{'1', '2', '3', '4', '5', '6',\}$. The outcomes are countable and finite in number. Hence, the sample space is called countably finite sample space.

Example 1.5.2: Toss a fair coin until a head is obtained. Determine if the sample space is continuous or discrete.

We may have to toss the coin any number of times before a head is obtained. Thus, the possible outcomes are H, TH, TTH, TTTH,...

The outcomes are countable but infinite in number. The countably infinite sample space is $S = \{H, TH, TTH, ...\}$.

Example 1.5.3: Determine the sample space of output of a radio receiver at any time and also if it is continuous or discrete.

Suppose the output voltage of a radio receiver at any time t is a value lying between –5 V and 5 V.

The associated sample space S is given by

$$S = \{s : s \in R, -5 \leq s \leq 5\} = [-5, 5]$$

Clearly S is a uncountable sample space and hence continuous.

1.5.4 Event

An event is the outcome or a combination of outcomes of an experiment. In other words, an event is a subset of the sample space.

For example, $\{H\}$ in the experiment of tossing a coin is an event and {sum is equal to 6} in the experiment of throwing a pair of dice is an event.

Occurrence of an Event

Suppose we throw a dice. Let E be the event of a perfect square number. Then $E = \{1, 4\}$. Suppose 3 appeared on the upper most face, then we say that the event E has not occurred. E occurs only when 1 or 4 appears on the upper most face. Therefore, whenever an outcome satisfies the conditions given in the event, we say that the event has occurred. In a random experiment, if E is the event of a sample space S and w is the outcome, then we say the event E has occurred if $w \in E$.

Types of Event

Simple event: If an event has one element of the sample space, then it is called a simple or elementary event.

Example: Consider the experiment of throwing a dice, there are 6 events and all events have single element.

Compound event: If an event has more than one sample points, the event is called a compound event.

In the above example of throwing a dice, $E = \{1, 4\}$ is a compound event.

Null event (ϕ): The null event, denoted by ϕ, is an event with no sample points. Thus $\phi = \overline{S}$.

Algebra of Events

In a random experiment, considering S (the sample space) as the universal set, let A, B and C be the events of S. We can define union, intersection and complement of events and their properties on S, which is similar to those in set theory.

i. $A \cup B$, $B \cap A$, A' are some events of random experiment.

ii. $A - B$ is an event, which is same as "A but not B".

iii. $A \cup B = B \cup A$, $A \cap B = B \cap A$

iv. $A \cup (B \cup C) = (A \cup B) \cup C$, $A \cap (B \cap C) = (A \cap B) \cap C$

v. $A \cup (B \cap C) = (A \cup B) \cap (A \cup C)$, $A \cap (B \cup C) = (A \cap B) \cup (A \cap C)$

vi. $(A \cup B)' = A' \cap B'$, $(A \cap B)' = A' \cup B'$

Complement of an event: The complement of an event E with respect to S is the set of all the elements of S which are not in E. The complement of E is denoted by E' or E^c.

Note: In an experiment if E has not occurred then E' has occurred.

Intersection of events: The intersection of two events, A and B, is the set of all outcomes which belong to A as well as B. It is denoted by $A \cap B$ or simply (AB). The intersection of A and B is also referred to as a joint event A and B. This concept can be generalized to the case of intersection of three or more events.

Union of events: The union of two events A and B, denoted by $A \cup B$ (also written as $(A + B)$ or (A or B)) is the set of all outcomes which belong to A or B or both. This concept can be generalized to the union of more than two events.

Mutually exclusive events: Two events associated with a random experiment are said to be mutually exclusive, if both cannot occur together in the same trial.

In the experiment of throwing a dice, the events $A = \{1, 4\}$ and $B = \{2, 5, 6\}$ are mutually exclusive events. In the same experiment, the events $A = \{1, 4\}$ and $C = \{2, 4, 5, 6\}$ are not mutually exclusive because, if 4 appears on the dice, then it is favorable to both events A and C. The definition of mutually exclusive events can also be extended to more than two events. We say that more than two events are mutually exclusive, if the happening of one of these, rules out the happening of all other events. The events $A = \{1, 3\}$, $B = \{5, 6\}$ and $C = \{2, 4\}$ are mutually exclusive in connection with the experiment of throwing a single dice.

Let $E_1, E_2, \ldots, E_n$ be n events associated with a random experiment. They are said to be pairwise mutually exclusive, if $E_i \cap E_j = \phi$ for all i, j and $i \neq j$.

Mutually exhaustive events: For a random experiment, let $E_1, E_2, E_3, \ldots, E_n$ be the subsets of the sample space S. $E_1, E_2, E_3, \ldots, E_n$ forms a set of exhaustive events if $E_1 \cup E_2 \cup E_3 \cup \ldots \cup E_n = S$.

A set of events $E_1, E_2, E_3, \ldots, E_n$ of S are said to be mutually exclusive and exhaustive events if

$$E_1 \cup E_2 \cup E_3 \cup \ldots \cup E_n = S \text{ and}$$

$$E_i \cap E_j = \phi \text{ and } i \neq j.$$

Equally likely outcomes: The outcomes of a random experiment are said to be equally likely, if each one of them has an equal chance of occurrence.

Example: The outcomes of an unbiased coin are equally likely.

1.6 PROBABILITY OF AN EVENT

The probability of an event has been defined in several ways. Two of the most popular definitions are the *relative frequency* definition, and the *classical* definition.

1.6.1 The Relative Frequency Definition

Suppose that a random experiment is repeated n times. If an event A occurs n_A times, then the probability of A, denoted by $P(A)$, is defined as

$$P(A) = \lim_{n\to\infty}\left(\frac{n_A}{n}\right) \quad ...(1.6)$$

where $\left(\frac{n_A}{n}\right)$ represents the fraction of occurrence of A in n trials.

1.6.2 The Classical Definition

The relative frequency definition given above has empirical flavor. In the classical approach, the probability of an event A is found without experimentation. This is done by counting the total number N of the possible outcomes of the experiment. If N_A of those outcomes are favorable to the occurrence of the event A, then

$$P(A) = \left(\frac{N_A}{N}\right) \quad ...(1.7)$$

where, it is assumed that all outcomes are *equally likely*.

Axiomatic Definition of Probability

Whatever may be the definition of probability, we require the probability measure (to the various events on the sample space) to obey the following postulates or axioms:

$$\left.\begin{array}{l} \text{i. } 0 \le P(A) \le 1 \\ \text{ii. } P(S) = 1 \\ \text{iii. If } A_1, A_2,..., A_n \text{ are mutually exclusive events, then} \\ \quad P(A_1 \cup A_2 \cup ... \cup A_n) = P(A_1) + P(A_2) + \cdots + P(A_n) \end{array}\right\} \quad ...(1.8)$$

1.6.3 Theorems of Probability

Theorem 1: Addition rule of probability

If A and B are any two events, then $P(A \cup B) = P(A) + P(B) - P(A \cap B)$

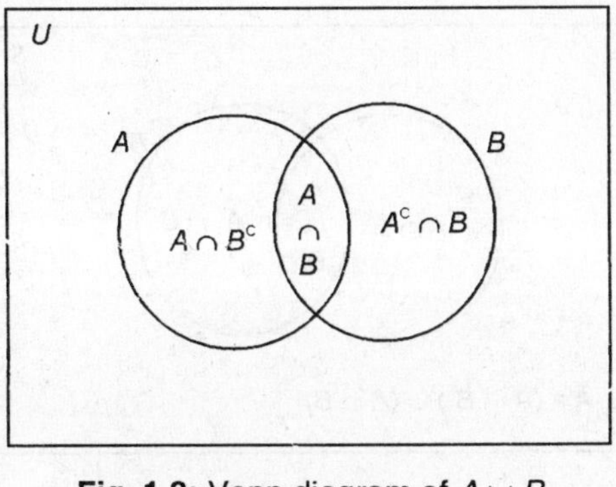

Fig. 1.3: Venn diagram of $A \cup B$

Proof: We have $A \cup B = A \cup (A^c \cap B)$ and $A \cap (A^c \cap B) = \phi$ (from the Venn diagram)

$$\therefore \quad P(A \cup B) = P(A) + P(A^c \cap B) \qquad ...(1.9)$$

($\because$ A and $A^c \cap B$ are mutually exclusive)

We have $B = (A \cap B) \cup (A^c \cap B)$ and $(A \cap B) \cap (A^c \cap B) = \phi$

$$\therefore \quad P(B) = P(A \cap B) + P(A^c \cap B)$$

$$\therefore \quad P(A^c \cap B) = P(B) - P(A \cap B) \qquad ...(1.10)$$

Substituting Eq. (1.10) in Eq. (1.9), we get

$$P(A \cup B) = P(A) + P(B) - P(A \cap B)$$

Note: If $A \cap B = \phi$ then $P(A \cup B) = P(A) + P(B)$

i.e. if A and B are mutually exclusive events, then

$$P(A \cup B) = P(A) + P(B)$$

Theorem 2: $P(A^c) = 1 - P(A)$

Proof: $A \cap A^c = \phi$ and $A \cup A^c = S$

$$P(A \cup A^c) = P(S)$$

$$\Rightarrow \quad P(A) + P(A^c) = 1$$

$$P(A) = 1 - P(A^c).$$

Theorem 3: $P(\phi) = 0$

Proof: The proof follows from the above Theorem 2.

$$P(\phi^c) = 1 - P(\phi)$$

$$\Rightarrow \quad P(\phi) = 1 - P(\phi^c)$$

$$= 1 - P(S) \qquad (\because \phi^c = S)$$

$$= 1 - 1 = 0$$

Theorem 4: Given that two events A and B are statistically independent. Show that:

i. A is independent of B^c

ii. A^c is independent of B

iii. A^c is independent of B^c

Proof: i. $A = (A \cap B^c) \cup (A \cap B)$

$$P(A) = P(A \cap B^c) + P(A \cap B)$$

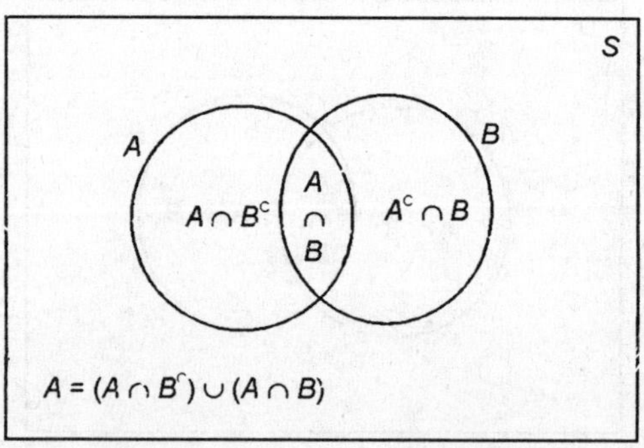

Fig. 1.4: Venn diagram of $A \cap B$

If A and B are independent,

$$P(A \cap B) = P(A) \cdot P(B)$$

$$P(A) = P(A \cap B^c) + P(A) \cdot P(B)$$

$$\begin{aligned} P(A \cap B^c) &= P(A) - P(A) \cdot P(B) \\ &= P(A)\,(1 - P(B)) \\ &= P(A) \cdot P(B^c) \end{aligned}$$

Hence A is independent of B^c.

ii. $$B = (A \cap B) \cup (A^c \cap B)$$

$$P(B) = P(A \cap B) + P(A^c \cap B)$$

$$P(A^c \cap B) = P(B) - P(A \cap B)$$

$$P(A^c \cap B) = P(B) - P(A) \cdot P(B)$$

(If A and B are independent, $P(A \cap B) = P(A) \cdot P(B)$)

$$\begin{aligned} &= P(B)(1 - P(A)) \\ &= P(B) \cdot P(A^c) \end{aligned}$$

Hence A^c and B are independent.

iii. $$(A \cup B)^c = S - (A \cup B)$$

$$\left[P(A \cup B)^c\right] = P(S) - P(A \cup B)$$

$$\begin{aligned} \left[P(A \cup B)^c\right] &= 1 - P(A) - P(B) + P(A \cap B) \\ &= 1 - P(A) - P(B) + P(A) \cdot P(B) \\ &= 1 - P(A) - P(B)\,(1 - P(A)) \\ &= (1 - P(A))\,(1 - P(B)) \\ &= P(A^c)\,P(B^c). \end{aligned}$$

Hence A^c and B^c are independent.

Theorem 5:

$$P(A \cup B \cup C) = P(A) + P(B) + P(C) - P(A \cap B) - P(B \cap C) - P(C \cap A) + P(A \cap B \cap C)$$

Proof: We know that

$$P(A \cup B \cup C) = P((A \cup B) \cup C)$$

$$P(A \cup B) = P(A) + P(B) - P(A \cap B)$$

$$\begin{aligned} \therefore\ P((A \cup B) \cup C) &= P(A \cup B) + P(C) - P((A \cup B) \cap C) \\ &= P(A) + P(B) + P(C) - P(A \cap B) - P((A \cup B) \cap C) \\ &= P(A) + P(B) + P(C) - P(A \cap B) - P\{(A \cap C) \cup (B \cap C)) \\ &= P(A) + P(B) + P(C) - P(A \cap B) - (P(A \cap C) + P(B \cap C) - P(A \cap B \cap C)) \end{aligned}$$

$$P(A \cup B \cup C) = P(A) + P(B) + P(C) -- P(A \cap B) - (B \cap C) - P(C \cap A) + P(A \cap B \cap C).$$

Theorem 6: If $B \subset A$, prove that $P(B) \leq P(A)$.

Proof: B and AB are mutually exclusive events such that $B \cup A\overline{B} = A$

$$P(B \cup AB^c) = P(A)$$

i.e. $$P(B) + P(AB^c) = P(A)$$

$$P(B) \leq P(A)$$

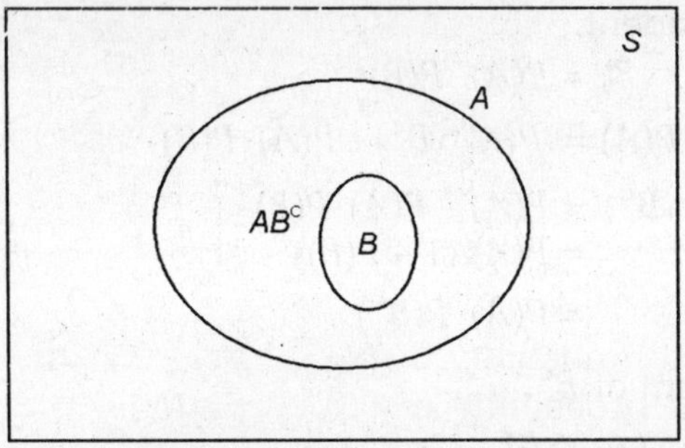

Fig. 1.5: Venn diagram of $B \cup AB^c = A$

1.7 MATHEMATICAL MODEL OF EXPERIMENTS

A real experiment is defined mathematically by three factors:

i. Assignment of a sample space
ii. Definition of events of interest
iii. Making probability assignments to events such that the axioms are satisfied.

Example 1.7.1: Three coins are tossed. Find the probability of getting:

i. Exactly two heads
ii. At least one tail
iii. Three heads

Assignment of sample space S:

$$S = \{HHH, HHT, HTH, HTT, THH, THT, TTH, TTT\}$$

Definition of events of our interest:

Let us take an event A as getting exactly two heads,
event B as getting at least one tail and
event C as getting three heads

$$n(A) = 3$$
$$n(B) = 7$$
$$n(C) = 1$$
$$P(A) = \frac{3}{8}$$
$$P(B) = \frac{7}{8}$$
$$P(C) = \frac{1}{8}.$$

1.8. JOINT AND CONDITIONAL PROBABILITY

For two events A and B, the common elements form the event $A \cap B$.

1.8.1 Joint Probability

The probability $P(A \cap B)$ is called the *joint probability* for two events A and B which intersect in the sample space. A study of Venn diagram will readily show that

$$P(A \cup B) = P(A) + P(B) - P(A \cap B) \qquad ...(1.11)$$

Equivalently,

$$P(A \cup B) = P(A) + P(B) - P(A \cap B) \leq P(A) + P(B) \qquad ...(1.12)$$

In other words, the probability of the union of two events never exceeds the sum of the event probabilities.

1.8.2 Conditional Probability

A very useful concept in probability theory is that of *conditional probability*, denoted $P(B/A)$; it represents the probability of B occurring, given that A has occurred. In a real world random experiment, it is quite likely that the occurrence of the event B is very much influenced by the occurrence of the event A. To give a simple example, let a bowl contain 3 resistors and 1 capacitor. The occurrence of the event 'the capacitor on the second draw' is very much dependent on what has been drawn at the first instant. Such dependencies between the events are brought out using the notion of conditional probability. The conditional probability $P(B/A)$ can be written in terms of the joint probability $P(AB)$ and the probability of the event $P(A)$. This relation can be arrived at by using either the relative frequency definition of probability or the classical definition. Using the former, we have

$P(B/A)$ = Probability of occurrence of event B when the event A as already occurred.

$$P(B/A) = \frac{\text{No. of cases favourable to } A \text{ which are also favourable to } B}{\text{No. of cases favourable to } A} \qquad ...(1.13)$$

$$\therefore \quad P(B/A) = \frac{\text{No. of cases favourable to } A \cap B}{\text{No. of cases favourable to A}} \qquad ...(1.14)$$

Also,

$$P(B/A) = \frac{\dfrac{\text{No. of cases favourable to } A \cap B}{\text{No. of cases in the sample space}}}{\dfrac{\text{No. of cases favourable to } A}{\text{No. of cases in the sample space}}}$$

$$\therefore \quad P(B/A) = \frac{P(A \cap B)}{P(A)} \qquad ...(1.15)$$

According to relative frequency probability definition,

$$P(AB) = P(A \cap B) = \lim_{n \to \infty}\left(\frac{n_{AB}}{n}\right)$$

$$P(A) = \lim_{n \to \infty}\left(\frac{n_A}{n}\right)$$

where n_{AB} is the number of times AB occurs in n repetitions of the experiment.

As $P(B/A)$ refers to the probability of B occurring, given that A has occurred, we have conditional probability

$$P(B/A) = \lim_{n \to \infty}\left(\frac{n_{AB}}{n_A}\right)$$

$$P(B/A) = \lim_{n\to\infty}\left(\frac{\frac{n_{AB}}{n}}{\frac{n_A}{n}}\right) = \frac{P(AB)}{P(A)}, P(A) \neq 0 \qquad ...(1.16)$$

$$P(AB) = P(A \cap B) = P(B/A)\,P(A) \qquad ...(1.17)$$

Interchanging the role of A and B, we have

$$P(A/B) = \frac{P(AB)}{P(B)}, P(B) \neq 0 \qquad ...(1.18)$$

Equations (1.16) and (1.17) can be written as

$$P(A) = P(B/A)\,P(A) = P(A/B)\,P(B) \qquad ...(1.19)$$

In view of Eq. (1.19), we can also write Eq. (1.16) as

$$P(B/A) = \frac{P(B)P(A/B)}{P(A)}, P(A) \neq 0 \qquad ...(1.20)$$

Similarly $$P(A/B) = \frac{P(A)P(B/A)}{P(B)}, P(B) \neq 0 \qquad ...(1.21)$$

Example 1.8.1: A card is drawn from an ordinary deck and we are told that it is red. What is the probability that the card is greater than 2 but less than 9.

Solution: Let A be the event of getting a card greater than 2 but less than 9.

B be the event of getting a red card. We have to find the probability of A given that B has occurred, that is, we have to find $P(A/B)$.

In a deck of cards, there are 26 red cards and 26 black cards.

$$n(B) = 26$$

Among the red cards, the number of outcomes which are favorable to A are 12, that is, $n(A \cap B) = 12$

$$\therefore \qquad P(A/B) = \frac{n(A \cap B)}{n(B)} = \frac{12}{26} = \frac{6}{13}$$

Example 1.8.2: A pair of dice is thrown. If it is known that one dice shows a 4, what is the probability that:

(a) The other dice shows a 5

(b) The total of both the dice is greater than 7.

Solution: Let A be the event that one dice shows up 4. Then the outcomes which are favourable to A are

(4, 1), (4, 2), (4, 3), (4, 4), (4, 5), (4, 6), (1, 4), (2, 4), (3, 4), (5, 4), (6, 4)

$$n(B) = 11$$

(a) Let B be the event of getting a 5 on one of the dice. Then the outcomes which are favourable to both A and B are (4, 4) and (4, 5)

$$n(A \cap B) = 2$$

$$\therefore \qquad P(B/A) = \frac{n(A \cap B)}{n(A)} = \frac{2}{11}$$

(b) Let C be the event of getting a total greater than 7 on both the dice.

The outcomes which are favourable to both C and A are (4, 4) (4, 5), (4, 6), (5, 4), (6, 4)

$$n(A \cap C) = 5$$

$$\therefore \quad P(C/A) = \frac{n(A \cap C)}{n(A)} = \frac{5}{11}$$

Note that in the above example, $P(B)$ and $P(B/A)$ are different.

$\therefore$ $P(B/A)$ is $\frac{2}{11}$ whereas $P(B) = \frac{11}{36}$

Similarly, $P(C)$ and $P(C/A)$ are different.

Example 1.8.3: In a housing colony 70% of the houses are well planned and 60% of the houses are well planned and well built. Find the probability that an arbitrarily chosen house in this colony is well built given that it is well planned.

Solution: Let A be the event that the house is well planned and B be the event that the house is well built.

$$P(A) = 0.7$$

$$P(A \cap B) = 0.6$$

Probability that a house selected is well built given that it is well planned,

$$P(B/A) = \frac{P(A \cap B)}{P(A)} = \frac{0.6}{0.7} = \frac{6}{7}$$

1.9 STATISTICAL INDEPENDENCE

Another useful probabilistic concept is that of statistical independence. Suppose two events A and B have nonzero probabilities of occurrence, that is, assume $P(A) \neq 0$ and $P(B) \neq 0$. We call the events *statistically independent* if the probability of occurrence of one event is not affected by the other event. Mathematically,

$$P(B/A) = P(B) \qquad ...(1.22)$$

Similarly

$$P(A/B) = P(A) \qquad ...(1.23)$$

Equation (1.23), then

$$P(AB) = P(A \cap B) = P(A)\,P(B) \qquad ...(1.24)$$

Equation (1.24) can be used to define the statistical independence of two events. Note that if A and B are independent, then $P(AB) = P(A \cap B) = P(A) \cdot P(B)$, whereas if they are disjoint, then $P(AB) = 0$. The notion of statistical independence can be generalized to the case of more than two events. A set of k events $A_1, A_2, ..., A_k$ are said to be statistically independent if and only if the probability of every intersection of k or fewer events equal the product of the probabilities of its constituents.

1.10 TOTAL PROBABILITY AND BAYE'S THEOREM

1.10.1 Total Probability

Theorem: (Law of total probability)

If $B_1, B_2, B_3, ..., B_n$ are mutually exclusive and exhaustive events of the sample space S, then for any event A of S,

$$P(A) = P(B_1) \times P(A/B_1) + P(B_2) \times P(A/B_2) + P(B_3) \times (A/B_3) + \cdots + P(B_n) \times P(A/B_n)$$

$$= \sum_{i=1}^{n} P(B_i)P(A/B_i)$$

Proof: From the Venn diagram, we have

$$S = B_1 \cup B_2 \cup B_3 \cup \ldots \cup B_n \qquad \text{...(1.25)}$$

We know that

$$A = S \cap A \qquad \text{...(1.26)}$$

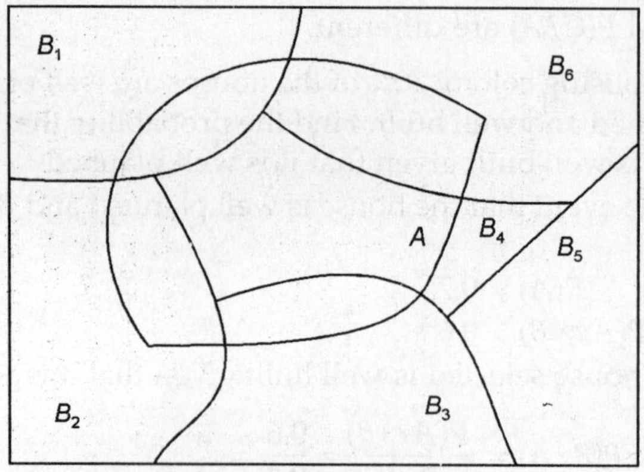

Fig. 1.6: Venn diagram showing *n*-mutually exclusive events B_n and another event A

Substituting Eq. (1.25) in Eq. (1.26),

$$A = (B_1 \cap A) + (B_2 \cap A) + (B_3 \cap A) + \cdots + (B_n \cap A)$$

where $B_1 \cap A, B_2 \cap A, B_3 \cap A, \ldots, B_n \cap A$ are mutually exclusive events.

$$\Rightarrow \; P(A) = P(B_1 \cap A) + P(B_2 \cap A) + P(B_3 \cap A) + \cdots + P(B_n \cap A)$$

$$P(A) = P(B_1) \times P(A/B_1) + P(B_2) \times P(A/B_2) + P(B_3) \times P(A/B_3) + \cdots + P(B_n) \times P(A/B_n)$$

$$P(A) = \sum_{i=1}^{n} P(B_i)P(A/B_i). \qquad \text{...(1.27)}$$

1.10.2 Baye's Theorem

Let S be a sample space. $B_1, B_2, B_3, \ldots, B_n$ are mutually exclusive and exhaustive events such that $P(B_i) \neq 0$ for all i.

Then for any event A which is a subset of

$$S = B_1 \cup B_2 \cup B_3 \cup \ldots \cup B_n \text{ and } P(A) > 0$$

we have,

$$P(B_i/A) = \frac{P(B_i) \times P(A/B_i)}{\sum_{i=1}^{n} P(B_i)P(A/B_i)} \text{ for all } i = 1, 2, 3, \ldots, n \qquad \text{...(1.28)}$$

Proof: We have $S = B_1 \cup B_2 \cup B_3 \cup \ldots \cup B_n$ and $B_i \cap B_j = \phi$ for $i \neq j$

Since $A \subseteq S$ and $A = S \cap A$

$$= A \cap (B_1 \cup B_2 \cup B_3 \cup ... \cup B_n)$$
$$= (A \cap B_1) + (A \cap B_2) + (A \cap B_3) + \cdots + (A \cap B_n)$$
$$\therefore \quad P(A) = P(A \cap B_1) + P(A \cap B_2) + P(A \cap B_3) + \cdots + P(A \cap B_n)$$
$$[\because (A \cap B_i) \cap (A \cap B_j) = \phi \quad \text{for } i \neq j]$$
$$P(A) = P(B_1) \times P(A/B_1) + P(B_2) \times P(A/B_2) + P(B_3) \times P(A/B_3) + \cdots + P(B_n) \times P(A/B_n)$$

$$P(A) = \sum_{i=1}^{n} P(B_i)P(A/B_i) \quad ...(1.29)$$

Also,
$$P(A \cap B_i) = P(A)\, P(B_i/A) = P(B_i)\, P(A/B_i)$$

$$P(B_i/A) = \frac{P(A \cap B_i)}{P(A)} \quad ...(1.30)$$

From Eqs. (1.29) and (1.30),

$$\Rightarrow \quad P(B_i/A) = \frac{P(B_i)P(A/B_i)}{\sum_{i=1}^{n} P(B_i)P(A/B_i)} \quad ...(1.31)$$

Example 1.10.1: Let S be the sample space which is the population of adults in a small town who have completed the requirement for a college degree. The population is categorized according to sex and employment status and is as follows:

	Employed	*Unemployed*	*Total*
Male	460	40	500
Female	140	260	400
Total	600	300	900

One of these individual is to be selected for a tour throughout the country. Knowing that the individual chosen is employed, what is the probability that the individual is a man? Suppose that we are now given the additional information that 36 of those employed and 12 of those unemployed are the members of the rotary club, what is the probability of the event A that the selected individual is a member of the rotary club?

Solution: Let M be the event that a man is selected and E be the event that the individual selected is employed.

Using the reduced sample space, we have

$$P(M/E) = \text{Probability of a man who is employed} = \frac{460}{600} = \frac{23}{30}$$

Also, we have $P(M/E) = \dfrac{P(M \cap E)}{P(E)}$

From the original sample space,

$$\text{probability of a person who is employed} = P(E) = \frac{600}{900} = \frac{2}{3}$$

$$P(U) = \frac{300}{900} = \frac{1}{3}$$

$$P(M \cap E) = \frac{460}{900} = \frac{23}{45}$$

$$\Rightarrow \quad P(M/E) = \frac{P(M \cap E)}{P(E)}$$

$$P(M/E) = \frac{23}{45} \times \frac{3}{2} = \frac{23}{30}$$

A is the event that the selected individual is a member of the rotary club.

$$P(A/E) = \frac{36}{600}$$

$$P(A/U) = \frac{12}{300}$$

According to the total probability theorem,

$$P(A) = P(E) \times P(A/E) + P(U) \times P(A/U)$$

$$P(A) = \frac{2}{3} \times \frac{3}{50} + \frac{1}{3} \times \frac{1}{25}$$

$$P(A) = \frac{4}{75}$$

Example 1.10.2: In a binary communication channel, A is the input and B is the output. $P(A) = 0.4$, $P(B/A) = 0.9$ and $P(\bar{B}/\bar{A}) = 0.6$. Find $P(A/B)$ and $P(A/\bar{B})$.

Solution: We know that

$$P(B/A) = \frac{P(A \cap B)}{P(A)}$$

$$0.9 = \frac{P(A \cap B)}{0.4}$$

$$P(A \cap B) = 0.36$$

Given that $\quad P(\bar{B}/\bar{A}) = 0.6$

$$\frac{P(\bar{A} \cap \bar{B})}{P(\bar{A})} = 0.6$$

$$P(\bar{A} \cap \bar{B}) = (0.6) \times (0.6) \qquad [\because P(\bar{A}) = 0.6]$$

$$P(\bar{A} \cap \bar{B}) = 0.36 \qquad [\because \bar{A} \cap \bar{B} = \overline{(A \cup B)}]$$

$$P(A \cup B) = 0.64$$

We know that

$$P(A \cup B) = P(A) + P(B) - P(A \cap B)$$
$$0.64 = 0.4 + P(B) - 0.36$$
$$P(B) = 0.6$$

$$P(A/B) = \frac{P(A \cap B)}{P(B)} = \frac{0.36}{0.6} = 0.6$$

By the law of total probability,

$$P(A) = P(A/B)P(B) + P(A/\bar{B})P(\bar{B})$$

$$0.4 = (0.6)(0.6) + P(A/\bar{B})(0.4)$$

$$P(A/\bar{B})(0.4) = 0.4 - 0.36$$

$$P(A/\bar{B}) = \frac{0.04}{0.4} = 0.1.$$

Example 1.10.3: In a bolt factory, 25%, 35% and 40% of the total is manufactured by machines A, B and C, out of which 5%, 4% and 2% are respectively defective. If the bolt drawn is found to be defective, what is the probability that it is manufactured by the machine A, machine B, machine C?

Solution: Given $P(A) = 0.25$, $P(B) = 0.35$ and $P(C) = 0.4$

Let D be the event of getting a defective bolt.

$$P(D/A) = 0.05, P(D/B) = 0.04, P(C) = 0.02$$

$$\therefore \quad P(A/D) = \frac{P(A)P(D/A)}{P(D)} = \frac{P(A)P(D/A)}{P(A)P(D/A) + P(B)P(D/B) + P(C)P(D/C)}$$

$$= \frac{(0.25)(0.05)}{(0.25)(0.05) + (0.35)(0.04) + (0.4)(0.02)}$$

$$= \frac{0.0125}{0.0345} = \frac{125}{345}$$

$$P(B/D) = \frac{P(B)P(D/B)}{P(A)P(D/A) + P(B)P(D/B) + P(C)P(D/C)}$$

$$= \frac{(0.35)(0.04)}{(0.25)(0.05) + (0.35)(0.04) + (0.4)(0.02)}$$

$$= \frac{0.0140}{0.0345} = \frac{140}{345}$$

$$P(C/D) = \frac{P(C)P(D/C)}{P(A)P(D/A) + P(B)P(D/B) + P(C)P(D/C)}$$

$$= \frac{(0.4)(0.02)}{(0.25)(0.05)+(0.35)(0.04)+(0.4)(0.02)}$$

$$= \frac{(0.008)}{(0.25)(0.05)+(0.35)(0.04)+(0.4)(0.02)} = \frac{8}{345}.$$

EXAMPLES

Example 1.1: A dice is rolled, find the probability that an even number is obtained.

Solution: Let us first write the sample space S of the experiment.

$$S = \{1, 2, 3, 4, 5, 6\}$$

Let E be the event "an even number is obtained" and write it down.

$$E = \{2, 4, 6\}$$

We now use the formula of the classical probability,

$$P(E) = n(E)/n(S) = 3/6 = 1/2$$

Example 1.2: Two coins are tossed, find the probability that two heads are obtained. Note that each coin has two possible outcomes H (head) and T (tail).

Solution: The sample space S is given by.

$$S = \{(H, T), (H, H), (T, H), (T, T)\}$$

Let E be the event "two heads are obtained"

$$E = \{(H, H)\}$$

We use the formula of the classical probability,

$$P(E) = n(E)/n(S) = 1/4.$$

Example 1.3: Which of these numbers cannot be a probability?

(a) – 0.00001 (b) 0.5 (c) 1.001 (d) 0 (e) 1 (f) 20%

Solution: Probability is always greater than or equal to 0 and less than or equal to 1. Hence, only (a) and (c) above cannot represent probabilities as –0.00001 is less than 0 and 1.001 is greater than 1.

Example 1.4: Two dice are rolled; find the probability that the sum is

(a) equal to 1 (b) equal to 4 (c) less than 13.

Solution: (a) The sample space S of two dice is shown below:

$$S = \begin{Bmatrix} (1,1), (1,2), (1,3), (1,4), (1,5), (1,6) \\ (2,1), (2,2), (2,3), (2,4), (2,5), (2,6) \\ (3,1), (3,2), (3,3), (3,4), (3,5), (3,6) \\ (4,1), (4,2), (4,3), (4,4), (4,5), (4,6) \\ (5,1), (5,2), (5,3), (5,4), (5,5), (5,6) \\ (6,1), (6,2), (6,3), (6,4), (6,5), (6,6) \end{Bmatrix}$$

Let E be the event "sum equal to 1". There are no outcomes which correspond to a sum equal to 1, hence

$$P(E) = n(E)/n(S) = 0/36 = 0$$

(b) Three possible outcomes give a sum equal to 4: $E = \{(1, 3), (2, 2), (3, 1)\}$

Hence $$P(E) = n(E)/n(S) = 3/36 = 1/12$$

(c) All possible outcomes, $E = S$, give a sum less than 13, hence

$$P(E) = n(E)/n(S) = 36/36 = 1.$$

Example 1.5: A dice is rolled and a coin is tossed, find the probability that the dice shows an odd number and the coin shows a head.

Solution: The sample space S of the experiment described is as follows:

$$S = \{(1, H), (1, T), (2, H), (2, T), (3, H), (3, T), (4, H), (4, T), (5, H),(5, T), (6, H),(6, T)\}$$

Let E be the event, "the dice shows an odd number and the coin shows a head". Event E may be described as follows:

$$E = \{(1,H),(3,H),(5,H)\}$$

The probability $P(E)$ is given by

$$P(E) = n(E)/n(S) = 3/12 = 1/4.$$

Example 1.6: A card is drawn at random from a deck of 52 cards. Find the probability of getting the 3 of diamond.

Solution: The sample space S of the experiment is shown below:

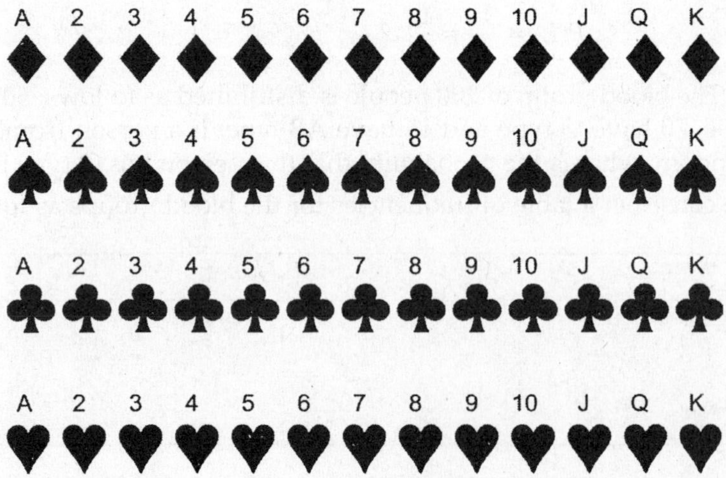

Let E be the event "getting the 3 of diamond". An examination of the sample space shows that there is one "3 of diamond" so that $n(E) = 1$ and $n(S) = 52$. Hence, the probability of event E occurring is given by

$$P(E) = n(E)/n(S) = 1/52.$$

Example 1.7: A card is drawn at random from a deck of 52 cards. Find the probability of getting a queen.

Solution: The sample space S of the experiment is shown above in Example 1.6.

Let E be the event "getting a queen". An examination of the sample space shows that there are 4 queens so that $n(E) = 4$ and $n(S) = 52$. Hence, the probability of event E occuring is given by

$$P(E) = n(E)/n(S) = 1/13.$$

Example 1.8: A jar contains 3 red marbles, 7 green marbles and 10 white marbles. If a marble is drawn from the jar at random, what is the probability that this marble is white?

Solution: We first construct a table of frequencies that gives the marble color distributions as follows:

Color	*Frequency*
Red	3
Green	7
White	10

We now use the empirical formula of probability,

$$P(E) = \frac{\text{Frequency of white color}}{\text{Total frequency}}$$

$$P(E) = \frac{10}{20} = \frac{1}{2}.$$

Example 1.9: The blood group of 200 people is distributed as follows: 50 have A type, 65 have B type, 70 have O type and 15 have AB type. If a person from this group is selected at random, what is the probability that this person has O type blood?

Solution: We construct a table of frequencies for the blood groups as follows:

Group	*Frequency*
A	50
B	65
O	70
AB	15

We use the empirical formula of the probability,

$$P(E) = \frac{\text{Frequency of O blood}}{\text{Total frequencies}}$$

$$P(E) = \frac{70}{200} = 0.35.$$

Example 1.10: In an experiment of picking up a resistor with same likelihood of being picked up for the events; A as "draw a 47 Ω resistor", B as "draw a resistor with 5% tolerance" and C as "draw a 100 Ω resistor" from a box containing 100 resistors having resistance and tolerance as shown in table below. Determine joint and conditional probabilities.

Resistance /	*Tolerance*		
	5%	*10%*	*Total*
22	10	14	24
47	28	16	44
100	24	8	32
Total	62	38	100

Solution:

$$P(A) = P(47\,\Omega) = \frac{44}{100} = \frac{11}{25}$$

$$P(B) = P(5\%) = \frac{62}{100} = \frac{31}{50}$$

$$P(C) = P(100\,\Omega) = \frac{32}{100} = \frac{8}{25}$$

Joint probabilities:

$$P(A \cap B) = P(47\,\Omega \cap 5\%) = \frac{28}{100} = \frac{7}{25}$$

$$P(B \cap C) = P(5\% \cap 100\,\Omega) = \frac{24}{100} = \frac{6}{25}$$

$$P(C \cap A) = P(100\,\Omega \cap 47\,\Omega) = 0$$

Conditional probabilities:

$$P(A/B) = \frac{P(A \cap B)}{P(B)} = \frac{7/25}{31/50} = \frac{14}{31}$$

$$P(B/C) = \frac{P(B \cap C)}{P(C)} = \frac{6/25}{8/25} = \frac{3}{4}$$

$$P(C/A) = \frac{P(C \cap A)}{P(A)} = 0.$$

Example 1.11: A binary communication channel carries data as one of the two types of signals denoted by 0 and 1. Owing to noise, a transmitted 0 is sometimes received as 1 and a transmitted 1 is sometimes received as a 0. For a given channel, assume a probability of 0.95 that a transmitted 0 is correctly received as a 0 and a probability of 0.90 that a transmitted 1 is received as a 1. Further, assume a probability of 0.40 of transmitting a 0. If a signal is sent, determine the probability that:

i. A 1 is received
ii. A 1 is transmitted given that a 1 is received
iii. A 0 is received
iv. A 0 is transmitted given that a 0 is received
v. Probability of errors.

Solution:

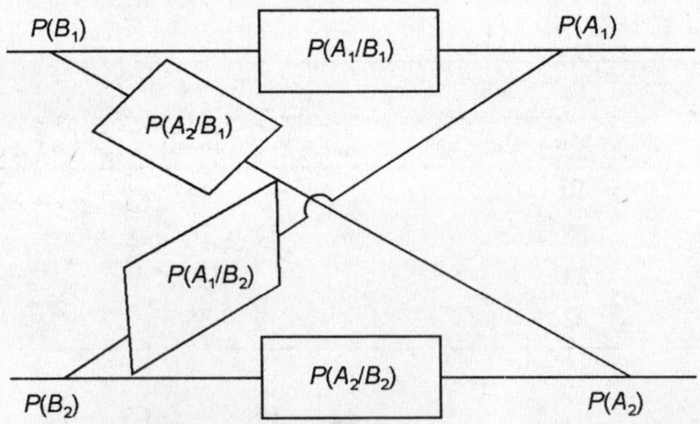

Let event B be the input of the channel and
event A be the output of the channel.
Assume B_1 if the input of the channel is zero.
Assume B_2 if the input of the channel is one.
Assume A_1 if the output of the channel is zero.
Assume A_2 if the output of the channel is one.

(a) Probability that a '1' is received is

$$P(A_2) = P(A_2/B_2)\,P(B_2) + P(A_2/B_1)\,P(B_1)$$
$$= 0.6 \times 0.9 + 0.4 \times 0.05$$
$$= 0.54 + 0.02 = 0.56$$

(b) Probability that a '1' was transmitted given that '1' was received is

$$P(B_2/A_2) = \frac{P(A_2/B_2)P(B_2)}{P(A_2)}$$
$$= \frac{0.6 \times 0.9}{0.56} = \frac{27}{28}$$

(c) Probability that a '0' is received is

$$P(A_1) = P(A_1/B_2)\,P(B_2) + P(A_1/B_1)\,P(B_1)$$
$$= 0.6 \times 0.1 + 0.4 \times 0.95$$
$$= 0.06 + 0.38 = 0.44$$

(d) Probability that a '0' was transmitted given that '0' was received is

$$P(B_1/A_1) = \frac{P(A_1/B_1)P(B_1)}{P(A_1)}$$
$$= \frac{0.4 \times 0.9}{0.44} = 0.87$$

(e) Here error means getting output is '1', even input is '0'.
Similarly getting output is '0', even input is '1'.
Getting output is '1', when input is '0',

$$P(B_1/A_2) = \frac{P(A_2/B_1)P(B_1)}{P(A_2)} = \frac{0.4 \times 0.05}{0.56} = 0.035$$

Getting output is '0', when input is '1',

$$P(B_2/A_1) = \frac{P(A_1/B_2)P(B_2)}{P(A_1)} = \frac{0.1 \times 0.6}{0.44} = 0.136.$$

Example 1.12: Roll a dice in which all faces are equally likely. What is the probability of each outcome? Find the probabilities of the events: (a) roll 4 or higher, (b) roll an even number, and (c) roll the square of an integer.

Solution: The probability of each outcome is

$$P(i) = \frac{1}{6}$$

where $i = 1, 2, 3, 4, 5, 6$

The probabilities of the three events are

(a) P (roll 4 or higher) $= P(4) + P(5) + P(6) = \frac{3}{6} = \frac{1}{2}$

(b) P(roll an even number) $= P(2) + P(4) + P(6) = \frac{3}{6} = \frac{1}{2}$

(c) P (roll the square of an integer) $= P(1) + P(4) = \frac{2}{6} = \frac{1}{3}$.

Example 1.13: Four persons are to be selected from a group of 12 people, 7 of whom are women.

(a) What is the probability that the first and third selected are women?

(b) What is the probability that three of those selected are women?

(c) What is the (conditional) probability that the first and third selected are women, given that three of those selected are women?

Solution: (a) Probability that first and third selected are women is

$$P(W_1W_3) = P(W_1W_2^c W_3) + P(W_1W_2W_3)$$

$$= \left(\frac{7}{12}\cdot\frac{5}{11}\cdot\frac{6}{10}\right) + \left(\frac{7}{12}\cdot\frac{6}{11}\cdot\frac{5}{10}\right) = \frac{7}{22}$$

(b) The probability that three of those selected are women is

$$P(W_1W_2W_3) = \frac{{}^7C_3}{{}^{12}C_3} = \frac{21}{55}$$

(c) Probability that the first and third selected are women, given that three of those selected are women

$$\frac{P(W_1W_3)}{P(W_1W_2W_3)} = \frac{7/22}{21/55} = \frac{7 \times 55}{22 \times 21} = \frac{385}{462} = 0.826.$$

Example 1.14: Five boxes of random access memory chips have 100 units per box. They have respectively one, two, three, four, and five defective units. A box is selected at random, on an equally likely basis, and a unit is selected at random from there. It is defective. What are the (conditional) probabilities that the unit was selected from each of the boxes?

Solution: Let the five events be H_1, H_2, H_3, H_4, H_5.

$$P(H_1) = P(H_2) = P(H_3) = P(H_4) = P(H_5) = \frac{1}{5}$$

$$P(D/H_1) = \frac{1}{100}, P(D/H_2) = \frac{2}{100}, P(D/H_3) = \frac{3}{100},$$

$$P(D/H_4) = \frac{4}{100}, P(D/H_5) = \frac{5}{100}$$

$$P(D) = P(D/H_1)\, P(H_1) + P(D/H_2)\, P(H_2) + P(D/H_3)\, P(H_3) + P(D/H_4)\, P(H_4) + P(D/H_5)\, P(H_5)$$

$$= \frac{1}{100}\frac{1}{5} + \frac{2}{100}\frac{1}{5} + \frac{3}{100}\frac{1}{5} + \frac{4}{100}\frac{1}{5} + \frac{5}{100}\frac{1}{5} = \frac{15}{500} = \frac{3}{100}$$

$$P(H_1/D) = \frac{P(D/H_1)P(H_1)}{P(D)} = \frac{1/500}{3/100} = \frac{1}{15}$$

$$P(H_2/D) = \frac{P(D/H_2)P(H_2)}{P(D)} = \frac{2/500}{3/100} = \frac{2}{15}$$

$$P(H_3/D) = \frac{P(D/H_3)P(H_3)}{P(D)} = \frac{3/500}{3/100} = \frac{3}{15} = \frac{1}{5}$$

$$P(H_4/D) = \frac{P(D/H_4)P(H_4)}{P(D)} = \frac{4/500}{3/100} = \frac{4}{15}$$

$$P(H_5/D) = \frac{P(D/H_5)P(H_5)}{P(D)} = \frac{5/500}{3/100} = \frac{5}{15} = \frac{1}{3}.$$

Example 1.15: A bag contains 10 white and 3 black balls. Another bag contains 3 white and 5 black balls. Two balls are drawn at random from the first bag and placed in the second bag and then 1 ball is taken at random from the latter. What is the probability that it is a white ball?

Solution: The two balls transferred may be both white or both black or one white and one black.

Let event B_1 be drawing 2 white balls from the first bag,
event B_2 be drawing 2 black balls from the first bag,
event B_3 be drawing 1 white and 1 black ball from the first bag.

Clearly B_1, B_2, B_3 are mutually exclusive events.

$$P(B_1) = \frac{^{10}C_2}{^{13}C_2} = \frac{15}{26}$$

$$P(B_2) = \frac{^{3}C_2}{^{13}C_2} = \frac{1}{26}$$

$$P(B_3) = \frac{^{10}C_1 \times {}^{3}C_1}{^{13}C_2} = \frac{10}{26}$$

Assume that the two balls transferred from bag one are both white, then total balls in second bag are 10, and white balls are 5.

$$P(A/B_1) = P\,(\text{drawing a white ball/2 balls have been transferred})$$
$$= \frac{5}{10}$$

Assume that the two balls transferred from bag one are both black, then total balls in second bag are 10, and white balls are 3.

$$P(A/B_2) = P\,(\text{drawing a white ball/2 balls have been transferred})$$
$$= \frac{3}{10}$$

Assume that the two balls transferred from bag one are such that one ball is white and another is black, then total balls in second bag are 10, and white balls are 4.

$$P(A/B_3) = \text{P(drawing a white ball/2 balls have been transferred)}$$
$$= \frac{4}{10}$$

Total probability getting white ball is

$$P(A) = P(A/B_1)\,P(B_1) + P(A/B_2)\,P(B_2) + P(A/B_3)\,P(B_3)$$
$$= \frac{5}{10}\frac{15}{26} + \frac{3}{10}\frac{1}{26} + \frac{4}{10}\frac{10}{26} = \frac{59}{130}.$$

Example 1.16: Given a class of students, what is the probability that two students in the class have the same birthday?

Solution: Let $k \le 365$ be the number of students in the class.

Then the number of possible birthdays = $365 \cdot 365 \cdot \ldots \cdot 365$ (k times) = 365^k

The no. of cases with each of the k students having a different birthday is

$$= {}^{365}P_k = 365 \cdot 364 \ldots (365 - k + 1)$$

Therefore, the probability of common birthday $= 1 - \dfrac{{}^{365}P_k}{365^k}$

No. of persons	*Probability*
2	0.0027
10	0.1169
15	0.4114
25	0.5687
40	0.8912
60	0.9941
80	0.9999

Example 1.17: Out of 50 houses, 25 houses have refrigerator and 30 houses have vacuum cleaner. Find how houses have both refrigerator and vacuum cleaner.

Solution: Let P = {Houses having refrigerator}

O = {Houses having vacuum cleaner}

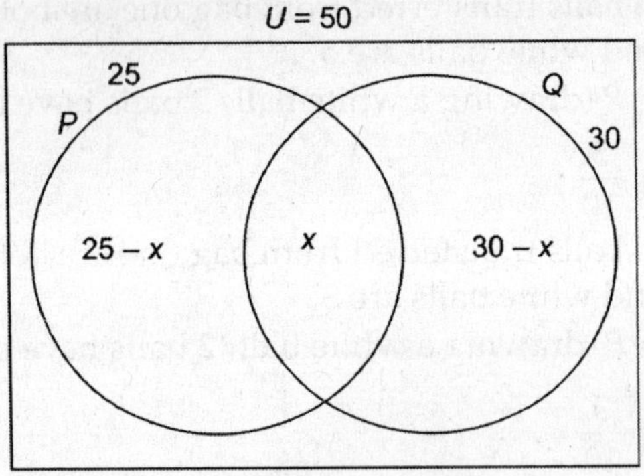

$$P \cap Q = \{\text{Houses having both refrigerator and vacuum cleaner}\}$$

Let $n(P \cap Q) = x$

From the Venn diagram:

$$25 - x + x + 30 - x = 50$$
$$55 - x = 50$$
$$-x = 50 - 55$$
$$x = 5$$

Hence, the number of houses having both refrigerator and vacuum cleaner = 6.

Example 1.18: Verify De Morgan's law $(A \cup B)' = A' \cap B'$ using Venn diagrams.

Solution: From Fig. 1.7(ii) and (v), we find that $(A \cup B)' = A' \cap B'$.

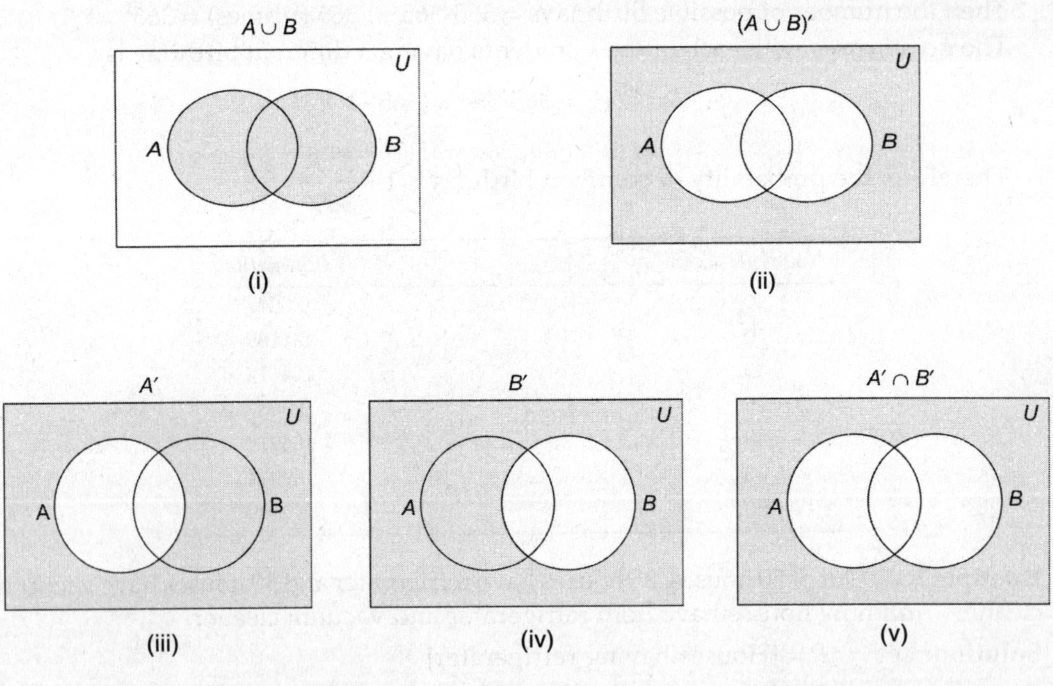

Fig. 1.7

Example 1.19: Verify the De Morgan's law $(A \cap B)' = A' \cup B'$, using Venn diagrams.

Solution: From Fig. 1.8(ii) and (v), we find that $(A \cap B)' = A' \cup B'$,

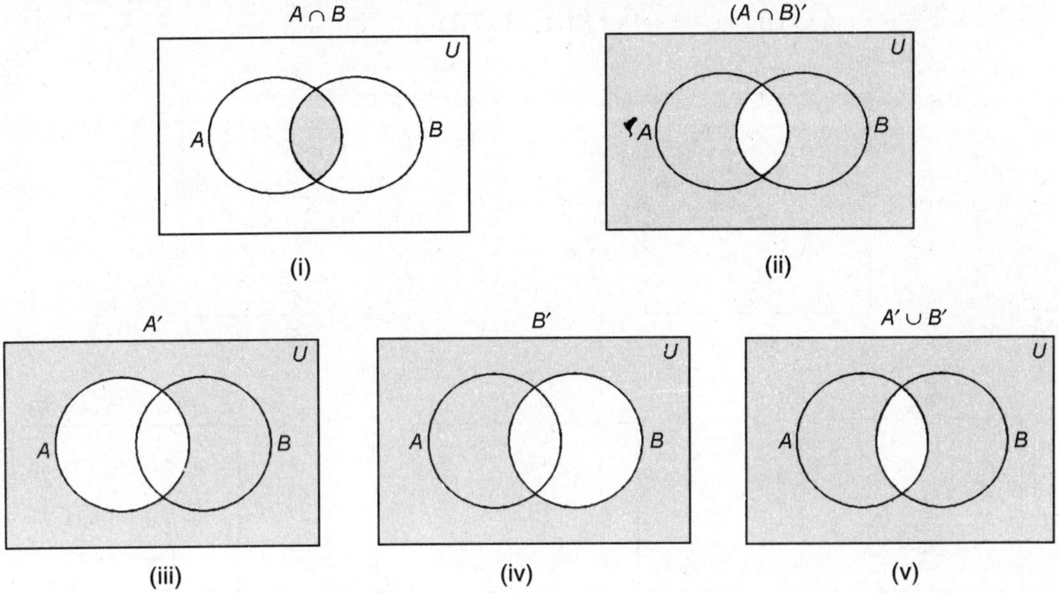

Fig. 1.8

Example 1.20: Verify the De Morgan's law $A - (B \cup C) = (A - B) \cap (A - C)$ using Venn diagrams.

Solution: From Fig. 1.9(ii) and (v), we find that $A - (B \cup C) = (A - B) \cap (A - C)$

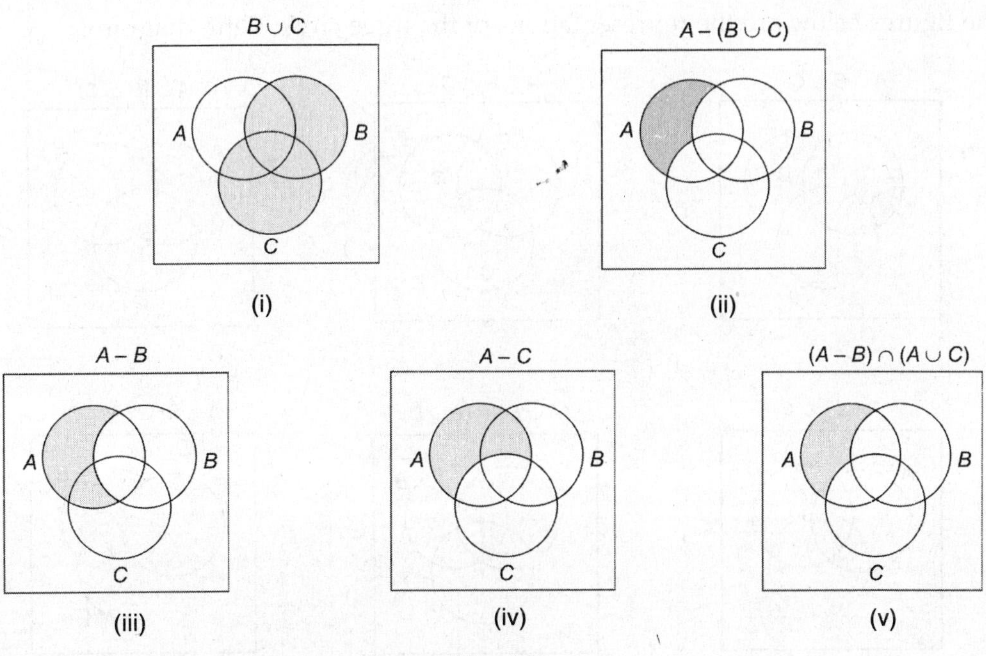

Fig. 1.9

Example 1.21: Verify the law $A \cup (B \cap C) = (A \cup B) \cap (A \cup C)$ using tri Venn diagrams template.

Solution: Using tri Venn diagrams, we find from Fig. 1.10(ii) and (v) that

$$A \cup (B \cap C) = (A \cup B) \cap (A \cup C)$$

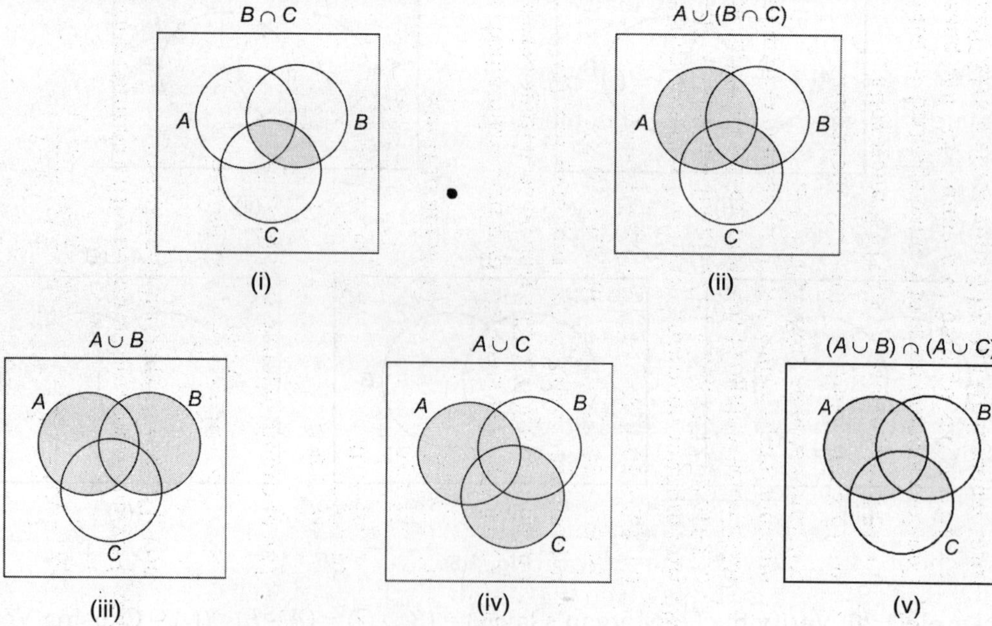

Fig. 1.10

Hence the proof.

The figures below are the representations of the three circle Venn diagrams:

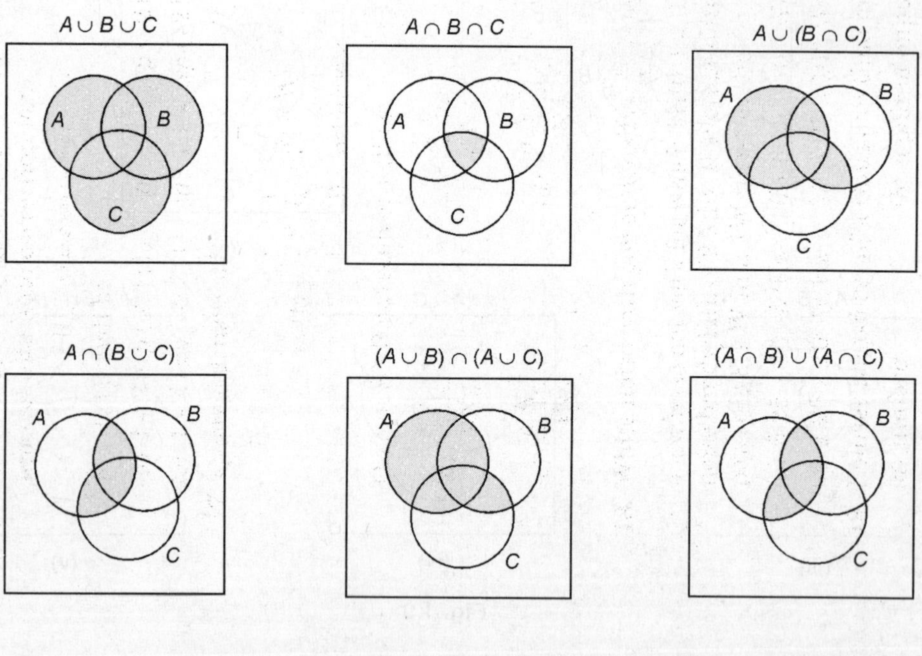

EXERCISE

1. Consider the universal set $S = \{0, 1, 2, 3, 4, 5, 6, 7, 8, 9\}$. Define two subsets $A = \{x : x \leq 8\}$ and $B = \{1, 3, 5, 7, 9\}$. Find: i. A^c ii. A/B iii. $A \cap B$ iv. $A \cup B$
2. Use Venn diagram to show that:
   i. $[A/(A \cap B) \cup B] = A \cup B$
   ii. $(A \cup B \cup C)/(A \cap B \cap C) = (A^c \cap B) \cup (B^c \cap C) \cup (C^c \cap A)$
3. State whether the following statements about the subsets A, B and C of a universal set S are true (T) or false (F).
   (a) $A \cup B \cup S = A \cap B$
   (b) $A \cup B = (A^c \cap B) \cup (A \cap B^c) \cup (A \cap B)$
   (c) $A/B = B/A$
   (d) $A = (A \cap B^c) \cup (A \cap B)$
   (e) $A/B \subseteq A$
   (f) $S^c = \{\phi\}$
4. A maths teacher gave her class two tests. 25% of the class passed both the tests and 42% of the class passed the first test. What percent of those who passed the first test also passed the second test?
5. The probability that it is Friday and that a student is absent is 0.03. Since there are 5 school days in a week, the probability that it is Friday is 0.2. What is the probability that a student is absent given that today is Friday?
6. At Kennedy Middle School, the probability that a student takes technology and Spanish is 0.087. The probability that a student takes technology is 0.68. What is the probability that a student takes Spanish given that the student is taking technology?
7. At a middle school, 18% of all students play football and basketball and 32% of all students play football. What is the probability that a student plays basketball given that the student plays football?
8. In New York state, 48% of all teenagers own a skateboard and 39% of all teenagers own a skateboard and roller blades. What is the probability that a teenager owns roller blades given that the teenager owns a skateboard?
9. In New England, 84% of the houses have a garage and 65% of the houses have a garage and a back yard. What is the probability that a house has a backyard given that it has a garage?
10. A spinner has 4 equal sectors colored yellow, blue, green and red. After spinning the spinner, what is the probability of landing on each color?
11. A single dice is rolled. What is the probability of each outcome? What is the probability of getting an even number and an odd number?
12. A glass jar contains 6 red, 5 green, 8 blue and 3 yellow marbles. If a single marble is chosen at random from the jar, what is the probability of choosing (i) a red marble (ii) a green marble (iii) a blue marble (iv) a yellow marble?
13. Choose a number at random from 1 to 5. What is the probability of each outcome? What is the probability that the number chosen is even? What is the probability that the number chosen is odd?

14. (a) What is an event and explain discrete and continuous events with an example.
    (b) Discuss joint and conditional probability.
    (c) Determine the probability of a card being either red or a queen.
15. (a) Is probability relative frequency of occurrence of some event? Explain with an example.
    (b) Define an event and explain discrete and continuous event with an example.
    (c) One card is drawn from a regular deck of 52 cards. What is the probability of the card being either red or a king?
16. (a) Discuss joint and conditional probability.
    (b) When are two events said to be mutually exclusive? Explain with an example?
    (c) Determine the probability of the card being either red or a king when one card is drawn from a regular deck of 52 cards.
17. (a) Define and explain the following with an example:
       i. Equally likely events
       ii. Exhaustive events
       iii. Mutually exclusive events
    (b) Give the classical definition of probability.
    (c) Find the probability of three half-rupee coins falling all heads up when tossed simultaneously.
18. A batch of 50 items contains 10 defective items. Suppose 10 items are selected at random and tested. What is the probability that exactly 5 of the items tested are defective?
19. Two boxes are selected randomly. The first box contains 2 black balls and 3 white balls. Second box contains 4 white and 6 black balls. What is the probability of drawing a white ball?
20. Three newspapers A, B and C are published in a city and a survey of readers indicates the following: 20% read A, 16% read B and 14% read C, 8% read A and B, 5% read A and C, 2% read A, B and C.

    For one adult chosen at random, compute the probability that:
    i. He reads none of the papers
    ii. He reads exactly one of papers
    iii. He reads at least A and B if it is known that he reads at least one of the papers published.

2

Random Variable, Distribution and Density Functions

2.0 INTRODUCTION

Chapter 1 begins with a physical model of an experiment. An experiment consists of a procedure and observations. The set of all possible observations, S is the sample space of the experiment. S is the beginning of the mathematical probability model. In addition to S, the mathematical model includes a rule for assigning numbers between 0 and 1 to sets A in S. Thus, for every $A \subset S$, the model gives us a probability $P(A)$, where $0 \leq P(A) \leq 1$.

In this chapter and for most of the remaining part of the book, we will examine probability models that assign numbers to the outcomes in the sample space. In application of probability, we are often concerned with numerical values which are random in nature. For example, we may consider the number of customers arriving at a service station at a particular interval of time or the transmission time of a message in a communication system. These random quantities may be considered as real-valued function on the sample space. Such a real-valued function is called a real random variable and plays an important role in describing random data. We shall introduce the concept of random variables in the following sections.

2.1 RANDOM VARIABLES

The outcome of an experiment need not be a number, for example, the outcome when a coin is tossed can be 'head' or 'tail'. However, we often want to represent outcomes as numbers. A random variable is a function that associates a unique numerical value with every outcome of an experiment. The value of the random variable will vary from trial to trial as the experiment is repeated.

Definition: A random variable is a function that maps the sample space S to real numbers. Therefore, when we assign probabilities to the sample space, we automatically assign probabilities that the random variables takes up a particular value. We shall represent a random variable by a capital letter (such as W, X, or Y) and any particular value of a random variable by a lowercase letter (such as w, x, or y).

Conditions for a Function to be a Random Variable

1. Every point in S must correspond to only one value of the random variable.
2. The set $\{X \leq x\}$ shall be an event for any real number x. The satisfaction of this condition is not a trouble in practical problems.

3. The probabilities of the events $\{X = \infty\}$ and $\{X = -\infty\}$ equal 0, i.e.
$P\{X = \infty\} = 0$ and $P\{X = -\infty\} = 0$

This condition does not prevent from being either $-\infty$ or ∞ for some values of s; it only requires that the probability of the set of that s be zero.

There are three types of random variable: *discrete, continuous and mixed.*

A *discrete random variable* is one which may take only a countable number of distinct values such as 0, 1, 2, 3, 4,.... Discrete random variables are usually (but not necessarily) counts. If a random variable can take only a finite number of distinct values, then it must be discrete. Examples of discrete random variables are the number of children in a family, the Friday attendance in a college, the number of defective light bulbs in a box of ten.

A *continuous random variable* is one which takes an infinite number of possible values. Continuous random variables are usually measurements. Consider modeling the distribution of age at which an animal dies, i.e. age of death measured perfectly with all the decimals and no rounding, is a continuous random variable, e.g. age of death could be 27.324574 years. Other examples of continuous random variables include time until the occurrence of the next earthquake in India, the lifetime of an electronic component, the annual rainfall in East Africa, etc.

A *mixed random variable* is one for which some of its values are discrete and some are continuous. It is the least important type of random variable, but it occurs in some problems of practical significance.

2.2 CUMULATIVE DISTRIBUTION FUNCTION (CDF)

The cumulative distribution function is also called the probability distribution function or the distribution function. All random variables (discrete and continuous) have a cumulative distribution function.

The CDF is the accumulation of the probability of all the outcomes up to a given value. In other words, it is a function giving the probability that the random variable X is less than or equal to x, for every value x.

Formally, the cumulative distribution function $F_X(x)$ is defined as follows:

For a discrete random variable, the cumulative distribution function is found by summing up the probabilities.

$$F_X(x) = P(X \le x) \quad \text{for all} \quad -\infty < x < \infty \qquad ...(2.1)$$

For a continuous random variable, the cumulative distribution function is the integraton of its probability density function.

$$F_X(x) = \int_{-\infty}^{x} f_X(x)\,dx \qquad ...(2.2)$$

Properties of Cumulative Distribution Function

i. $F_X(-\infty) = 0$

According to the definition of CDF,

$$F_X(x) = P(X \le x)$$
$$F_X(-\infty) = P(X \le -\infty)$$

But we know that $P(X \le -\infty) = 0$

Hence $F_X(-\infty) = 0$

ii. $F_X(\infty) = 1$

According to the definition of CDF,

$$F_X(x) = P(X \le x)$$
$$F_X(\infty) = P(X \le \infty)$$

But we know that $P(X \le \infty) = 1$

Hence $F_X(\infty) = 1$

iii. $0 \le F_X(x) \le 1$

From above two properties, the maximum and minimum values of CDF are 1 and 0 respectively.

Hence $0 \le F_X(x) \le 1$

iv. $F_X(x_1) \le F_X(x_2)$ if $x_1 < x_2$

Because $F_X(x_2)$ is considering as many or more events in the sample space as $F_X(x_1)$,

$F_X(x_1) \le F_X(x_2)$

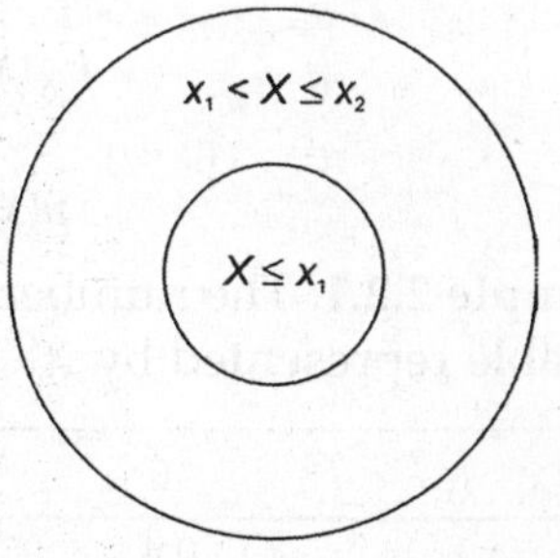

Fig. 2.1: Venn diagram

v. $F_X(x_2) - F_X(x_1) = P(x_1 < X \le x_2)$ if $x_2 > x_1$

From Fig. 2.1, we know that

$$X \le x_2 = (X \le x_1) \cup (x_1 < X \le x_2)$$

Apply probability on both sides,

$$P(X \le x_2) = P\{(X \le x_1) \cup (x_1 < X \le x_2)\}$$
$$P(X \le x_2) = P(X \le x_1) + P(x_1 < X \le x_2)$$
$$P(x_1 < X \le x_2) = P(X \le x_2) - P(X \le x_1)$$
$$P(x_1 < X \le x_2) = F_X(x_2) - F_X(x_1)$$

If X is a discrete random variable, its distribution function $F_X(x)$ must have a stair step form, such as shown in Fig. 2.2. The amplitude of a step will equal the probability

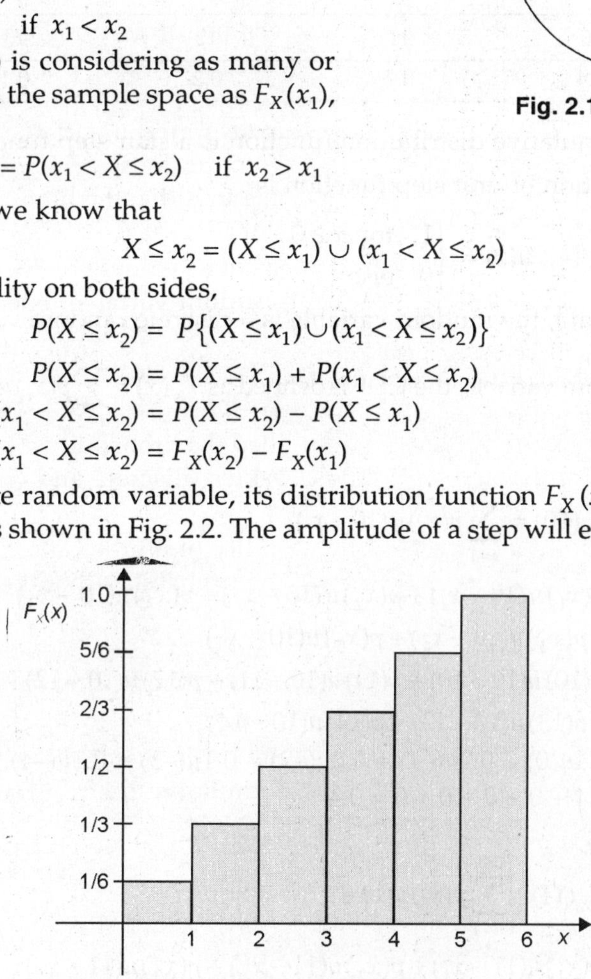

Fig. 2.2: Discrete cumulative distribution function

of occurrence of the value of X where the step occurs. If the values of X are denoted by x_i, we may write $F_X(x)$ as

$$F_X(x) = \sum_{i=1}^{N} P\{X = x_i\} u(x - x_i)$$

where $u(.)$ is the step function defined by

$$u(x) = \begin{cases} 1 & x \geq 0 \\ 0 & x < 0 \end{cases}$$

In shortened notation,

$$F_X(x) = \sum_{i=1}^{N} p(x_i)u(x - x_i)$$

$$p(x_i) = P\{X = x_i\}.$$

Example 2.2.1: The number of patients seen in the ER in any given hour is a random variable represented by X. The probability distribution for X is

$X = x_i$	10	11	12	13	14
$P(X = x_i)$	0.4	0.2	0.2	0.1	0.1

Show that the cumulative distribution function is a stair step function.

Solution: The definition of unit step function is

$$u(x) = \begin{cases} 1 & \text{for } x \geq 0 \\ 0 & \text{else} \end{cases}$$

From the above data, the random variable is a discrete random variable.

For discrete random variable, the CDF is defined as $F_X(x) = \sum_{i=1}^{N} p(x_i)u(x - x_i)$ and $N = 5$.

For $x = 10$

$$\begin{aligned} F_X(x) = F_X(10) &= \sum_{i=1}^{5} p(x_i)u(10 - x_i) \\ &= p(x_1)u(10 - x_1) + p(x_2)u(10 - x_2) + p(x_3)u(10 - x_3) \\ &\quad + p(x_4)u(10 - x_4) + p(x_5)u(10 - x_5) \\ &= p(10)u(10 - 10) + p(11)u(10 - 11) + p(12)u(10 - 12) \\ &\quad + p(13)u(10 - 13) + p(14)u(10 - 14) \\ &= 0.4u(0) + 0.2u(-1) + 0.2u(-2) + 0.1u(-3) + 0.1u(-4) \\ &= 0.4 + 0 + 0 + 0 + 0 = 0.4 \end{aligned}$$

For $x = 11$

$$\begin{aligned} F_X(x) = F_X(11) &= \sum_{i=1}^{5} p(x_i)u(11 - x_i) \\ &= p(x_1)u(11 - x_1) + p(x_2)u(11 - x_2) + p(x_3)u(11 - x_3) \\ &\quad + p(x_4)u(11 - x_4) + p(x_5)u(11 - x_5) \\ &= p(10)u(11 - 10) + p(11)u(11 - 11) + p(12)u(11 - 12) \\ &\quad + p(13)u(11 - 13) + p(14)u(11 - 14) \\ &= 0.4u(1) + 0.2u(0) + 0.2u(-1) + 0.1u(-2) + 0.1u(-3) \\ &= 0.4 + 0.2 + 0 + 0 + 0 = 0.6 \end{aligned}$$

For $x = 12$

$$
\begin{aligned}
F_X(x) = F_X(12) &= \sum_{i=1}^{5} p(x_i)u(12-x_i) \\
&= p(x_1)u(12-x_1)+p(x_2)u(12-x_2)+p(x_3)u(12-x_3) \\
&\quad +p(x_4)u(12-x_4)+p(x_5)u(12-x_5) \\
&= p(10)u(12-10)+p(11)u(12-11)+p(12)u(12-12) \\
&\quad +p(13)u(12-13)+p(14)u(12-14) \\
&= 0.4u(2) + 0.2u(1) + 0.2u(0) + 0.1u(-1) + 0.1u(-2) \\
&= 0.4 + 0.2 + 0.2 + 0 + 0 = 0.8
\end{aligned}
$$

For $x = 13$

$$
\begin{aligned}
F_X(x) = F_X(13) &= \sum_{i=1}^{5} p(x_i)u(13-x_i) \\
&= p(x_1)u(13-x_1)+p(x_2)u(13-x_2)+p(x_3)u(13-x_3) \\
&\quad +p(x_4)u(13-x_4)+p(x_5)u(13-x_5) \\
&= p(10)u(13-10)+p(11)u(13-11)+p(12)u(13-12) \\
&\quad +p(13)u(13-13)+p(14)u(13-14) \\
&= 0.4u(3) + 0.2u(2) + 0.2u(1) + 0.1u(0) + 0.1u(-1) \\
&= 0.4 + 0.2 + 0.2 + 0.1 + 0 = 0.9
\end{aligned}
$$

For $x = 14$

$$
\begin{aligned}
F_X(x) = F_X(14) &= \sum_{i=1}^{5} p(x_i)u(14-x_i) \\
&= p(x_1)u(14-x_1)+p(x_2)u(14-x_2)+p(x_3)u(14-x_3) \\
&\quad +p(x_4)u(14-x_4)+p(x_5)u(14-x_5) \\
&= p(10)u(14-10)+p(11)u(14-11)+p(12)u(14-12) \\
&\quad +p(13)u(14-13)+p(14)u(14-14) \\
&= 0.4u(4) + 0.2u(3) + 0.2u(2) + 0.1u(1) + 0.1u(0) \\
&= 0.4 + 0.2 + 0.2 + 0.1 + 0.1 = 1
\end{aligned}
$$

Hence, for a discrete random variable the distribution function is a stair step function as shown in Fig. 2.3.

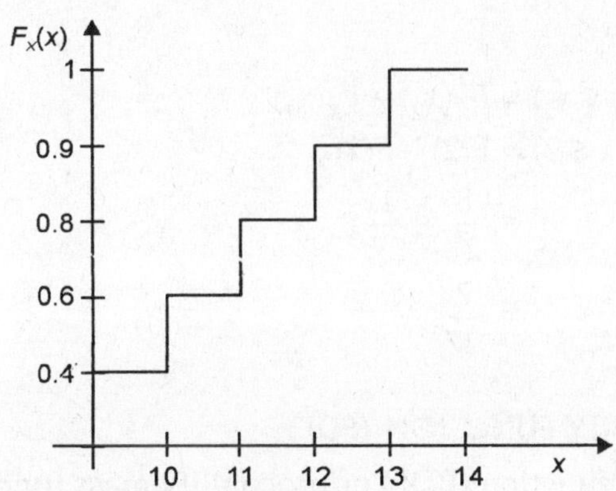

Fig. 2.3

Example 2.2.2: Find the distribution function for the random variable X whose density function is

$$f_X(x) = \begin{cases} \frac{1}{9}x^2 & 0 \le x < 3 \\ 0 & \text{else} \end{cases}$$

Find $P(1 < x \le 2)$.

Solution: $F_X(x) = \int_{-\infty}^{x} f_X(x)dx$

If $x < 0$ then $F_X(x) = 0$

If $0 \le x < 3$ then $F_X(x) = \int_0^x f_X(x)dx$

$$F_X(x) = \int_0^x \frac{1}{9}x^2 dx$$

$$= \frac{x^3}{27}$$

If $x \ge 3$ then $F_X(x) = \int_0^3 f_X(x)dx + \int_3^{\infty} f_X(x)dx$

$$F_X(x) = \int_0^3 \frac{1}{9}x^2 dx + \int_3^{\infty} 0\,dx$$

$$= \left.\frac{x^3}{27}\right|_0^3 = 1$$

Hence $$F_X(x) = \begin{cases} 0 & x \le 0 \\ \frac{x^3}{27} & 0 \le x < 3 \\ 1 & x \ge 3 \end{cases}$$

We know that

$$P(x_1 < x \le x_2) = F_X(x_2) - F_X(x_1)$$
$$P(1 < x \le 2) = F(2) - F(1)$$
$$= \frac{8}{27} - \frac{1}{27}$$
$$= \frac{7}{27}.$$

2.3 PROBABILITY DENSITY FUNCTION (PDF)

The probability density function (PDF) or probability mass function of X is a function which allocates probabilities. Simply, it is a function which tells you the probability of certain events occurring. The probability density function (PDF) is denoted by $f_X(x)$ and expressed as the derivative of the distribution function.

Mathematically, it is expressed as

$$f_X(x) = \frac{dF_X(x)}{dx} \quad ...(2.3)$$

Properties of Probability Density Function

i. $f_X(x) \geq 0$ for all x in the sample space.

ii. $\int_{-\infty}^{\infty} f_X(x)dx = 1$

$$\int_{-\infty}^{\infty} f_X(x)dx = \int_{-\infty}^{\infty} \frac{dF_x(x)}{dx}dx$$

$$= F_X(x)\Big|_{-\infty}^{\infty}$$

$$= F_X(\infty) - F_X(-\infty)$$

$$= 1 - 0 = 1$$

iii. $F_X(x) = \int_{-\infty}^{x} f_X(\xi)d\xi$

$$\int_{-\infty}^{x} f_X(\xi)d\xi = \int_{-\infty}^{x} \frac{d}{d\xi}F_X(\xi)d\xi$$

$$= F_X(\xi)\Big|_{-\infty}^{x}$$

$$= F_X(x) - F_X(-\infty)$$

$$= F_X(x) - 0$$

$$= F_X(x)$$

Hence $F_X(x) = \int_{-\infty}^{x} f_X(\xi)d\xi$

iv. $P(x_1 < X \leq x_2) = \int_{x_1}^{x_2} f_X(x)dx$

$$\int_{x_1}^{x_2} f_X(x)dx = \int_{x_1}^{x_2} \frac{dF_X(x)}{dx}dx = F_X(x)\Big|_{x_1}^{x_2} = F_X(x_2) - F_X(x_1)$$

But according to fifth property of CDF, $P(x_1 < X \leq x_2) = F_X(x_2) - F_X(x_1)$

Hence $P(x_1 < X \leq x_2) = \int_{x_1}^{x_2} f_X(x)dx$

If X is discrete random variable, the cumulative distribution function is

$$F_X(x) = \sum_{i=1}^{N} p(x_i)u(x - x_i)$$

Probability density function

$$f_X(x) = \frac{dF_X(x)}{dx}$$

$$f_X(x) = \frac{d}{dx}\left(\sum_{i=1}^{N} p(x_i)u(x-x_i)\right)$$

$$= \sum_{i=1}^{N} p(x_i)\frac{d}{dx}\{u(x-x_i)\}$$

$$= \sum_{i=1}^{N} p(x_i)\,\delta(x-x_i) \qquad \because \frac{du(x)}{dx} = \delta(x)$$

$$f_X(x) = \sum_{i=1}^{N} p(x_i)\delta(x-x_i).$$

Example 2.3.1: The number of patients seen in the ER in any given hour is a random variable represented by X. The probability distribution for X is given in table below.

$X = x_i$	10	11	12	13	14
$P(X = x_i)$	0.4	0.2	0.2	0.1	0.1

Plot the graph for probability density function.

Solution: The definition of unit step function is

$$\delta(x) = \begin{cases} 1 & \text{for } x = 0 \\ 0 & \text{else} \end{cases}$$

From the above data, it is clear that the random variable is a discrete random variable. For discrete random variable, the PDF is defined as

$$f_X(x) = \sum_{i=1}^{N} p(x_i)\delta(x-x_i) \text{ and } N = 5.$$

For $x = 10$

$$f_X(x) = f_X(10) = \sum_{i=1}^{5} p(x_i)\delta(10-x_i)$$

$$= p(x_1)\delta(10-x_1) + p(x_2)\delta(10-x_2) + p(x_3)\delta(10-x_3) + p(x_4)\delta(10-x_4) + p(x_5)\delta(10-x_5)$$

$$= p(10)\delta(10-10) + p(11)\delta(10-11) + p(12)\delta(10-12) + p(13)\delta(10-13) + p(14)\delta(10-14)$$

$$= 0.4\delta(0) + 0.2\delta(-1) + 0.2\delta(-2) + 0.1\delta(-3) + 0.1\delta(-4)$$

$$= 0.4 + 0 + 0 + 0 + 0 = 0.4$$

For $x = 11$

$$\begin{aligned}
f_X(x) &= f_X(11) = \sum_{i=1}^{5} p(x_i)\delta(11-x_i) \\
&= p(x_1)\delta(11-x_1) + p(x_2)\delta(11-x_2) + p(x_3)\delta(11-x_3) \\
&\quad + p(x_4)\delta(11-x_4) + p(x_5)\delta(11-x_5) \\
&= p(10)\delta(11-10) + p(11)\delta(11-11) + p(12)\delta(11-12) \\
&\quad + p(13)\delta(11-13) + p(14)\delta(11-14) \\
&= 0.4\delta(1) + 0.2\delta(0) + 0.2\delta(-1) + 0.1\delta(-2) + 0.1\delta(-3) \\
&= 0 + 0.2 + 0 + 0 + 0 = 0.2
\end{aligned}$$

For $x = 12$

$$\begin{aligned}
f_X(x) &= f_X(12) = \sum_{i=1}^{5} p(x_i)\delta(12-x_i) \\
&= p(x_1)\delta(12-x_1) + p(x_2)\delta(12-x_2) + p(x_3)\delta(12-x_3) \\
&\quad + p(x_4)\delta(12-x_4) + p(x_5)\delta(12-x_5) \\
&= p(10)\delta(12-10) + p(11)\delta(12-11) + p(12)\delta(12-12) \\
&\quad + p(13)\delta(12-13) + p(14)\delta(12-14) \\
&= 0.4\delta(2) + 0.2\delta(1) + 0.2\delta(0) + 0.1\delta(-1) + 0.1\delta(-2) \\
&= 0 + 0 + 0.2 + 0 + 0 = 0.2
\end{aligned}$$

For $x = 13$

$$\begin{aligned}
f_X(x) &= f_X(13) = \sum_{i=1}^{5} p(x_i)\delta(13-x_i) \\
&= p(x_1)\delta(13-x_1) + p(x_2)\delta(13-x_2) + p(x_3)\delta(13-x_3) \\
&\quad + p(x_4)\delta(13-x_4) + p(x_5)\delta(13-x_5) \\
&= p(10)\delta(13-10) + p(11)\delta(13-11) + p(12)\delta(13-12) \\
&\quad + p(13)\delta(13-13) + p(14)\delta(13-14) \\
&= 0.4\delta(3) + 0.2\delta(2) + 0.2\delta(1) + 0.1\delta(0) + 0.1\delta(-1) \\
&= 0 + 0 + 0 + 0.1 + 0 = 0.1
\end{aligned}$$

For $x = 14$

$$\begin{aligned}
f_X(x) &= f_X(14) = \sum_{i=1}^{5} p(x_i)\delta(14-x_i) \\
&= p(x_1)\delta(14-x_1) + p(x_2)\delta(14-x_2) + p(x_3)\delta(14-x_3) \\
&\quad + p(x_4)\delta(14-x_4) + p(x_5)\delta(14-x_5) \\
&= p(10)\delta(14-10) + p(11)\delta(14-11) + p(12)\delta(14-12) \\
&\quad + p(13)\delta(14-13) + p(14)\delta(14-14) \\
&= 0.4\delta(4) + 0.2\delta(3) + 0.2\delta(2) + 0.1\delta(1) + 0.1\delta(0) \\
&= 0 + 0 + 0 + 0 + 0.1 = 0.1
\end{aligned}$$

The graph for probability density function is shown in Fig. 2.4.

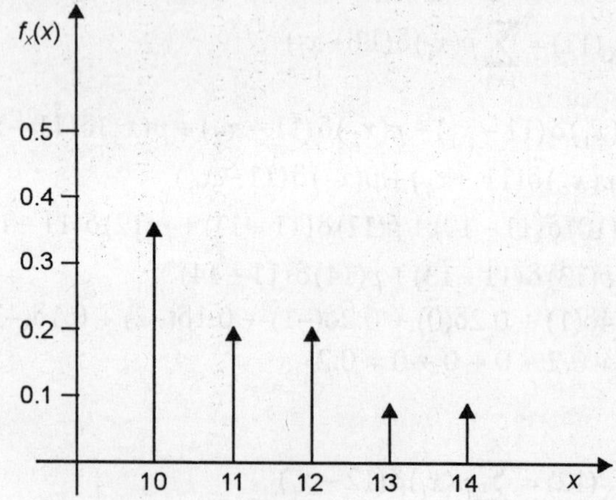

Fig. 2.4

Example 2.3.2: Determine if the following function is a valid distribution function or not.

$$F_X(x) = \begin{cases} 1 - e^{-\frac{x}{2}} & x \geq 0 \\ 0 & x < 0 \end{cases}$$

Solution: $F_X(x)$ is a valid distribution function only if $f_X(x)$ is valid density function.

$$f_X(x) = \frac{dF_X(x)}{dx} = -\left(-\frac{1}{2}\right)e^{-\frac{x}{2}} = \left(\frac{1}{2}\right)e^{-\frac{x}{2}}$$

The condition for valid density function is

$$\int_{-\infty}^{\infty} f_X(x)dx = 1$$

$$\int_{-\infty}^{\infty} f_X(x)dx = \int_{-\infty}^{0} f_X(x)dx + \int_{0}^{\infty} f_X(x)dx$$

$$= \int_{-\infty}^{0} 0\,dx + \int_{0}^{\infty}\left(\frac{1}{2}\right)e^{-\frac{x}{2}}dx$$

$$= \left(\frac{1}{2}\right)\frac{e^{-\frac{x}{2}}}{-\frac{1}{2}}\Bigg|_0^{\infty}$$

$$= -e^{-\infty} + e^0 = 1$$

Hence, the density function is valid and so is the distribution function.

Note: If the density function is valid, then the distribution function is also valid.

2.4 SOME IMPORTANT DISTRIBUTION AND DENSITY FUNCTIONS

i. Discrete distributions:
   (a) Bernoulli distribution
   (b) Binomial distribution
   (c) Poisson distribution

ii. Continuous distributions:
   (a) Uniform distributions
   (b) Exponential distribution
   (c) Normal distribution
   (d) Rayleigh distribution

2.4.1 Discrete Distributions

Bernoulli Distribution

The *Bernoulli trials process,* named after Jacob Bernoulli, is one of the simplest yet most important random variable in probability. This distribution best describes all situations where a "trial" is made resulting in either "success" or "failure," such as when tossing a coin, or when modeling the success or failure of a surgical procedure. The Bernoulli distribution is defined as:

$$f_X(x) = p^x (1-p)^{1-x} \qquad \text{for } x = 0, 1 \qquad \text{...(2.4)}$$

where p is the probability that a particular event (e.g. success) will occur.

A sequence of *Bernoulli trials* satisfies the following assumptions:

i. Each trial has two possible outcomes, in the language of reliability called *success* and *failure.*
ii. The trials are independent. Intuitively, the outcome of one trial has no influence over the outcome of another trial.
iii. On each trial, the probability of success is p and the probability of failure is $1-p$ where $p \in [0, 1]$ is the *success parameter* of the process.

Example 2.4.1: Suppose that a student takes a multiple choice test. The test has 20 questions, each of which has 4 possible answers (only one correct). If the student blindly guesses the answer to each question, do the questions form a sequence of Bernoulli trials? If so, identify the trial outcomes and the parameter p.

Solution: Yes. The outcomes are *correct* and *incorrect* and $p = \frac{1}{4}$.

Example 2.4.2: A fair coin is tossed. Let the variable x take values 1 and 0 according to whether the toss results in head or tail. Then, X is a Bernoulli variable with parameter $p = \frac{1}{2}$. Here, X denotes the number of heads obtained in the toss.

Probability of success $= \frac{1}{2}$ and the probability of failure $= \frac{1}{2}$.

Example 2.4.3: In a single throw of a dice, the outcome "5" is called a success and any other outcome is called a failure, then the successive throws of a dice will contain Bernoulli trials. The probability of success $= \frac{1}{6}$ and the probability of failure $= \frac{5}{6}$.

Binomial Distribution

Binomial distribution is a discrete probability distribution. A binomial experiment is one that possesses the following properties:

i. The experiment consists of n repeated trials.

ii. Each trial results in an outcome that may be classified as a **success** or a **failure** (hence the name **binomial**).

iii. The probability of success, denoted by p, remains constant from trial to trial and repeated trials are independent.

The number of successes X in n trials of a binomial experiment is called a **binomial random variable** and it is denoted by $B(n, p)$.

The probability distribution of the random variable X is called a **binomial distribution**, and is given by

$$F_X(x) = \sum_{k=0}^{n} {}^nC_k\, p^k\, q^{(n-k)} u(x-k) \qquad \text{...(2.5)}$$

The density function is defined as

$$f_X(x) = \sum_{k=0}^{n} {}^nC_k\, p^k\, q^{(n-k)} \delta(x-k) \qquad \text{...(2.6)}$$

where n = the number of trials

p = the probability of success in a single trial

q = the probability of failure in a single trial (i.e. $q = 1 - p$).

Note: ${}^nC_k = \dfrac{n!}{k!(n-k)!}$

Example 2.4.4: Consider an exam that contains 10 multiple choice questions with 4 possible choices for each question, only one of which is correct.

i. What is the probability that the student gets no answer correct?

ii. What is the probability that the student gets two answers correct?

iii. What is the probability that the student fails in the test (i.e. to have less than 6 correct answers)?

Solution: Suppose a student is to select the answer for every question randomly. Let X be the number of questions the student answers correctly. Then, X has a binomial distribution with parameters $n = 10$, $p = 0.25$ and $q = 0.75$.

$$P(X = k) = {}^nC_k\, p^k\, q^{(n-k)}$$

i. $P(X = 0) = {}^{10}C_0\, p^0\, q^{(10-0)}$

$$= \frac{10!}{0!(10-0)!}(0.25)^0\,(0.75)^{10}$$

$$= (0.75)^{10}$$

$$= 0.0563$$

ii. $P(X=2) = {}^{10}C_2 p^2 q^{(10-2)}$

$$= \frac{10!}{2!(10-2)!}(0.25)^2(0.75)^8$$

$$= 45\,(0.25)^2\,(0.75)^8 = 0.2816$$

iii. $P(X \le 5) = \sum_{i=0}^{5} P(X = x_i)$

$$= P(X=x_1)+P(X=x_2)+P(X=x_3)+P(X=x_4)+P(X=x_5)$$

$$= 0.0563 + 0.1877 + 0.2816 + 0.2503 + 0.1460 + 0.0584 = 0.9803.$$

Example 2.4.4: Hospital records show that out of patients suffering from a certain disease, 75% of them die. What is the probability that out of 6 randomly selected patients, 4 will recover?

Solution: This is a binomial distribution because there are only 2 outcomes (the patient will live or die).

Let X = number of patients who recover.

Here, $n = 6$ and $k = 4$. Let $p = 0.25$ (success, i.e. they live), $q = 0.75$ (failure, i.e. they die).

The probability that 4 will recover is:

$$P(X) = {}^nC_k\, p^k\, q^{(n-k)}$$

$$P(X) = {}^6C_4\,(0.25)^4\,(0.75)^{(6-4)}$$

$$= {}^6C_4\,(0.25)^4\,(0.75)^{(6-4)}$$

$$= 15 \times 2.1973 \times 10^{-3}$$

$$= 0.0329595.$$

Example 2.4.5: A (blindfolded) marksman finds that on an average he hits the target 4 times out of 5. If he fires 4 shots, what is the probability of:

(a) more than 2 hits?

(b) at least 3 misses?

Solution: Here, $n = 4$, $p = 0.8$, $q = 0.2$.

Let X = number of hits

k_0 = no hits, k_1 = 1 hit, k_2 = 2 hits, etc.

(a) $P(X) = P(k=3) + P(k=4)$

$$P(X) = {}^nC_k\, p^k\, q^{(n-k)}$$

$$P(X) = {}^4C_3\,(0.8)^3\,(0.2)^{(4-3)} + {}^4C_4\,(0.8)^4\,(0.2)^{(4-4)}$$

$$= 4(0.8)^3\,(0.2) + (0.8)^4$$

$$= 0.8192$$

(b) 3 misses means 1 hit, and 4 misses means 0 hits.

$$P(X) = P(k=1) + P(k=0)$$

$$P(X) = {}^nC_k\, p^k\, q^{(n-k)}$$

$$P(X) = {}^4C_1\,(0.8)^1\,(0.2)^{(4-1)} + {}^4C_0\,(0.8)^0\,(0.2)^{(4-0)}$$

$$= 4(0.8)^1\,(0.2)^3 + (0.2)^4 = 0.0272.$$

Example 2.4.6: A manufacturer of metal pistons finds that on an average, 12% of his pistons are rejected because they are either oversize or undersize. What is the probability that a batch of 10 pistons will contain:

(a) not more than 2 rejects?

(b) at least 2 rejects?

Solution: Let X = number of rejected pistons (in this case, success means rejection)

Here, $n = 10, p = 0.12, q = 0.88$.

(a) No rejects, i.e. $k = 0$

$$P(X) = {}^nC_k\, p^k\, q^{(n-k)}$$

$$P(X) = {}^{10}C_0\,(0.12)^0\,(0.88)^{(10-0)} = 0.2785$$

One reject, i.e. $k = 1$

$$P(X) = {}^{10}C_1\,(0.12)^1\,(0.88)^{(10-1)} = 0.37977$$

Two rejects, i.e. $k = 2$

$$P(X) = {}^{10}C_2\,(0.12)^2\,(0.88)^{(10-2)} = 0.23304$$

So the probability of getting not more than 2 rejects is

$$P(X \le 2) = 0.2785 + 0.37977 + 0.23304 = 0.89131$$

(b) We could work out all the cases for X = 2, 3, 4,..., 10, but it is much easier to proceed as follows:

Probability of at least 2 rejects

$$= 1 - P(X \le 1)$$
$$= 1 - P(k = 0) + P(k = 1)$$
$$= 1 - (0.2785 + 0.3797)$$
$$= 0.34173.$$

Poisson Distribution

The Poisson distribution was developed by the French mathematician Simeon Denis Poisson in 1837. The Poisson distribution is another family of distribution that arises in a great number of business situations. It is usually applicable in situations where random events occur at a certain rate over a period of time.

We can apply Poisson distribution to counting applications, like the hourly number of customers arriving at a railway booking counter, the daily number of emergency calls in a district, the daily number of accidents on a particular stretch of highway and monthly demands for a particular product.

Like binomial distribution, Poisson distribution arises when a set of canonical assumptions are reasonably valid. These are:

i. The number of events that occur in any time interval is independent of the number of events in any other disjoint interval. Here, time interval is the standard example of an exposure variable and other interpretations are possible, e.g. error rate per page of a book.

ii. The distribution of number of events in an interval is the same for all intervals of the same size.

iii. For a small time interval, the probability of observing an event is proportional to the length of the interval. The proportionality constant corresponds to the rate at which the event occurs.

iv. The probability of observing two or more events in an interval approaches zero as the interval becomes smaller.

Under the above assumptions, let λ be the rate at which events occur, t be the length of a time interval, and X be the total number of events in that time interval. Then, X is called a Poisson random variable and the probability distribution of X is called the Poisson distribution.

Let $\mu = \lambda t$, then μ can be interpreted as the average, or mean, number of events in an interval of length t.

The Poisson random variable X has the density and distribution functions as follows:

$$f_X(x) = e^{-\mu}\sum_{k=0}^{\infty}\frac{\mu^k}{k!}\delta(x-k) \qquad \text{...(2.7)}$$

$$F_X(x) = e^{-\mu}\sum_{k=0}^{\infty}\frac{\mu^k}{k!}u(x-k) \qquad \text{...(2.8)}$$

where $\mu > 0$ is a real constant. When plotted, these functions appear quite similar to those for the binomial variable. In fact, if $n \to \infty$ and $p \to 0$ for the binomial case in such a way that $np = \mu$, a constant, Poisson case results.

Example 2.4.7: A life insurance salesman sells on an average 3 life insurance policies per week. Use Poisson's law to calculate the probability that in a given week, he will sell:

(a) some policies

(b) 2 or more policies but less than 5 policies

(c) assuming that there are 5 working days per week, what is the probability that in a given day he will sell one policy?

Solution: Here $\mu = 3$

(a) "Some policies" means "1 or more policies". We can work this out by finding 1 minus the "zero policies" probability, i.e.

$$P(X > 0) = 1 - P(k = 0)$$

Now
$$P(X) = e^{-\mu}\frac{\mu^k}{k!}$$

so
$$P(X) = e^{-3}\frac{3^0}{0!} = 4.9787 \times 10^{-2}$$

$\therefore$
$$\begin{aligned} P(X > 0) &= 1 - P(k = 0) \\ &= 1 - 4.9787 \times 10^{-2} \\ &= 0.95021 \end{aligned}$$

(b)
$$P(2 \le X < 5) = P(k = 2) + P(k = 3) + P(k = 4)$$

$$P(X) = e^{-3}\frac{3^2}{2!} + e^{-3}\frac{3^3}{3!} + e^{-3}\frac{3^4}{4!} = 0.61611$$

(c) Average number of policies sold per day $= \frac{3}{5} = 0.6$

So on a given day, $P(X) = e^{-0.6}\frac{(0.6)^1}{1!} = 0.32929.$

Example 2.4.8: Vehicles pass through a junction on a busy road at an average rate of 300 per hour.

(a) Find the probability that none passes in a given minute.

(b) What is the expected number passing in two minutes?

(c) Find the probability that this expected number actually pass through in a given two-minute period.

Solution: The average number of cars per minute is $\mu = \frac{300}{60} = 5$

(a) $P(k = 0) = e^{-5}\frac{(5)^0}{0!} = 6.7379 \times 10^{-3}$

(b) $5 \times 2 = 10$

(c) Now, with $\mu = 10$, we have $P(k = 10) = e^{-10}\frac{(10)^{10}}{10!} = 0.12511.$

2.4.2 Continuous Distributions

Uniform Probability Density

The uniform distribution is the simplest example of a continuous probability distribution. The uniform or rectangular distribution has random variable X restricted to a finite interval $[a, b]$ and $f(x)$ has constant density over the interval. The PDF and its corresponding CDF are

$$f_X(x) = \begin{cases} \frac{1}{b-a} & a \le x \le b \\ 0 & \text{elsewhere} \end{cases} \quad ...(2.9)$$

$$F_X(x) = \begin{cases} 0 & x < a \\ \frac{x-a}{b-a} & a \le x < b \\ 1 & x \ge b \end{cases} \quad ...(2.10)$$

The plots for the PDF and CDF of a uniform random variable are shown in Fig. 2.5. We can create many other random variables of interest, by performing a transformation on this uniform random variable.

Applications: Some continuous random variables in the physical, management, and biological sciences have approximately uniform probability distributions.

1. An example of a uniform random variable would be the phase of a radio frequency sinusoid in a communication system. Although the transmitter knows the phase of the sinusoid, the receiver may have no information about the phase. In this case, the phase at the receiver could be modeled as a random variable uniformly distributed over the interval $[0, 2\pi]$.

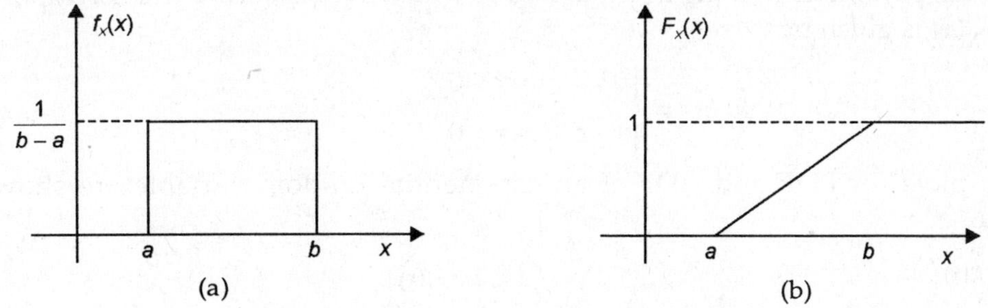

Fig. 2.5: (a) Probability density function and (b) Cumulative distribution function of a uniform random variable

2. Most computer random number generators will generate a random variable that closely approximates a uniform random variable over the interval (0, 1).
3. Suppose we are counting events that have a Poisson distribution, such as telephone calls coming into a switch board. If it is known that exactly one such event has occurred in a given interval, say (0, *t*), then the actual time of occurrence is distributed uniformly over this interval.
4. The errors introduced in the round-off process in quantization are uniformly distributed. If a signal is discritized into steps of Δ, then the quantization error is uniformly distributed between $\frac{-\Delta}{2}$ and $\frac{\Delta}{2}$.

Example 2.4.9: The arrival of customers at a certain checkout counter follow Poisson distribution. It is known that during a given 30-minute period, one customer arrived at the counter. Find the probability that the customer arrived during the last 5 minutes of the 30-minute period.

Solution: As just mentioned, the actual time of arrival follows a uniform distribution over the interval (0, 30). If X denotes the arrival time, then

$$P(25 \le X \le 30) = \int_{25}^{30} \frac{1}{30-0} dx = \frac{1}{30} x|_{25}^{30} = \frac{30-25}{30} = \frac{1}{6}$$

Example 2.4.10: Suppose the amount of gasoline sold daily at a service station is uniformly distributed with a minimum of 2,000 gallons and a maximum of 5,000 gallons. What is the probability that daily sales will fall between 2,500 gallons and 3,000 gallons?

Solution: $P(2500 \le X \le 3000) = \int_{2500}^{3000} \frac{1}{5000-2000} dx = \frac{1}{3000} x|_{2500}^{3000} = \frac{3000-2500}{3000} = \frac{1}{6}$

Exponential Probability Density Function

Exponential distributions are widely used in queueing theory, reliability theory, and communication theory. The PDF of an exponential distribution is given by

$$f_X(x) = \begin{cases} 0 & x \le 0 \\ \lambda e^{-\lambda x} & x > 0 \end{cases} \qquad ...(2.11)$$

where the parameter λ is the average rate at which events occur. The corresponding CDF $F_X(x)$ is given by

$$F_X(x) = \begin{cases} 0 & x \le 0 \\ 1 - e^{-\lambda x} & x > 0 \end{cases} \qquad ...(2.12)$$

The plots for PDF and CDF of an exponential random variable are shown in Fig. 2.6.

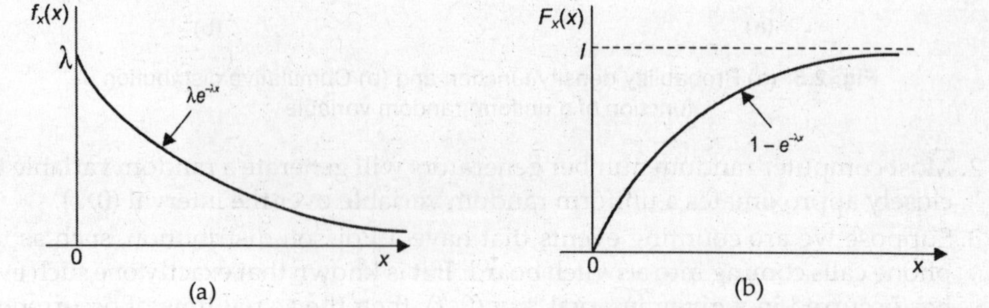

Fig. 2.6: Exponential random variable: (a) Density function (b) Distribution function

Applications: The time between two consecutive occurrences of independent events (*waiting time*) can be modeled by the exponential RV. For example, the exponential distribution gives probability density function of the time between two successive counts in Poisson distribution. Exponential random variables are commonly encountered in the study of queueing systems. The exponential random variable is often used to model lifetimes, such as how long a cell phone call lasts or how long it takes a computer network to transmit a message from one node to another.

An important fact is that when time between random "events" follow the exponential distribution with rate λ, then the total number of events in a time period of length t follows the Poisson distribution with parameter λt.

Example 2.4.11: Assume that the length of phone calls made at a particular telephone booth is exponentially distributed with a mean of 4 minutes. If you arrive at the telephone booth just when Ravi was about to make a call, find the following:

(a) The probability that you will wait for more than 5 minutes before Ravi is done with the call.

(b) The probability that Ravi's call will last between 2 and 6 minutes.

Solution: Let X be a random variable that denotes the length of calls made at the telephone booth. Since the mean length of calls is $1/\lambda = 4$, the PDF of X is given by

$$f_X(x) = \lambda e^{-\lambda x} = \frac{1}{4} e^{-x/4}$$

(a) The probability that you will wait for more than 5 minutes is the probability that X is greater than 5 minutes, which is given by

$$P(X > 5) = \int_5^\infty \frac{1}{4} e^{-x/4}\, dx = [-e^{-x/4}]_5^\infty = e^{-5/4} = 0.2865$$

(b) The probability that the call lasts between 2 and 6 minutes is given by

$$P(2 \le X \le 6) = \int_2^6 \frac{1}{4} e^{-x/4}\, dx = \left[-e^{-x/4}\right]_2^6 = e^{-2/4} - e^{-6/4} = 0.3833.$$

Gaussian (Normal) Distribution

The normal distribution, also called the Gaussian distribution, plays a basic role in error calculus. The normal distribution is a bell-shaped, continuous and symmetrical distribution (the coefficient of skew is zero). This is the most important distribution of all because of the *central limit theorem*, which states that the sum (or average) of a large number of independent random variables is approximately normal, no matter what the original distribution looks like.

A random variable X called Gaussian density function has the form

$$f_X(x) = \frac{1}{\sqrt{2\pi\sigma^2}} e^{-(x-\mu)^2/2\sigma^2} \qquad \text{...(2.13)}$$

where μ is the average (mean) value and σ stands for the standard deviation.

The CDF of the Gaussian random variable is

$$F_X(x) = \frac{1}{\sqrt{2\pi\sigma^2}} \int_{-\infty}^{x} e^{-(\xi-\mu)^2/2\sigma^2} d\xi \qquad \text{...(2.14)}$$

The plots for PDF and CDF of a Gaussian random variable are shown in Fig. 2.7

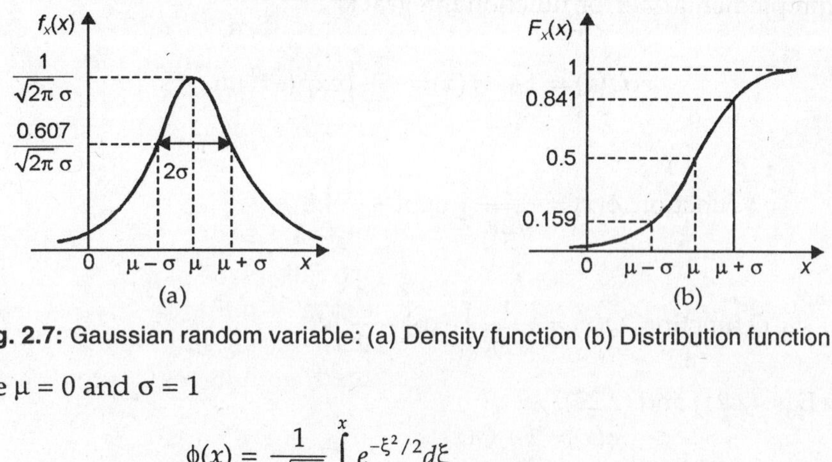

Fig. 2.7: Gaussian random variable: (a) Density function (b) Distribution function

Assume μ = 0 and σ = 1

$$\phi(x) = \frac{1}{\sqrt{2\pi}} \int_{-\infty}^{x} e^{-\xi^2/2} d\xi \qquad \text{...(2.15)}$$

This is called normalized or standard Gaussian distribution $N(0, 1)$.

From the definition of the Gaussian distribution,

$$F_X(x) = \frac{1}{\sqrt{2\pi\sigma^2}} \int_{-\infty}^{x} e^{-(\xi-\mu)^2/2\sigma^2} d\xi$$

Let $z = (\xi - \mu)/\sigma$ and $dz = d\xi/\sigma$. Then $d\xi = \sigma dz$ and the CDF of X becomes

$$F_X(x) = \frac{1}{\sqrt{2\pi}} \int_{-\infty}^{(x-\mu)/\sigma} e^{-(z)^2/2} dz \qquad \text{...(2.16)}$$

Thus, with the above transformation, X becomes a standard normal random variable. The above integral cannot be evaluated in closed form. It is usually evaluated numerically through the function $\phi(x)$, which is defined as follows:

$$\phi(x) = \frac{1}{\sqrt{2\pi}} \int_{-\infty}^{x} e^{-\xi^2/2} d\xi$$

Thus, the CDF of X is given by

$$F_X(x) = \frac{1}{\sqrt{2\pi}} \int_{-\infty}^{(x-\mu)/\sigma} e^{-(z)^2/2} dz = \phi\left(\frac{x-\mu}{\sigma}\right) \quad ...(2.17)$$

The values of $\phi(x)$ are usually given for non-negative values of x. For negative values of x, $\phi(x)$ can be obtained from the following relationship:

$$\phi(-x) = 1 - \phi(x) \quad ...(2.18)$$

Whether the Gaussian CDF is to be evaluated by using a table or a program, the required CDF must be converted into one of the few commonly used standard forms. A few of these common forms are as follows:

i. Error function integral,

$$erf(x) = \frac{2}{\sqrt{\pi}} \int_{-\infty}^{x} \exp(-t^2) dt \quad ...(2.19)$$

ii. The complementary error function integral is

$$erfc(x) = 1 - erf(x) = \frac{2}{\sqrt{\pi}} \int_{x}^{\infty} \exp(-t^2) dt \quad ...(2.20)$$

$$\phi \text{ function, } \phi(x) = \frac{1}{\sqrt{2\pi}} \int_{-\infty}^{x} \exp\left(-\frac{t^2}{2}\right) dt \quad ...(2.21)$$

$$Q \text{ function, } Q(x) = \frac{1}{\sqrt{2\pi}} \int_{x}^{\infty} \exp\left(-\frac{t^2}{2}\right) dt \quad ...(2.22)$$

From Eqs. (2.21) and (2.22),

$$\phi(x) = 1 - Q(x) \quad ...(2.23)$$

If table of $\phi(x)$ is not given, $Q(x)$ is defined as

$$Q(x) = \left[\frac{1}{(1-a)x + a\sqrt{x^2 + b}}\right] \frac{e^{-x^2/2}}{\sqrt{2\pi}} \quad ...(2.24)$$

where $a = 0.339$ and $b = 5.510$.

Some important features of Gaussian distribution are as follows:

i. Each normal distribution can be differentiated and represented by its mean and standard deviation.

ii. The highest point on the curve is the mean median and mode.

iii. It can theoretically have values from negative infinity to positive infinity.

Some of the applications of Gaussian random variables can be seen in the following cases:

i. Physical attributes of a large population such as height and weight.
ii. Exam scores of a large class of students.
iii. Errors in measurement or production processes can often be approximated by a normal distribution. In particular, since the macroscopic noise current measured in a circuit results from the sum of forces of many independent collisions on an atomic scale, noise current is well described by the Gaussian density. For this reason, Gaussian random variables are the noise model of choice in electronic communication and control systems.

Example 2.4.11: Suppose the time required to build a motor cycle is normally distributed with a mean of 50 minutes and a standard deviation of 10 minutes. What is the probability for the assembly time of a computer to be between 40 and 60 minutes?

Solution: We wish to compute $P(40 < X \leq 60)$. To do this, we first rewrite the event of interest into a form that is in terms of a standardized variable $Z = (X - 50)/10$, as follows:

$$F_X(x) = \phi\left(\frac{x-\mu}{\sigma}\right)$$

$$P(40 < X \leq 60) = F_X(60) - F_X(40) = \phi\left(\frac{60-50}{10}\right) - \phi\left(\frac{40-50}{10}\right)$$

$$= \phi(1) - \phi(-1)$$

$$\text{But } \phi(-x) = 1 - \phi(x)$$

$$= \phi(1) - 1 + \phi(1)$$

$$= 0.8413 - 1 + 0.8413 = 0.6826$$

Rayleigh Probability Density Function

The original derivation of this density function (by Lord Rayleigh in 1880) was applied to the envelope of the sum of many sine waves of different frequencies.

The Rayleigh density and distribution functions are:

$$f_X(x) = \begin{cases} \frac{2}{b}(x-a)e^{-(x-a)^2/b} & x \geq a \\ 0 & x < a \end{cases} \qquad \text{...(2.25)}$$

$$F_X(x) = \begin{cases} 1 - e^{-(x-a)^2/b} & x \geq a \\ 0 & x < a \end{cases} \qquad \text{...(2.26)}$$

The density function is plotted as shown in Fig. 2.6. Note that the maximum value of the density function is at $a+\sqrt{\frac{b}{2}}$, but the density function is not symmetrical about this maximum point.

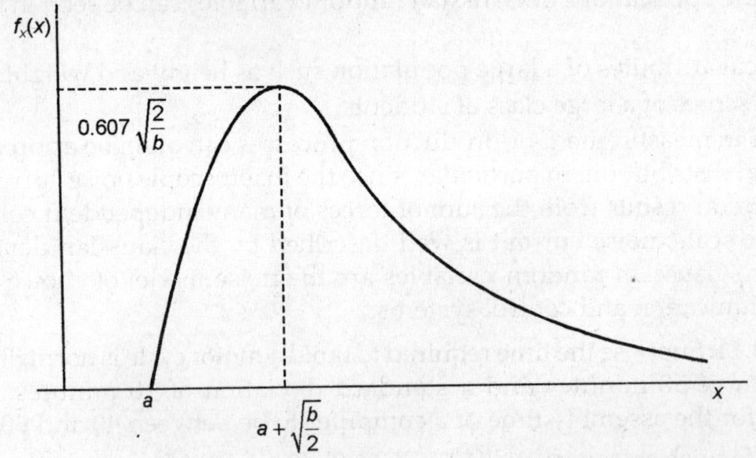

Fig. 2.8: Rayleigh density function

Relation between the Rayleigh Distribution and the Gaussian Distribution

A Rayleigh RV is related to a Gaussian RV as follows: If $X_1 \sim N(0, \sigma^2)$ and $X_2 \sim N(0, \sigma^2)$ are independent, then the envelope $X = \sqrt{X_1^2 + X_2^2}$ has the Rayleigh distribution with the parameter σ.

Applications of the Rayleigh RV

i. The Rayleigh probability density function arises in several different physical situations. For example, the peak values (i.e. the envelope) of a random voltage or current having a Gaussian probability density function will follow the Rayleigh density function.
ii. The Rayleigh distribution is frequently used to model wave heights in oceanography, and in communication theory to describe hourly median and instantaneous peak power of received radio signals. It has been used to model the frequency of different wind speeds over a year at wind turbine sites.
iii. Modeling the *root mean square error.*
iv. Models magnitude variations of a signal (called fading) in a wireless communication channel, assuming there is no significant line-of-sight (LoS) propagation.
v. Modeling the envelope of a signal with two *orthogonal components* as in the case of a signal of the form: $s(t) = X_1 \cos \omega t + Y_1 \sin \omega t$.

2.5 CONDITIONAL DISTRIBUTION AND DENSITY FUNCTION

We know that, for two events A and B where $P(B) \neq 0$, the conditional probability of A given B had occurred is

$$P\left(\frac{A}{B}\right) = \frac{P(A \cap B)}{P(B)} \qquad ...(2.27)$$

In this section we extend the same conditional probability concept to random variables.

Conditional Distribution Function

Let A be identified as the event $\{X \le x\}$ for the variable X. The resulting probability $P\{X \le x/B\}$ is defined as the conditional distribution function of X, which we denote by

$$F_X(x/B) = P\{X \le x/B\} = \frac{P\{X \le x, B\}}{P(B)} \qquad ...(2.28)$$

The probability distribution function $P\{X \le x, B\}$ is the joint probability of the events $\{X \le x\}$ and B, that is, the intersection of the event $\{X \le x\}$ and event B.

Mathematically, it is expressed as

$$P\{X \le x, B\} = P\{X \le x \cap B\} \qquad ...(2.29)$$

Hence Eq. (2.28) can be rewritten as

$$F_X(x/B) = P\{X \le x/B\} = \frac{P\{X \le x \cap B\}}{P(B)} \qquad ...(2.30)$$

Properties of Conditional Distribution Function

All the properties of ordinary distribution apply to $F_X(x/B)$.

i. $F_X(-\infty/B) = P\{X \le -\infty/B\} = \dfrac{P\{X < -\infty \cap B\}}{P(B)} = \dfrac{0}{P(B)} = 0$

ii. $F_X(\infty/B) = P\{X \le \infty/B\} = \dfrac{P\{X \le \infty \cap B\}}{P(B)} = \dfrac{P(B)}{P(B)} = 1$

iii. $0 \le F_X(x/B) \le 1$

iv. $F_X(x_1/B) = F_X(x_2/B)$, if $x_1 < x_2$, it states that it is a nondecreasing function.

v. $P\{x_1 < X \le x_2/B\} = F_X(x_2/B) - F_X(x_1/B)$.

Conditional Density Function

The conditional density function is defined as

$$f_X(x/B) = \frac{dF_X(x/B)}{dx} \qquad ...(2.31)$$

Properties of Conditional Density Function

i. $f_X(x/B) \ge 0$

ii. $\displaystyle\int_{-\infty}^{\infty} f_X(x/B)dx = 1$

iii. $F_X(x/B) \displaystyle\int_{-\infty}^{x} f_X(\xi/B)d\xi$

iv. $P\{x_1 < X \le x_2/B\} = \displaystyle\int_{-\infty}^{\infty} f_X(x/B)dx.$

2.6 METHODS OF DEFINING CONDITIONING EVENT

Let B be identified as the event $\{X \le b\}$ for the variable X, where b is some real number, $\{-\infty < b \le \infty\}$.

From the definition of conditional distribution,

$$F_X(x/B) = F_X\{x/X \le b\} = P\{X \le x/X \le b\} = \frac{P\{X \le x \cap X \le b\}}{P(X \le b)} \quad ...(2.32)$$

For all the events $\{X \le b\}$, for which $P(X \le b) \ne 0$.
Two cases must be considered

Case 1. When $b \le X$
The event $\{X \le b\}$ is a subset of the event $\{X \le x\}$, so $\{X \le x\} \cap \{X \le b\} = \{X \le b\}$

$$F_X\{x/X \le b\} = \frac{P\{X \le x \cap X \le b\}}{P(X \le b)} = \frac{P(X \le b)}{P(X \le b)} = 1 \quad ...(2.33)$$

Case 2. When $X < b$
The event $\{X \le x\}$ is a subset of the event $\{X \le b\}$, so $\{X \le x\} \cap \{X \le b\} = \{X \le x\}$

$$F_X\{x/X \le b\} = \frac{P\{X \le x \cap X \le b\}}{P(X \le b)} = \frac{P(X \le x)}{P(X \le b)} = \frac{F_X(x)}{F_X(b)} \quad ...(2.34)$$

By combining Eqs. (2.33) and (2.34),

$$F_X\{x/X \le b\} = \begin{cases} \dfrac{F_X(x)}{F_X(b)} & x < b \\ 1 & b \le x \end{cases} \quad ...(2.35)$$

Similarly,
$$f_X\{x/X \le b\} = \begin{cases} \dfrac{f_X(x)}{F_X(b)} = \dfrac{f_X(x)}{\int_{-\infty}^{x} F_X(b)\,dx} & x < b \\ 1 & b \le x \end{cases} \quad ...(2.36)$$

From Eq. (2.35), the conditional distribution function is never smaller than normal distribution function.

$$F_X\{x/X \le b\} \ge F_X(x) \qquad \text{for } x < b \quad ...(2.37)$$

Similarly,
$$f_X\{x/X \le b\} \ge f_X(x) \qquad \text{for } x < b \quad ...(2.38)$$

EXAMPLES

Example 2.1: The number of calls received in a telephone exchange follows a Poisson distribution with an average of 8 calls per minute. What is the probability that in one minute duration, the following happens:

i. No call is received
ii. Exactly 5 calls are received
iii. More than 2 calls are received.

Solution: Let X be the random variable representing the number of calls received.
According to Poisson distribution

$$P_X(k) = \frac{e^{-\lambda}\lambda^k}{k!}, \text{ where } \lambda = 8.$$

i. Probability that no call is received is $P_X(0) = \dfrac{e^{-8}8^0}{0!} = 0.000335$

ii. Probability that exactly 5 calls are received is $P_X(5) = \dfrac{e^{-8}8^5}{5!} = 0.0916$

iii. Probability that more than 2 calls are received is

$$P_X(x > 2) = 1 - \sum_{k=0}^{2} P_X(k) = 1 - e^{-8}\left(\frac{8^0}{0!} + \frac{8^1}{1!} + \frac{8^2}{2!}\right) = 0.9862.$$

Example 2.2: Find the probability of the event $\{X \le 5.5\}$ for a Gaussian random variable having $\mu = 4$ and $\sigma = 2.5$?

Solution: $\dfrac{x-\mu}{\sigma} = \dfrac{5.5-4}{2.5} = 0.6$

$\because$ $F_X(x) = \phi\left(\dfrac{x-\mu}{\sigma}\right)$ where $\phi(\,)$ is standard Gaussian distribution function

$$P\{X \le 5.5\} = F_X(5.5) = \phi(0.6) = 0.7257.$$

Example 2.3: Assume that the height of an airplane at some location above the ground is a gaussian random variable X with $\mu = 1800$ m and $\sigma = 400$ m. Find the probability that airplane will be higher than 2500 m?

Solution: $P(X > 2500) = 1 - P(X \le 2500)$

$$= 1 - F_X(2500) = 1 - \phi\left(\frac{2500-1800}{400}\right)$$

$$= 1 - \phi(1.75) = 1 - 0.9599 = 0.0401.$$

Example 2.4: In a certain junior Olympics, long jump distances are well approximated by a Gaussian distribution for which $\mu = 30$ m and $\sigma = 5$ m. In a qualifying round, contestants must throw farther than 25 m to qualify. In the main event the record throw is 40 m.

(a) What is the probability of being disqualified in the qualifying round?

(b) In the main event, what is the probability that the record will be broken?

Solution: (a) Probability of being qualified is $P(X > 25)$.

Probability of being not qualified is

$$P(X \le 25) = F_X(25)$$

$$= \phi\left(\frac{25-30}{5}\right)$$

$$= \phi\left(\frac{-5}{5}\right) = \phi(-1)$$

$$= 1 - \phi(1) = 1 - 0.8413 = 0.1587$$

(b) Probability that the record will be broken is $P(X > 40)$.

$$P(X > 40) = 1 - P(X \le 40)$$

$$1 - F_X(40) = 1 - \phi\left(\frac{40-30}{5}\right)$$

$$= 1 - \phi\left(\frac{10}{5}\right) = 1 - \phi(2) = 1 - 0.9772 = 0.0228.$$

Example 2.5: A random variable X is Gaussian with $\mu = 0$ and $\sigma = 1$.
(a) What is the probability that $|X| > 2$?
(b) What is the probability that $X > 2$?

Solution: (a)
$$\begin{aligned} P\{|X| > 2\} &= P\{X < -2\} + P\{X > 2\} \\ &= 1 - P\{X \le 2\} + P\{X < -2\} \\ &= 1 - F_X(2) + F_X(-2) \\ &= 1 - F_X(2) + 1 - F_X(2) \\ &= 2 - 2F_X(2) \\ &= 2 - 2(0.9772) = 0.0456 \end{aligned}$$

(b) $P\{X > 2\} = 1 - P\{X \le 2\} = 1 - F_X(2) = 1 - 0.9772 = 0.0228.$

Example 2.6: Certain military radar is set up at a remote site with no repair facilities. If the radar is known to have a mean-time-between-failure (MTBF) of 200 hr, find the probability that repair is still in operation one week later when picked up for maintenance and repair.

Solution: $\lambda = \dfrac{1}{\text{MTBF}} = \dfrac{1}{200}$, $T = 7(24) = 168$ hr (1 week).

$P(0 \text{ failure in 1 week}\} = e^{-\lambda T} = e^{-168/200} = 0.4317.$

Example 2.7: A certain large city averages three murders per week and occurrences follow a Poisson distribution.
(a) What is the probability that there will be five or more murders in a given week?
(b) On an average, how many weeks a year can this city expect to have no murders?
(c) How many weeks per year can the city expect the number of murders per week to equal or exceed the average number per week?

Solution: (a) $\lambda = 3$ murders/week, $t = 1$ week and $\mu = 3$

The probability that there will be five or more murders in a given week is

$$P\{X \ge 5\} = 1 - P\{X \le 4\} = 1 - e^{-3}\sum_{k=0}^{\infty} \frac{3^k}{k!}$$

$$= 1 - e^{-3}\left(\frac{3^0}{0!} + \frac{3^1}{1!} + \frac{3^2}{2!} + \frac{3^3}{3!} + \frac{3^4}{4!}\right)$$

$$= 1 - e^{-3}(1 + 3 + 4.5 + 4.5 + 3.75) = 0.1847$$

(b) Probability of no murder in a given week is

$$P\{X = 0\} = e^{-3}\frac{3^0}{0!} = 0.04978$$

We know that the number of weeks in a year is 52.

On average number of weeks a year have no murders = 0.04978(52) = 2.5889.

(c) $P\{X \ge 3\} = 1 - P\{X < 3\} = 1 - P\{X \le 2\} = 1 - e^{-3}\sum_{k=0}^{2} \frac{3^k}{k!}$

$$= 1 - e^{-3}\left(\frac{3^0}{0!} + \frac{3^1}{1!} + \frac{3^2}{2!}\right)$$

$$= 1 - e^{-3}(1 + 3 + 4.5) = 1 - e^{-3}(8.5) = 0.576$$

The average number of weeks in a year having equal or more than three murders $= 0.57619(52) = 29.99$.

Example 2.8: Assume that an electronic manufacturer made transistors with a probability of 0.05 of being inoperable when new. A person purchases ten transistors for laboratory use.

(a) What is the probability that exactly one of the ten transistors is inoperable?

(b) What is the probability that all ten transistors are functional?

(c) Determine the probability that one or more transistors are not operable?

Solution: The experiment has only two outcomes, operable and inoperable so that it is a binomial random variable.

The binomial distribution is defined as

$$F_X(x) = \sum_{k=0}^{n} (p)^k (q)^{n-k}\, {}^nC_k\, u(x-k)$$

where p = the probability that a transistor is inoperable

n = total number of transistors

(a) P (exactly one transistor is inoperable)

$$= {}^{10}C_1 (0.05)^1 (0.95)^{10-1}$$

$$= (0.05)\,(0.95)^9\,(10) = 0.315$$

(b) P (all transistors are operable) = P(zero transistors are inoperable)

$$= {}^{10}C_0 (0.05)^0 (0.95)^{10-0}$$

$$= (0.95)^{10} = 0.5987$$

(c) P(one or more transistors inoperable) = 1 – P (zero transistors are inoperable)

$$= 1 - 0.5987 = 0.4013.$$

Example 2.9: Assume that lifetime of a system expressed in weeks is a Rayleigh random variable X for which

$$f_X(x) = \begin{cases} \dfrac{x}{200} e^{-(x^2/400)} & x \geq 0 \\ 0 & x < 0 \end{cases}$$

(a) What is the probability that the system will not last a full week?

(b) What is the probability that the system's lifetime will exceed one year?

Solution: (a) $P(x \leq 1) = \int_{-\infty}^{1} f_X(x)dx = \int_{-\infty}^{0} 0\,dx + \int_{0}^{1} \frac{x}{200} e^{-(x^2/400)} dx$

$$= \int_{0}^{1} \frac{x}{200} e^{-(x^2/400)} dx$$

Assume $\dfrac{x^2}{400} = k \Rightarrow \dfrac{x}{200}dx = dk$

$$= \int_0^{1/400} e^{-(k)}dk = \frac{e^{-(k)}}{-1}\Big|_0^{1/400} = 1 - e^{-(1/400)} = 0.0024$$

$$\text{(b) } P(x > 52) = 1 - P(x \le 52) = 1 - \int_{-\infty}^{52} f_X(x)dx = 1 - \int_{-\infty}^{0} 0\,dx - \int_0^{52} \frac{x}{200}e^{-(x^2/400)}dx$$

$$= 1 - \int_0^{52} \frac{x}{200}e^{-(x^2/400)}dx$$

Assume $\dfrac{x^2}{400} = k \Rightarrow \dfrac{x}{200}dx = dk$

$$= 1 - \int_0^{(52)^2/400} e^{-(k)}dk = 1 - \frac{e^{-(k)}}{-1}\Big|_0^{(52)^2/400} = 1 - 1 + e^{-\{(52)^2/400\}} = 0.0012.$$

Example 2.10: The number of cars arriving at a certain bank drive-in window during any 10 min period is a Poisson random variable X with μ = 2. Find:
(a) The probability that more than three cars will arrive during any 10 min period.
(b) The probability that no car will arrive.

Solution: μ = 2 and T = 10 min

(a) P(more than three cars arrive) = $P(X > 3) = 1 - P(X \le 3)$

$$= 1 - e^{-2}\sum_{k=0}^{\infty}\frac{2^k}{k!} = 1 - e^{-2}\left(\frac{2^0}{0!} + \frac{2^1}{1!} + \frac{2^2}{2!} + \frac{2^3}{3!}\right)$$

$$= 1 - e^{-2}(1 + 2 + 2 + 1.33) = 1 - 0.857 = 0.1428$$

(b) P(zero cars arrive) = $P(X = 0) = e^{-2}\dfrac{2^0}{0!} = e^{-2} = 0.1353.$

Example 2.11: Prove that $F_X(x/a < X \le b) = \begin{cases} 0 & x < a \\ \dfrac{F_x(x) - F_x(a)}{F_x(b) - F_x(a)} & a < x \le b \\ 1 & b \le x \end{cases}$

Solution: We know that $F_X(x/a < X \le b) = \dfrac{P(X \le x \cap a < X \le b)}{P(a < X \le b)}$

If $x < a$ $\qquad F_X(x/a < X \le b) = \dfrac{P(\phi)}{P(a < X \le b)} = 0$

If $a \le x < b$ $\qquad F_X(x/a < X \le b) = \dfrac{P(a \le X < b)}{P(a < X \le b)} = \dfrac{F_x(x) - F_x(a)}{F_x(b) - F_x(a)}$

If $x \ge b$ $\qquad F_X(x/a < X \le b) = \dfrac{P(a \le X \le b)}{P(a < X \le b)} = 1$

By combining above three conditions,

$$F_X(x/a < X \le b) = \begin{cases} 0 & x < a \\ \dfrac{F_x(x) - F_x(a)}{F_x(b) - F_x(a)} & a < x \le b \\ 1 & b \le x \end{cases}$$

Example 2.12: Assume that the lifetime of a laboratory research plant is defined by Rayleigh density with $a = 0$ and $b = 30$. If for some clinical reasons it is known that the plant will live at most 25 weeks, what is the probability it will live 15 weeks or less?

Solution: We know that Rayleigh density function is defined as

$$f_X(x) = \begin{cases} \dfrac{2}{b}(x-a)e^{-(x-a)^2/b} & x \ge a \\ 0 & x < a \end{cases}$$

For the above problem, it is defined as

$$f_X(x) = \begin{cases} \dfrac{2}{30}(x)e^{-(x)^2/30} & 0 < x < 25 \\ 0 & \text{otherwise} \end{cases}$$

$$P(X \le 15) = F_X(15) = \int_{-\infty}^{25} f_X(x)dx = \int_{-\infty}^{0} 0dx + \int_{0}^{15} \frac{2}{30}(x)e^{-(x)^2/30}dx$$

$$= \int_{0}^{15} \frac{2}{30}(x)e^{-(x)^2/30}dx$$

$$\text{Let } \frac{x^2}{30} = u \Rightarrow \frac{x\,dx}{15} = du$$

$$= \int_{0}^{15} e^{-u}du = \left.\frac{e^{-u}}{-1}\right|_0^{15} = 1 - e^{-15} = 0.9999996.$$

Example 2.13: A random variable X has the density function

$$f_X(x) = \begin{cases} \dfrac{1}{4} & -2 < x < 2 \\ 0 & \text{elsewhere} \end{cases}$$

Find: (a) $P(X < 1)$ (b) $P(|X| > 1)$ (c) $P(2X + 3 > 5)$

Solution: (a) $P(X < 1) = \int_{-2}^{1} \frac{1}{4}dx = \frac{1}{4}\int_{-2}^{1} dx = \frac{1}{4}x|_{-2}^{1} = \frac{3}{4}$

(b) $$P(|X| > 1) = P(X > 1 \text{ or } X < -1) = \int_{-2}^{-1} \frac{1}{4}dx + \int_{1}^{2} \frac{1}{4}dx$$

$$= \frac{1}{4} + \frac{1}{4} = \frac{1}{2}$$

(c) $$P(2X + 3 > 5) = P(2X > 2) = P(X > 1) = \frac{1}{4}\int_{1}^{2} dx = \frac{1}{4}x|_1^2 = \frac{1}{4}.$$

Example 2.14: On a certain aircraft, the main control circuit of an autopilot fails with probability 0.4. A redundant backup circuit fails independently with probability 0.6. The airplane can fly if at least one of the circuits is functioning. Find the probability that the airplane cannot fly.

Solution: Let X be a random variable representing the main control circuit.

Let Y be a random variable representing the backup circuit.

The airplane cannot fly if both circuits fail.

From the above, $X = \begin{cases} 0 & \text{if the main control circuit is functional} \\ 1 & \text{if the main control circuit is not functional} \end{cases}$

$$P(X = 1) = 0.4$$

$$Y = \begin{cases} 0 & \text{if the backup circuit is functional} \\ 1 & \text{if the backup circuit is not functional} \end{cases}$$

$$P(Y = 1) = 0.6$$

$$P(X = 1, Y = 1) = P(\{X = 1\} \cap \{Y = 1\})$$

$$P(X = 1, Y = 1) = P(\{X = 1\})P(\{Y = 1\})$$

$$P(X = 1, Y = 1) = (0.4)(0.6) = 0.24.$$

EXERCISE

1. Find the distribution function for the random variable X whose density function is

$$f_X(x) = \begin{cases} \frac{1}{9}x^2 & 0 \le x < 3 \\ 0 & \text{elsewhere} \end{cases}$$

and also find $P(1 < X \le 2)$?

2. Determine if the following is a valid distribution function or not?

$$F_X(x) = \begin{cases} 1 - e^{-x/2} & x \ge 0 \\ 0 & \text{elsewhere} \end{cases}$$

3. The mileage that car owners get with certain kind of tires is a random variable having the probability density

$$f_X(x) = \begin{cases} \frac{1}{20}e^{-x/20} & \text{for } x > 0 \\ 0 & \text{elsewhere} \end{cases}$$

Find the probability that one of these tires will give mileage: (a) atmost 1000 miles (b) anywhere from 15000 to 24000 miles (c) at least 30000 miles.

4. A random variable X is known to be Gaussian with $\mu = 1.6$ and $\sigma = 0.4$. Find:

(a) $P(1.4 < X \le 2.0)$ (b) $P(-0.6 < (X - 1.6) \le 0.6)$

5. Determine PDF for the following distribution function.

$$F_X(x) = \begin{cases} 0 & x < 0 \\ x^2 & 0 \le x < \dfrac{1}{2} \\ 1 - \dfrac{3}{50}(3-x)^2 & \dfrac{1}{2} \le x < 3 \\ 1 & x \ge 3 \end{cases}$$

6. Suppose that depth of water, measured in meters, behind a dam is described by an exponential random variable having density function $f_X(x) = \dfrac{1}{13.5}u(x)e^{-x/13.5}$. There is an emergency overflow at the top of the dam that prevents the depth from exceeding 40.6 m. There is pipe placed at 32.0 m that feeds water to a hydroelectric generator.
   (a) What is the probability that water is wasted through emergency overflow?
   (b) Given that water is not wasted in overflow, what is the probability the generator will have water to drive it?
   (c) What is the probability that water will be too low to produce power?
7. The diameter X of an electric cable is a continuous random variable with PDF $f_X(x) = kx(1-x)$ for $0 \le x \le 1$. Find: (a) k (b) CDF (c) the value of a such that $P(X < a) = 2P(X > a)$.
8. The random variable X has the discrete variable in the set $\{-1, -0.5, 0.7, 1.5, 3\}$, the corresponding probabilities are assumed to be $\{0.1, 0.2, 0.1, 0.4, 0.2\}$. Plot its distribution function and state whether it is a discrete or continuous distribution function.
9. Assume automobile arrives at a gasoline station are Poisson and occur at an average rate of 50/hr. The station has only one gasoline pump. If a car is assumed to take one minute in obtaining fuel, what is the probability that a waiting line will occur at the pump?
10. A fair coin is tossed three times and faces showing up are observed.
    i. Write the sample space.
    ii. If X is the number of heads in each of the outcomes of the experiment, find the probability function.
    iii. Sketch the CDF and PDF.
11. A production line manufactures resistors that must satisfy a 10% tolerance.
    (a) If resistance is adequately described by a Gaussian random variable X for which $\mu = 1000\ \Omega$ and $\sigma = 40\ \Omega$, what fraction of the resistors is expected to be rejected?
    (b) If a machine is not properly adjusted, the product resistances change to the case where $\mu = 1050\ \Omega$ (5% shift). What fraction is rejected now?
12. Suppose height to the bottom of cloud is a Gaussian distribution for which $\mu = 5000$ m and $\sigma = 1500$ m. A person bets that cloud height tomorrow will fall in set $A = \{2000 \text{ m} < X \le 4000 \text{ m}\}$, while a second person bets that height will be satisfied by $B = \{3000 \text{ m} < X \le 5500 \text{ m}\}$. A third person bets they both are correct. Find the probability that each person will win the bet?
13. If X is a continuous random variable with uniform probability density function having $a = -2$ and $b = 5$, find $P(X < 0)$ and $P(X > 1 \mid X > 0)$.

3

Operations on One Random Variable

3.0 INTRODUCTION

In the last chapter, we have defined a random variable as a mapping from the sample space to the real function. In our study of random variables, we use the probability density function or the cumulative distribution function to provide a complete statistical description of the random variable. From these functions, we could, in theory, determine anything we might want to know about the random variable. In this chapter, we will introduce some operations on a random variable. Most of these operations are based on the concept of "expectation". We introduce the concepts of expectation, moments, functions that give moments and transformation of a random variable. We also present methods for generating random variables.

Particularly, this chapter will focus on the concept of expectation.

Let us assume a set of data $X_1, X_2, \ldots, X_N$. We know that the arithmetic average (or arithmetic mean) is given by

$$\bar{X} = \frac{X_1 + X_2 + \cdots + X_n}{N} \quad \text{...(3.1)}$$

When the above data occur with different frequencies, we usually assign weights $a_1, a_2, \ldots, a_N$ to them and the so-called weighted arithmetic mean becomes

$$\bar{X} = \frac{a_1 X_1 + a_2 X_2 + \cdots + a_N X_N}{a_1 + a_2 + \cdots + a_N} \quad \text{...(3.2)}$$

The average is a value that is representative of a set of data and tends to lie centrally within a set of data that is arranged according to their magnitudes. Thus, it is usually called a measure of central tendency.

The term expectation is used for the process of averaging when a random variable is involved. It is a number used to locate the "center" of the distribution of a random variable. In many situations, we are primarily interested in the central tendency of a random variable.

Another measure of central tendency of a random variable is its variance, which measures the degree to which a random variable is spread out. This chapter deals with how the expectation (or mean or average) and variance of a random variable can be computed.

3.1 EXPECTATION AND VARIANCE

Expected value is also called statistical average, mean, and first moment. The term expectation is used for the process of averaging when a random variable is involved. The expectation of a random variable is one measure of the location and center of its probability density function (PDF).

If X is a random variable, then the expectation (or expected value or mean) of X, denoted by $E[X]$, is defined by

$$E[X] = \begin{cases} \sum_i x_i P(x_i) & X \text{ is discrete} \\ \int_{-\infty}^{\infty} x\, f_X(x)\,dx & X \text{ is continuous} \end{cases} \quad ...(3.3)$$

Thus, the expected value of X is a weighted average of the possible values that X can take, where each value is weighted by the probability that X takes that value. The expected value of X is sometimes denoted by $\bar{X}$.

3.1.1 Properties of Expectation

Assume that X is a continuous random variable.

(a) If k is a constant, then $E[kX] = kE[X]$.

Proof:

$$E[kX] = \int_{-\infty}^{\infty} kxf_X(x)\,dx \qquad \because \; E[X] = \int_{-\infty}^{\infty} xf_X(x)\,dx$$

$$= k\int_{-\infty}^{\infty} xf_X(x)\,dx = kE[X]$$

Hence $\quad E[kX] = kE[X] \quad ...(3.4a)$

(b) If X and Y are two random variables, then

$$E[X+Y] = E[X] + E[Y]$$

Proof: Assume X and Y are independent random variables.

According to the definition of expectation,

$$E[X+Y] = \int_{-\infty}^{\infty}\int_{-\infty}^{\infty} (x+y) f_{XY}(x,y)\,dx\,dy$$

$$E[X+Y] = \int_{-\infty}^{\infty}\int_{-\infty}^{\infty} xf_{XY}(x,y)\,dx\,dy + \int_{-\infty}^{\infty}\int_{-\infty}^{\infty} yf_{XY}(x,y)\,dx\,dy$$

$$= \int_{-\infty}^{\infty}\int_{-\infty}^{\infty} xf_X(x)\,f_Y(y)\,dx\,dy + \int_{-\infty}^{\infty}\int_{-\infty}^{\infty} yf_X(x)\,f_Y(y)\,dx\,dy$$

$\because$ X and Y are independent, $f_{XY}(x,y) = f_X(x)\,f_Y(y)$

$$= \int_{-\infty}^{\infty} f_Y(y)dy \int_{-\infty}^{\infty} xf_X(x)\,dx + \int_{-\infty}^{\infty} f_X(x)\,dx \int_{-\infty}^{\infty} yf_Y(y)\,dy \qquad \because \int_{-\infty}^{\infty} f_X(x)\,dx = 1$$

$$= \int_{-\infty}^{\infty} xf_X(x)\,dx + \int_{-\infty}^{\infty} yf_Y(y)\,dy = E[X] + E[Y]$$

Hence $E[X+Y] = E[X] + E[Y] \quad ...(3.4b)$

(c) If X and Y are independent random variables, then

$$E[XY] = E[X]E[Y]$$

According to the definition of expectation,

$$E[XY] = \int_{-\infty}^{\infty}\int_{-\infty}^{\infty} (xy)\, f_{XY}(x,y)\,dx\,dy$$

$$E[XY] = \int_{-\infty}^{\infty} x f_X(x)\,dx \int_{-\infty}^{\infty} (y)\, f_Y(y)\,dy$$

$\because$ X and Y are independent, $f_{XY}(x, y) = f_X(x)\, f_Y(y) = E[X]E[Y]$

Hence $$E[XY] = E[X]E[Y] \qquad \text{...(3.4c)}$$

(d) $E[aX + b] = aE[X] + b$, when X is a random variable and a and b are constants.

Proof: $E[aX + b] = \int_{-\infty}^{\infty} ax f_X(x)\,dx + \int_{-\infty}^{\infty} b f_X(x)\,dx \qquad \because E[X] = \int_{-\infty}^{\infty} x f_X(x)\,dx$

$$= a\int_{-\infty}^{\infty} x f_X(x)\,dx + b\int_{-\infty}^{\infty} f_X(x)\,dx = aE[X] + b \qquad \because \int_{-\infty}^{\infty} f_X(x)\,dx = 1$$

Hence $E[aX + b] = aE[X] + b.$ ...(3.4d)

3.1.2 Expected Value of a Function of a Random Variable

We know that $E[X] = \int_{-\infty}^{\infty} x f_X(x)\,dx$

Let X is not a random variable, it is a function of x, i.e. $X = g(x)$.
Expectation of $g(x)$ is defined as

$$E[g(x)] = \begin{cases} \sum_i g(x_i)P(x_i) & g(x) \text{ is discrete} \\ \int_{-\infty}^{\infty} g(x) f_X(x)\,dx & g(x) \text{ is continuous} \end{cases} \qquad \text{...(3.5)}$$

Example 3.1.1: A random variable X is defined as follows:

$$X = \begin{cases} -2 & \text{with a probability } \frac{1}{3} \\ 3 & \text{with a probability } \frac{1}{2} \\ 1 & \text{with a probability } \frac{1}{6} \end{cases}$$

Find $E[X]$.

Solution: $E[X] = \sum_{i=1}^{N} x_i P(x_i) = \sum_{i=1}^{3} x_i P(x_i) = x_1 P(x_1) + x_2 P(x_2) + x_3 P(x_3)$

$$= -2\left(\frac{1}{3}\right) + 3\left(\frac{1}{2}\right) + 1\left(\frac{1}{6}\right) = \left(\frac{-4+9+1}{6}\right) = 1$$

Example 3.1.2: Find the mean value of the density function given below:

$$f_X(x) = \begin{cases} \frac{1}{b}e^{-(x-a)/b} & x > a \\ 0 & x < a \end{cases}$$

Solution: For continuous random variable, expectation is defined as

$$E[X] = \int_{-\infty}^{\infty} x f_X(x)dx$$

$$E[X] = \int_{-\infty}^{a} x \cdot 0\, dx + \int_{a}^{\infty} x\frac{1}{b}e^{-(x-a)/b}dx = \frac{1}{b}e^{a/b}\int_{a}^{\infty} xe^{-x/b}dx$$

$$= \frac{e^{a/b}}{b}\int_{a}^{\infty} xe^{-x/b}dx = \frac{e^{a/b}}{b}\left[x\frac{e^{-a/b}}{-1/b} - \frac{e^{-x/b}}{-1/b^2}\right]_a^{\infty}$$

$$= \frac{e^{a/b}}{b}[0 - 0 + b\cdot a\cdot e^{-a/b} + b^2e^{-a/b}]$$

$$= \frac{e^{a/b}}{b}[b\cdot a\cdot e^{-a/b} + b^2e^{-a/b}] = (a+b)$$

Example 3.1.3: A random variable X has the density function given below:

$$f_X(x) = \begin{cases} e^{-x} & \text{for } x \geq 0 \\ 0 & \text{else} \end{cases}$$

Find $E\left[e^{2x/3}\right]$.

Solution: $E\left[e^{2x/3}\right] = \int_0^{\infty} e^{2x/3}e^{-x}dx = \int_0^{\infty} e^{-1/3x}dx = -3e^{-1/3x}\Big|_0^{\infty} = 3.$

3.1.3 Conditional Expectation

We know that $E[X] = \int_{-\infty}^{\infty} x f_X(x)dx.$

If the density function is conditional density function, replace the density function with conditional density function $f_X(x/B)$, where B is any event defined on the sample space. It gives the "conditional expected" value of X, denoted by $E[x/B]$.

$$E[x/B] = \int_{-\infty}^{\infty} x f_X(x/B)dx$$

Let B be defined as $\quad B = \{x \leq b\} \quad -\infty < b < \infty$

Conditional density function is defined as

$$f_X(x/x \leq b) = \begin{cases} \dfrac{f_X(x)}{\int_{-\infty}^{b} f_X(x)dx} & x < b \\ 0 & x \geq b \end{cases}$$

Thus $$E[X/x \le b] = \frac{\int_{-\infty}^{b} x f_X(x)\,dx}{\int_{-\infty}^{b} f_X(x)\,dx} \quad \text{...(3.6)}$$

3.2 MOMENTS

An immediate application of the expected value of a function $g(.)$ of a random variable X is in calculating moments. Generally, there are two types of moments, viz. moments about the origin and moments about the mean.

Definition: Expectations of functions of a random variable are called moments. Four important moments that provide much information about the location and shape of a random variable are summarized in Fig. 3.1.

3.2.1 Moments About the Origin

Let $g(X) = X^n$ for $n = 0, 1, 2...$ be a function.

Expectation of $g(X)$ is defined as $E[g(X)] = E(X^n)$, which gives the nth moment about the origin of the random variable X. We denote the nth moment by m_n. Then

$$m_n = E(X^n) = \begin{cases} \sum_{i=1}^{N} x_i^n P(x_i) & X \text{ is discrete} \\ \int_{-\infty}^{\infty} x^n f_X(x)\,dx & X \text{ is continuous} \end{cases} \quad \text{...(3.7)}$$

From the above, it is clear that when $n = 1$, the first moment, $E[X]$, is the expected value of X.

3.2.2 Moments About the Mean

Moments about the mean value of random variable X is also called as central moments.

Let $g(X) = (X - \bar{X})^n$ for $n = 0, 1, 2,...$ is a function.

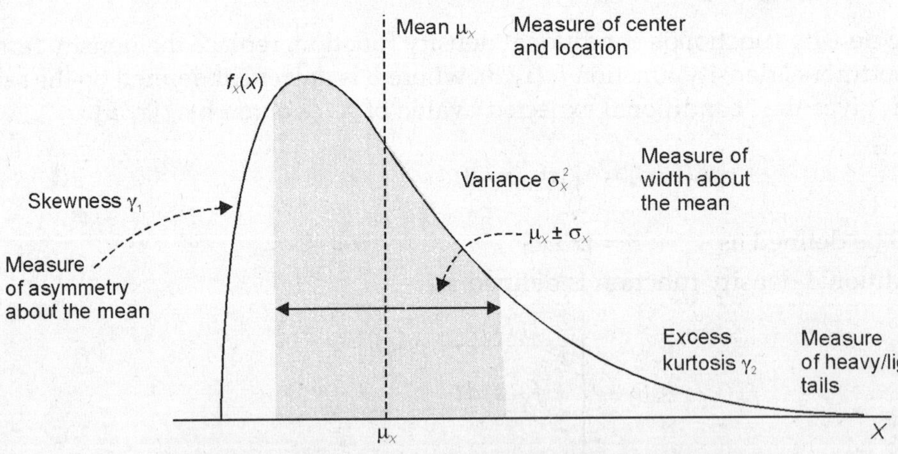

Fig. 3.1: Four important moments of random variable X

Expectation of $g(X)$ is defined as

$$E[g(X)] = E\left[(X-\bar{X})^n\right]$$

which gives the nth moment about the origin of the random variable X. We denote the nth moment by μ_n. Then,

$$\mu_n = E\left[(X-\bar{X})^n\right] = \overline{(X-\bar{X})^n}$$

$$= \begin{cases} \sum_{i=1}^{N} (x_i - \bar{X})^n P(x_i) & X \text{ is discrete} \\ \int_{-\infty}^{\infty} (x-\bar{X})^n f_X(x)dx & X \text{ is continuous} \end{cases} \qquad \text{...(3.8)}$$

For $n = 0$, $\mu_0 = 1$.
For $n = 1$, $\mu_1 = 0$.

Variance

The second central moment μ_2 is called *variance*. The variance of the random variable X is denoted by σ_X^2.

$$\sigma_X^2 = E\left[(X-\bar{X})^2\right] = \int_{-\infty}^{\infty} (x-\bar{X})^2 f_X(x)dx$$

$$= \int_{-\infty}^{\infty} (x^2 - 2x\bar{X} + \bar{X}^2) f_X(x)dx$$

$$= \int_{-\infty}^{\infty} x^2 f_X(x)dx - \int_{-\infty}^{\infty} 2x\bar{X} f_X(x)dx + \int_{-\infty}^{\infty} \bar{X}^2 f_X(x)dx$$

$$= E[X^2] - \bar{X}\int_{-\infty}^{\infty} 2x f_X(x)dx + \bar{X}^2 \int_{-\infty}^{\infty} f_X(x)dx$$

$$= E[X^2] - 2\bar{X}^2 + \bar{X}^2 = E[X^2] - (E[X])^2$$

$$\sigma_X^2 = E[X^2] - 2\bar{X}^2 + \bar{X}^2 = E[X^2] - (E[X])^2 \qquad \text{...(3.9)}$$

The positive square root of variance is called *standard variation*.

Properties:

i. Var(X + k) = Var(X), where X is the random variable and k is a constant.

Proof: $\text{Var}(X) = E\left[X^2\right] - (E[X])^2$

$$\text{Var}(X+k) = E\left[(X+k)^2\right] - (E[X+k])^2$$

$$= E\left[X^2 + 2Xk + k^2\right] - E[X]^2 - E[k]^2 - 2E[X]k$$

$$= E\left[X^2\right] + E[2Xk] + E\left[k^2\right] - E[X]^2 - E[k]^2 - 2E[X]k$$

$$= E\left[X^2\right] + 2kE[X] + E\left[k^2\right] - E[X]^2 - E[k]^2 - 2E[X]k = E\left[X^2\right] - (E[X])^2$$

$$\therefore \quad \text{Var}(X) = E\left[X^2\right] - (E[X])^2 \qquad \text{...(3.10a)}$$

ii. $Var(kX) = k^2 Var(X)$ where k is a constant.

Proof: We know that $Var(X) = E[X^2] - (E[X])^2$

$$Var(kX) = E[(kX)^2] - (E[kX])^2$$

$$= k^2 E[X^2] - k^2 (E[X])^2$$

$$= k^2 [E[X^2] - (E[X])^2]$$

$$= k^2 Var(X)$$

Hence $\quad Var(kX) = k^2 Var(X)$ ...(3.10b)

iii. $Var(X + Y) = Var(X) + Var(Y)$ where X and Y are two random variables.

We know that $Var(X) = E[X^2] - (E[X])^2$

$$Var(X+Y) = E[(X+Y)^2] - (E[X+Y]^2)$$

$$= E[(X^2 + 2XY + Y^2)] - E[X]^2 - E[Y]^2 - 2E[X]E[Y]$$

$$= E[X^2] + E[2XY] + E[Y^2] - E[X]^2 - E[Y]^2 - 2E[X]E[Y]$$

$$= E[X^2] + 2E[X]E[Y] + E[Y^2] - E[X]^2 - E[Y]^2 - 2E[X]E[Y]$$

$$= E[X^2] - E[X]^2 + E[Y^2] - E[Y]^2$$

$$= Var(X) + Var(Y)$$

Hence $\quad Var(X + Y) = Var(X) + Var(Y).$ ...(3.10c)

Skewness

The third central moment μ_3 is called *skewness*.

$$\mu_3 = E\left[(X - \overline{X})^3\right] = \int_{-\infty}^{\infty} (x - \overline{X})^3 f_X(x)dx$$

$$E\left[(X-\overline{X})^3\right] = E\left[X^3 - 3\overline{X}X^2 + 3X\overline{X}^2 - \overline{X}^3\right]$$

$$= E[X^3] - E[3\overline{X}X^2] + E[3\overline{X}^2X] - E[\overline{X}^3]$$

$$= E[X^3] - 3\overline{X}E[X^2] + 3\overline{X}^2E[X] - E[\overline{X}^3]$$

$$= \overline{X^3} - 3\overline{X}E[X^2] + 3\overline{X}^2\overline{X} - \overline{X}^3$$

$$= \overline{X^3} - 3\overline{X}(E[X^2] - \overline{X}^2) - \overline{X}^3$$

$$= \overline{X^3} - 3\overline{X}\sigma_X^2 - \overline{X}^3$$

$$\mu_3 = \overline{X^3} - 3\overline{X}\sigma_X^2 - \overline{X}^3 \quad ...(3.11)$$

Example 3.2.1: Let X have the exponential density function. Find the variance and skewness.

$$f_X(x) = \begin{cases} \dfrac{1}{b}e^{-(x-a)/b} & x \ge a \\ 0 & x < a \end{cases}$$

Solution: $\text{Var(X)} = \sigma_X^2 = E[X^2] - (E[X]^2)$

$$E[X] = \int_{-\infty}^{a} x \cdot 0\,dx + \int_{a}^{\infty} x\frac{1}{b}e^{-(x-a)/b}dx = \frac{1}{b}e^{a/b}\int_{a}^{\infty} xe^{-x/b}dx = (a+b)$$

$$E[X^2] = \int_{-\infty}^{a} x^2 \cdot 0\,dx + \int_{a}^{\infty} x^2\frac{1}{b}e^{-(x-a)/b}dx = \frac{1}{b}e^{a/b}\int_{a}^{\infty} x^2e^{-x/b}dx$$

$$= \frac{1}{b}e^{a/b}\left[\frac{x^2e^{-x/b}}{-1/b} - \left(\int_{a}^{\infty}\frac{2xe^{-x/b}}{-1/b}dx\right)\right] \qquad \left[\because \int uv\,dx = u\int v\,dx - \int u'\left[\int v\,dx\right]dx\right]$$

$$= \frac{1}{b}e^{a/b}\left[\frac{x^2e^{-x/b}}{-1/b} - \frac{2}{-1/b}\left(\frac{xe^{-x/b}}{-1/b} - \frac{e^{-x/b}}{-1/b^2}\right)\right]_a^{\infty}$$

$$= \frac{1}{b}e^{a/b}\left[\frac{x^2e^{-x/b}}{-1/b} - \frac{2xe^{-x/b}}{1/b^2} - \frac{e^{-x/b}}{1/b^3}\right]_a^{\infty} = \frac{1}{b}e^{a/b}\left[a^2be^{-a/b} + 2ab^2e^{-a/b} + 2b^3e^{-a/b}\right]$$

$$= \left[a^2 + 2ab + 2b^2\right]$$

$$\text{Var(X)} = \sigma_X^2 = E\left[X^2\right] - (E[X])^2$$

$$= \left[a^2 + 2ab + 2b^2\right] - (a+b)^2 = \left[a^2 + 2ab + 2b^2\right] - \left(a^2 + 2ab + b^2\right) = b^2$$

$$\text{Var(X)} = \sigma_X^2 = b^2$$

$$E[X^3] = \int_{-\infty}^{a} x^3 \cdot 0\,dx + \int_{a}^{\infty} x^3\frac{1}{b}e^{-(x-a)/b}dx = \frac{1}{b}e^{a/b}\int_{a}^{\infty} x^3e^{-x/b}dx$$

$$= \frac{1}{b}e^{a/b}\left[x^3e^{-x/b}(-b) - \int_{a}^{\infty}3x^2(-b)e^{-x/b}dx\right]_a^{\infty}$$

$$= \frac{1}{b}e^{a/b}\left[-bx^3e^{-x/b} + 3b\int_{a}^{\infty}x^2e^{-x/b}dx\right]_a^{\infty}$$

$$= \frac{1}{b}e^{a/b}\left[-bx^3e^{-x/b} + 3b\left\{x^2e^{-x/b}(-b) - \int_{a}^{\infty}2x(-b)e^{-x/b}dx\right\}\right]_a^{\infty}$$

$$= \frac{1}{b}e^{a/b}\left[-bx^3e^{-x/b} + 3b\left\{x^2e^{-x/b}(-b) + 2b\int_a^\infty xe^{-x/b}dx\right\}\right]_a^\infty$$

$$= \frac{1}{b}e^{a/b}\left[-bx^3e^{-x/b} - 3b^2x^2e^{-x/b} + 6b^2\int_a^\infty xe^{-x/b}dx\right]_a$$

$$= e^{a/b}\left[-x^3e^{-x/b} - 3bx^2e^{-x/b} + 6b\left\{xe^{-x/b}(-b) - \int_a^\infty (-b)e^{-x/b}dx\right\}\right]_a^\infty$$

$$= e^{a/b}\left[-x^3e^{-x/b} - 3bx^2e^{-x/b} - 6b^2\left\{xe^{-x/b} - e^{-x/b}(-b)\right\}\right]_a^\infty$$

$$= e^{a/b}\left[-x^3e^{-x/b} - 3bx^2e^{-x/b} - 6b^2xe^{-x/b} - 6b^3e^{-x/b}\right]_a^\infty$$

$$= \left[a^3 + 3ba^2 + 6b^2a + 6b^3\right]$$

$$= a^3 + 6b^3 + 3a^2b + 6ab^2$$

$$\mu_3 = \overline{X^3} - 3\overline{X}\sigma_X^2 - \overline{X}^3 = a^3 + 6b^3 + 3a^2b + 6ab^2 - 3(a+b)b^2 - (a+b)^3 = 2b^3.$$

3.3 THE MARKOV AND CHEBYSHEV INEQUALITIES

Many practical applications require knowledge of probabilities of the form $P(X > a)$ for some threshold a. For example, an optical communication receiver may decide an incoming bit is a 1 if the number X of photoelectrons detected exceeds a and decide 0 otherwise, or a could be the buffer size of an internet router, and packets would be dropped if the packet backlog X exceeds a. Unfortunately, it is often difficult to compute such probabilities when X is the output of a complicated system. The Markov and Chebyshev inequalities provide bounds on probabilities in terms of expectations that are more readily computable.

3.3.1 Markov Inequality

Often, for a random variable X that we are interested in, we want to know what is the probability that the value of the random variable X is far from its expectation. A generic answer to this, which holds for any non-negative random variable, is given by Markov inequality.

Theorem: Suppose that X is a random variable taking only non-negative values. Then, for any $a > 0$, we have

$$P(X \ge a) \le \frac{E[X]}{a}$$

Proof: Let $f_X(x)$ be the probability density function of X. It holds that

$$E[X] = \int_a^\infty x f_X(x)dx$$

$$= \int_0^\infty x f_X(x)dx \qquad \because X > 0$$

$$\geq \int_a^\infty x f_X(x)dx \qquad \text{for any } a > 0$$

$$\geq \int_a^\infty a f_X(x)dx \qquad \because x > a \text{ in the integrated region}$$

$$\geq a\int_a^\infty f_X(x)dx$$

$$E[X] \geq aP[X \geq a]$$

$$\therefore \quad P[X \geq a] \leq \frac{E[X]}{a} \qquad ...(3.12)$$

3.3.2 Chebyshev Inequality

Another answer to the question that what is the probability that the value of X is far from its expectation, is given by Chebyshev inequality, which works for any random variable (not necessarily a non-negative one).

Theorem: Let X be a random variable with finite mean $\mu = \overline{X} = m$ and variance σ_X^2, then for any $\in > 0$, we have

$$P\{|X - \overline{X}| \geq \in\} \leq \frac{\sigma_X^2}{\in^2}$$

Proof: $\sigma_X^2 = \int_{-\infty}^{\infty} (x-m)^2 f_X(x)dx$

$$= \int_{-\infty}^{m-\in} (x-m)^2 f_X(x)dx + \int_{m-\in}^{m+\in} (x-m)^2 f_X(x)dx + \int_{m+\in}^{\infty} (x-m)^2 f_X(x)dx$$

$$= \int_{m-\in}^{m+\in} (x-m)^2 f_X(x)dx > 0 \qquad \because \in > 0 \text{ so that } m + \in > m - \in$$

$$\sigma_X^2 \geq \int_{-\infty}^{m-\in} (x-m)^2 f_X(x)dx + \int_{m+\in}^{\infty} (x-m)^2 f_X(x)dx \qquad ...(3.13)$$

From the first term of Eq. (3.13), the upper limit of integral is $m - \in$, i.e.

$$x \leq m - \in$$

$$x - m \leq - \in$$

From the second term of Eq. (3.13), the lower limit of integral is $m + \in$, i.e.

$$x \geq m + \in$$

$$x - m \geq \in$$

$$\sigma_X^2 \geq \int_{-\infty}^{m-\epsilon} (-\epsilon)^2 f_X(x)dx + \int_{m+\epsilon}^{\infty} (\epsilon)^2 f_X(x)dx$$

$$\geq \epsilon^2 \left(\int_{-\infty}^{m-\epsilon} f_X(x)dx + \int_{m+\epsilon}^{\infty} f_X(x)dx \right)$$

$$\geq \epsilon^2 \int_{|x-m|\geq\epsilon}^{\infty} f_X(x)dx$$

$$\sigma_X^2 \geq \epsilon^2 P(|X-m|\geq\epsilon)$$

$$P(|X-m|\geq\epsilon) \leq \frac{\sigma_X^2}{\epsilon^2}. \qquad \text{...(3.14)}$$

Example 3.3.1: A post office handles, on average, 10,000 letters a day. What can be said about the probability that it will handle at least 15,000 letters tomorrow?

Solution: Let X be the number of letters to be handled tomorrow. Then $E[X] = 10,000$. Further, by Markov inequality, we get

$$P[X \geq a] \leq \frac{E[X]}{a}$$

$$P[X \geq 15000] \leq \frac{10000}{15000} = \frac{2}{3}.$$

Example 3.3.2: A post office handles 10,000 letters per day with a variance of 2,000 letters. What can be said about the probability that this post office handles between 8,000 and 12,000 letters tomorrow? What is the probability that more than 15,000 letters come in?

Solution: We have to find

$$P(8000 < X < 12000) = P(-2000 < X - 10000 < 2000)$$

$$= P(|X-10000| < 2000) = 1 - P(|X-10000| \geq 2000)$$

By Chebyshev inequality,

$$P(|X-m|\geq\epsilon) \leq \frac{\sigma_X^2}{\epsilon^2}$$

$$P(|X-10000|\geq 2000) \leq \frac{2000}{(2000)^2} = \frac{1}{2000} = 0.0005$$

Thus $\quad P(8000 < X < 12000) \geq 1 - 0.0005 = 0.9995$

The probability that more than 15000 letters come in is

$$P(X \geq 15000) = P(|X-10000| \geq 5000) \leq \frac{2000}{(5000)^2} = 0.00008$$

Recall that the old bound (no information about variance) was 2/3.

3.4 FUNCTIONS THAT GENERATE MOMENTS

There are two functions that allow moments to be calculated for a random variable X. They are:

i. Moment generating function
ii. Characteristic function

3.4.1 Moment Generating Function

Moment generating function is defined as

$$M_X(v) = E[e^{vx}]$$

where v is a real number, $-\infty < v < \infty$. Thus $M_X(v)$ is given by

$$M_X(v) = \int_{-\infty}^{\infty} f_X(x)e^{vx}dx \qquad ...(3.15)$$

The main advantage of the moment generating function drives its ability to give the moments. It gives n^{th} moment as

$$m_n = \left.\frac{d^n M_X(v)}{dv^n}\right|_{v=0} \qquad ...(3.16)$$

Proof: $$E[e^{vx}] = \int_{-\infty}^{\infty} f_X(x)e^{vx}dx$$

$$= \int_{-\infty}^{\infty}\left[1+\frac{vx}{1!}+\frac{(vx)^2}{2!}+\cdots+\frac{(vx)^n}{n!}+\cdots\right]f_X(x)dx$$

$$M_X(v) = \int_{-\infty}^{\infty}\left[1+\frac{vx}{1!}+\frac{(vx)^2}{2!}+\cdots+\frac{(vx)^n}{n!}+\cdots\right]f_X(x)dx$$

$$\frac{d^n[M_X(v)]}{dv^n} = \frac{d^n}{dv^n}\left[\int_{-\infty}^{\infty}\left\{1+\frac{vx}{1!}+\frac{(vx)^2}{2!}+\cdots+\frac{(vx)^n}{n!}+\cdots\right\}f_X(x)dx\right]$$

$$\frac{d^n[M_X(v)]}{dv^n} = \int_{-\infty}^{\infty}\left[\frac{d^n}{dv^n}\left\{1+\frac{vx}{1!}+\frac{(vx)^2}{2!}+\cdots+\frac{(vx)^n}{n!}+\cdots\right\}f_X(x)dx\right]$$

$$\frac{d^n[M_X(v)]}{dv^n} = \int_{-\infty}^{\infty}\left[\left\{0+0+0+\cdots+n!\frac{(x)^n}{n!}+(n+1)(n)(n-1)...(3)(2)\frac{(x)^{n+1}}{(n+1)!}v...\right\}f_X(x)dx\right]$$

$$\left.\frac{d^n[M_X(v)]}{dv^n}\right|_{v=0} = \int_{-\infty}^{\infty} x^n f_X(x)dx = m_n$$

$$m_n = \left.\frac{d^n[M_X(v)]}{dv^n}\right|_{v=0}$$

The main disadvantage of the moment generating function is that it may not exist for all random variables. In fact, $M_X(v)$ exists only if all the moments exist.

3.4.2 Characteristic Function

The characteristic function of a random variable X is defined by

$$\phi_X(\omega) = E\left[e^{j\omega x}\right] \quad \text{...(3.17)}$$

The nth moment is given by

$$m_n = (-j)^n \left.\frac{d^n[\phi_X(\omega)]}{d\omega^n}\right|_{\omega=0} \quad \text{...(3.18)}$$

Proof: $\phi_X(\omega) = E\left[e^{j\omega x}\right]$

$$E\left[e^{j\omega x}\right] = \int_{-\infty}^{\infty} f_X(x)e^{j\omega x}dx$$

$$= \int_{-\infty}^{\infty}\left[1+\frac{j\omega x}{1!}+\frac{(j\omega x)^2}{2!}+\cdots+\frac{(j\omega x)^n}{n!}+\cdots\right]f_X(x)dx$$

$$\phi_X(\omega) = \int_{-\infty}^{\infty}\left[1+\frac{j\omega x}{1!}+\frac{(j\omega x)^2}{2!}+\cdots+\frac{(j\omega x)^n}{n!}+\cdots\right]f_X(x)dx$$

$$\frac{d^n[\phi_X(\omega)]}{d\omega^n} = \frac{d^n}{d\omega^n}\left[\int_{-\infty}^{\infty}\left\{1+\frac{j\omega x}{1!}-\frac{(\omega x)^2}{2!}-j\frac{(\omega x)^3}{3!}+\cdots+(j)^n\frac{(\omega x)^n}{n!}+\cdots\right\}f_X(x)dx\right]$$

$$\frac{d^n[\phi_X(\omega)]}{d\omega^n} = \int_{-\infty}^{\infty}\left[\frac{d^n}{d\omega^n}\left[1+\frac{j\omega x}{1!}-\frac{(\omega x)^2}{2!}-j\frac{(\omega x)^3}{3!}+\cdots+(j)^n\frac{(\omega x)^n}{n!}+\cdots\right]f_X(x)dx\right]$$

$$\frac{d^n[\phi_X(\omega)]}{d\omega^n} = \int_{-\infty}^{\infty}\left[\left\{0+0+0+0+\cdots+(j)^n\frac{(x)^n}{n!}n!+(j)^{n+1}\frac{(x)^{n+1}}{(n+1)!}\omega(n+1)!\ldots\right\}f_X(x)dx\right]$$

$$\left.\frac{d^n[\phi_X(\omega)]}{d\omega^n}\right|_{\omega=0} = \int_{-\infty}^{\infty}(j)^n x^n f_X(x)dx = m_n$$

$$m_n = \frac{1}{(j)^n}\left.\frac{d^n[\phi_X(\omega)]}{d\omega^n}\right|_{\omega=0} = (-j)^n\left.\frac{d^n[\phi_X(\omega)]}{d\omega^n}\right|_{\omega=0}$$

The major advantage of using $\phi_X(\omega)$ to find moments is that $\phi_X(\omega)$ always exists, so the moments can always be found if $\phi_X(\omega)$ is known.

Note: It can be shown that the maximum magnitude of a characteristic function is unity and occurs at $\omega = 0$, i.e.

$$|\phi_X(\omega)| \le \phi_X(0) = 1$$

Example 3.4.1: Consider the random variable X which follows an exponential distribution:

$$f_X(x) = \begin{cases} \frac{1}{b} e^{-x/b} & x \geq 0 \\ 0 & \text{else} \end{cases}$$

Find the characteristic function and the first moment.

Solution: $\phi_X(\omega) = E[e^{j\omega x}]$

$$= \int_0^\infty \frac{1}{b} e^{-x/b} e^{j\omega x} dx$$

$$= \frac{1}{b} \int_0^\infty e^{-x\left(\frac{1}{b} - j\omega\right)} dx$$

$$= \frac{1}{b} \left. \frac{e^{-x\left(\frac{1}{b} - j\omega\right)}}{-\left(\frac{1}{b} - j\omega\right)} \right|_0^\infty$$

$$= \frac{1}{b} \frac{1}{\left(\frac{1}{b} - j\omega\right)} = \frac{1}{(1 - j\omega b)}$$

The nth moment is given by

$$m_n = (-j)^n \left. \frac{d^n [\phi_X(\omega)]}{d\omega^n} \right|_{\omega=0}$$

For $n = 1$,

$$m_1 = (-j)^1 \left. \frac{d[\phi_X(\omega)]}{d\omega} \right|_{\omega=0}$$

$$= (-j) \left. \frac{d\left(\frac{1}{1 - j\omega b}\right)}{d\omega} \right|_{\omega=0}$$

$$= (-j) \left. \frac{-1}{(1 - j\omega b)^2} (-jb) \right|_{\omega=0}$$

$$= \left. \frac{1}{(1 - j\omega b)^2} (b) \right|_{\omega=0}$$

$$m_1 = b$$

Example 3.4.2: If X represents the outcome when a fair dice is rolled, find the MGF of X and hence find $E[X]$ and Var(X).

Solution: Here X is a discrete random variable. The moment generating function for discrete random variable is

$$M_X(v) = \sum_i e^{vx_i} P_i$$

$$P_i = P(X=i) = \frac{1}{6}, \qquad \text{where } i = 1, 2, ..., 6$$

$$M_X(v) = \sum_{i=1}^{6} e^{vx_i} P_i = \frac{1}{6}\left(e^v + e^{2v} + e^{3v} + e^{4v} + e^{5v} + e^{6v}\right)$$

$$E[X] = \left.\frac{dM_X(v)}{dv}\right|_{v=0} = \frac{1}{6}\frac{d\left(e^v + e^{2v} + e^{3v} + e^{4v} + e^{5v} + e^{6v}\right)}{dv} = \frac{7}{2}$$

$$E[X^2] = \left.\frac{d^2 M_X(v)}{dv^2}\right|_{v=0} = \frac{1}{6}\frac{d\left(e^v + 4e^{2v} + 9e^{3v} + 16e^{4v} + 25e^{5v} + 36e^{6v}\right)}{dv} = \frac{91}{6}$$

$$\text{Var}(X) = E[X^2] - (E[X])^2 = \frac{91}{6} - \frac{49}{4} = \frac{35}{12}.$$

3.5 TRANSFORMATIONS OF A CONTINUOUS RANDOM VARIABLE

Consider a random variable X with a PDF and CDF given by $f_X(x)$ and $F_X(x)$ respectively. Define a new random variable Y such that $Y = T(X)$ for some function $T(\cdot)$. What is the PDF, $f_Y(y)$ (or CDF), of the new random variable? This problem is often encountered in the study of systems where the PDF for the input random variable X is known and the PDF for the output random variable Y needs to be determined. In such a case, we say that the input random variable has undergone a transformation.

Let us consider a new random variable Y generated by applying the following transformation to X:

$$Y = T(X) \qquad ...(3.19)$$

In general, X can be a discrete, continuous or mixed random variable and the transformation T can be linear, nonlinear, segmented, staircase, etc. Clearly, there are many cases to consider in a general study, depending on the form of X and T. In this book we shall consider only three cases: (i) X continuous, T continuous and either monotonically increasing or decreasing with X; (ii) X continuous and T continuous but nonmonotonic; (iii) X discrete and T continuous. Note that the transformation T is supposed to be continuous and differentiable in all the three cases.

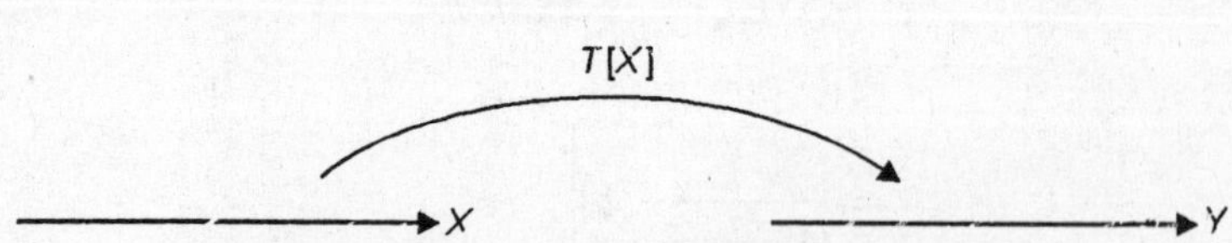

Fig. 3.2: Transformation of a random variable

3.5.1 Monotonic Transformation of a Continuous Random Variable

A monotonic function is a function that defines a one to one correspondence between the values X and Y.

- A transformation T is called monotonically increasing if $T(x_1) < T(x_2)$ for any $x_1 < x_2$.
- T is said to be monotonically decreasing when $T(x_1) > T(x_2)$ for any $x_1 < x_2$.

Monotonic Increasing Function

A monotonic increasing function is shown in Fig. 3.4a. We have

$$y_0 = T(x_0) \quad ...(3.20)$$

$$x_0 = T^{-1}(y_0) \quad ...(3.21)$$

Due to the one-to-one correspondence between X and Y, we can write

$$F_Y(y_0) = P\{Y \le y_0\} = P\{X \le x_0\} = F_X(x_0) \quad ...(3.22)$$

$$F_Y(y_0) = \int_{-\infty}^{y_0} f_Y(y_0)\,dy$$

$$= \int_{-\infty}^{x_0} f_X(x_0)\,dx \quad ...(3.23)$$

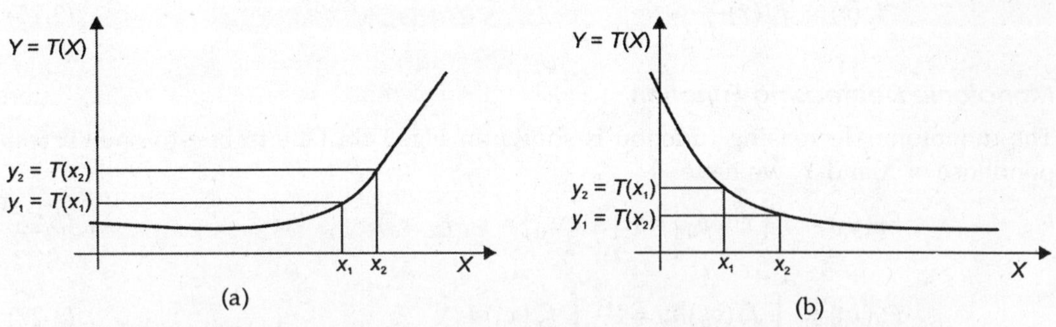

Fig. 3.3: (a) Monotonic increasing function (b) Monotonic decreasing function

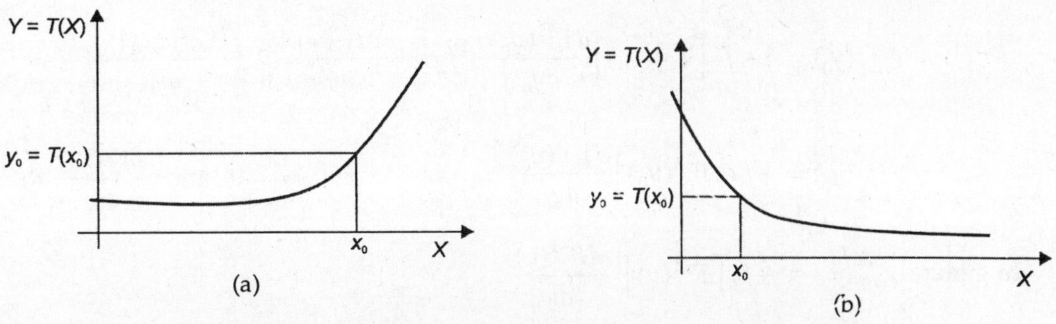

Fig. 3.4: Monotonic transformations: (a) Increasing function (b) Decreasing function

Differentiate both sides with respect to y_0 and $x_0 = T^{-1}(y_0)$ using Leibniz rule.

Leibniz rule

Let $\quad G(u) = \int_{\alpha(u)}^{\beta(u)} H(x,u)dx$

then $\quad \frac{dG(u)}{du} = H[\beta(u),u]\frac{d\beta(u)}{du} - H[\alpha(u),u]\frac{d\alpha(u)}{du} + \int_{\alpha(u)}^{\beta(u)} \frac{\partial H(x,u)}{\partial u}dx \quad \text{...(3.24)}$

$$f_Y(y_0) = f_X\left[T^{-1}(y_0)\right]\frac{dT^{-1}(y_0)}{dy_0} - f_X[-\infty]\frac{d(-\infty)}{dy_0} + \int_{-\infty}^{T^{-1}(y_0)} \frac{\partial f_X(x_0)}{\partial y_0}dx$$

$$f_Y(y_0) = f_X\left[T^{-1}(y_0)\right]\frac{dT^{-1}(y_0)}{dy_0} \qquad \because\ f_X[-\infty] = 0 \text{ and } \frac{\partial f_X(x_0)}{\partial y_0} = 0$$

In general, $f_Y(y) = f_X\left[T^{-1}(y)\right]\frac{dT^{-1}(y)}{dy}$

$$f_Y(y) = f_X(x)\frac{dx}{dy} \qquad \text{...(3.25)}$$

Monotonic Decreasing Function

The monotonic decreasing function is shown in Fig. 3.4b. Due to one-to-one correspondence of X and Y, we have

$$F_Y(y_0) = P\{Y \le y_0\} = P\{X \ge x_0\} = 1 - F_X(x_0) \qquad \text{...(3.26)}$$

$$F_Y(y_0) = \int_{-\infty}^{y_0} f_Y(y_0)dy = 1 - \int_{-\infty}^{x_0} f_X(x_0)dx \qquad \text{...(3.27)}$$

Differentiate both sides with respect to y_0 and $x_0 = T^{-1}(y_0)$ using Leibniz rule.

$$f_Y(y_0) = 1 - f_X\left[T^{-1}(y_0)\right]\frac{dT^{-1}(y_0)}{dy_0} + f_X[-\infty]\frac{d(-\infty)}{dy_0} - \int_{-\infty}^{T^{-1}(y_0)} \frac{\partial f_X(x_0)}{\partial y_0}dx$$

$$f_Y(y_0) = -f_X\left[T^{-1}(y_0)\right]\frac{dT^{-1}(y_0)}{dy_0} \qquad \because f_X[-\infty] = 0 \text{ and } \frac{\partial f_X(x_0)}{\partial y_0} = 0$$

In general, $f_Y(y_0) = -f_X\left[T^{-1}(y)\right]\frac{dT^{-1}(y)}{dy}$

$$f_Y(y) = -f_X(x)\frac{dx}{dy} \qquad \text{...(3.28)}$$

From Eqs. (3.27) and (3.28),

$$f_Y(y) = f_X(x)\left|\frac{dx}{dy}\right| \qquad ...(3.29)$$

3.5.2 Nonmonotonic Transformation of a Continuous Random Variable

In general, a transformation may not be monotonic; it can be both increasing and decreasing function as shown in Fig. 3.5. When the event $\{Y \le y_0\}$ corresponds to more than one event of the random variable X, then the transformation is called nonmonotonic transformation. For the values of y_0 shown in Fig. 3.5, the event $\{Y \le y_0\}$ corresponds to the event $\{X \le x_1$ and $x_2 \le X \le x_3\}$. Thus, the probability of the event $\{Y \le y_0\}$ now equals the probability of the event $\{x$ values yielding $Y \le y_0\}$, which we shall write as $\{x/Y \le y_0\}$. In other words,

$$F_Y(y_0) = P\{Y \le y_0\} = P\{x/Y \le y_0\} = \int_{x/Y \le y_0} F_X(x_0)$$

$$F_Y(y_0) = P\{Y \le y_0\} = P\{X \le x_1\} + P\{X \ge x_2\} + P\{X \le x_3\}$$

$$F_Y(y_0) = F_X(x_1) + 1 - F_X(x_2) + F_X(x_3)$$

$$F_Y(y_0) = P\{Y \le y_0\} = \int_{-\infty}^{x_1} f_X(x)\,dx + \int_{x_2}^{\infty} f_X(x)\,dx + \int_{-\infty}^{x_3} f_X(x)\,dx$$

$$\int_{-\infty}^{y_0} f_Y(y_0)\,dy = \int_{-\infty}^{x_1=T^{-1}(y_0)} f_X(x)\,dx + \int_{x_2=T^{-1}(y_0)}^{\infty} f_X(x)dx + \int_{-\infty}^{x_3=T^{-1}(y_0)} f_X(x)\,dx$$

By using Leibniz rule, differentiate with respect to y_0,

$$f_Y(y_0) = f_X\left[T^{-1}(y_0)\right]\frac{dT^{-1}(y_0)}{dy_0} - f_X\left[T^{-1}(y_0)\right]\frac{dT^{-1}(y_0)}{dy_0} + f_X\left[T^{-1}(y_0)\right]\frac{dT^{-1}(y_0)}{dy_0}$$

$$f_Y(y_0) = f_X(x_1)\frac{dx_1}{dy_0} - f_X(x_2)\frac{dx_2}{dy_0} + f_X(x_3)\frac{dx_3}{dy_0}$$

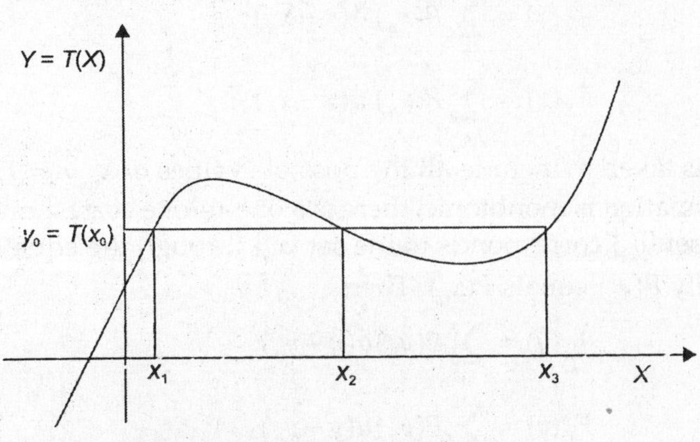

Fig. 3.5: Nonmonotonic transformation

In general, $$f_Y(y_0) = f_X(x_1)\left|\frac{dx_1}{dy_0}\right| + f_X(x_2)\left|\frac{dx_2}{dy_0}\right| + f_X(x_3)\left|\frac{dx_3}{dy_0}\right|$$

$$f_Y(y_0) = \sum_{i=1}^{n} f_X(x_i)\left|\frac{dx_i}{dy_0}\right| \qquad ...(3.30)$$

Example 3.5.1: Suppose X is a Gaussian random variable with mean, μ and variance, σ^2. A new random variable is formed according to $Y = aX + b$, where $a > 0$ (so that the transformation is monotonically increasing). Find the probability density function of Y.

Solution: From Eq. (3.25),

$$f_Y(y) = f_X(x)\frac{dx}{dy}$$

Since $\frac{dx}{dy} = \frac{1}{a}$ and $x = \frac{y-b}{a}$

$$\therefore \quad f_Y(y) = f_X(x)\frac{dx}{dy} = \frac{1}{a} f_X\left(\frac{y-b}{a}\right)$$

$$f_Y(y) = \frac{1}{a} f_X\left(\frac{y-1}{a}\right) = \frac{1}{a}\frac{1}{\sqrt{2\pi\sigma^2}} \exp\left(\frac{-\left(\frac{y-b}{a}-\mu\right)^2}{2\sigma^2}\right)$$

$$f_Y(y) = \frac{1}{\sqrt{2\pi(a\sigma)^2}} \exp\left(\frac{-[y-(a\mu+b)]^2}{2(a\sigma)^2}\right)$$

Note that the PDF of Y still has a Gaussian form. In this example, the transformation did not change the form of the PDF, it merely changed the mean and variance.

3.6 TRANSFORMATIONS OF A DISCRETE RANDOM VARIABLE

If X is a discrete random variable while $Y = T(X)$ is a continuous transformation, the problem is especially simple. Here

$$f_X(x) = \sum_n P(x_n)\,\delta(x - x_n) \qquad ...(3.31)$$

$$F_X(x) = \sum_n P(x_n)\,u(x - x_n) \qquad ...(3.32)$$

where the sum is taken to include all the possible values of x_n, $n = 1, 2,...$

If the transformation is monotonic, there is a one-to-one correspondence between X and Y so that a set $\{y_n\}$ corresponds to the set $\{x_n\}$ through the equation $y_n = T(x_n)$.

The probability $P(y_n)$ equals $P(x_n)$. Then

$$f_Y(y) = \sum_n P(y_n)\delta(y - y_n) \qquad ...(3.33)$$

$$F_Y(y) = \sum_n P(y_n)u(y - y_n) \qquad ...(3.34)$$

where $$y_n = T(x_n) \quad ...(3.35)$$
$$P(y_n) = P(x_n) \quad ...(3.36)$$

If T is not monotonic, the above procedure remains valid except that there now exists the possibility that more than one value x_n corresponds to a value y_n. In such a case, $P(y_n)$ will equal the sum of probabilities of the various x_n for which $y_n = T(x_n)$.

EXAMPLES

Example 3.1: Suppose X is a Gaussian random variable with zero mean and variance σ^2, together with a quadratic transformation, $Y = X^2$. Find PDF of random variable Y.

Solution: For any positive value of y, $y = x^2$ has two real roots, namely $x = \pm\sqrt{y}$ (for negative values of y, there are no real roots). $Y = X^2$ is a nonmonotonic transformation.

We know that for nonmonotonic transformation,

$$f_Y(y) = \sum_{i=1}^{n} f_X(x_i)\left|\frac{dx_i}{dy}\right|$$

$$\left|\frac{dx_i}{dy}\right| = \left|\frac{1}{\pm 2\sqrt{y}}\right| = \frac{1}{2\sqrt{y}}$$

$$f_Y(y) = f_X(x_1)\frac{dx_1}{dy} + f_X(x_2)\frac{dx_2}{dy} = \left(f_X\left(+\sqrt{y}\right)\frac{1}{2\sqrt{y}} + f_X\left(-\sqrt{y}\right)\frac{1}{2\sqrt{y}} \right)u(y)$$

For a zero mean Gaussian PDF, $f_X(x)$ is an even function so that $f_X\left(+\sqrt{y}\right) = f_X\left(-\sqrt{y}\right)$. Therefore,

$$f_Y(y) = \frac{1}{\sqrt{y}} f_X\left(+\sqrt{y}\right)u(y)$$

$$f_Y(y) = \frac{1}{\sqrt{2\pi y}} \exp\left(-\frac{y}{2}\right)u(y).$$

Example 3.2: Given the random variable X with density function

$$f_X(x) = \begin{cases} 2x & 0 < x < 1 \\ 0 & \text{else} \end{cases}$$

Find the PDF of $y = 8x^3$.

Solution: Since $y = 8x^3$ is a strictly monotonic increasing function in the interval (0, 1)

$$f_Y(y) = f_X(x)\frac{dx}{dy}$$

Since $\frac{dx}{dy} = \frac{1}{6}y^{-2/3}$ and $x = \sqrt[3]{\frac{y}{8}}$

$$f_Y(y) = f_X(x)\frac{dx}{dy} = y^{1/3}\frac{1}{6}y^{-2/3} = \frac{1}{6}y^{-1/3} \qquad 0 < y < 8.$$

Example 3.3: If the continuous random variable X has PDF

$$f_X(x) = \begin{cases} \frac{2}{9}(x+1) & -1 < x < 2 \\ 0 & \text{else} \end{cases}$$

find the PDF of $Y = X^2$.

Solution: The transformation function $y = x^2$ is not monotonic in the interval (–1, 2). Hence the interval is divided into two parts, i.e. (–1, 1) and (1, 2).

Since (–1, 1) is a symmetric interval, $f_Y(y)$ is found out by using

$$f_Y(y) = \sum_{i=1}^{n} f_X(x_i)\left|\frac{dx_i}{dy}\right| \quad \text{and } x_i = \pm\sqrt{y}$$

$$f_Y(y) = f_X(x_1)\frac{dx_1}{dy} + f_X(x_2)\frac{dx_2}{dy}$$

$$f_Y(y) = \frac{2}{9}(1+\sqrt{y})\frac{d\sqrt{y}}{dy} + \frac{2}{9}(1-\sqrt{y})\frac{d\sqrt{y}}{dy}$$

$$= \frac{2}{9}(1+\sqrt{y})\frac{1}{2\sqrt{y}} + \frac{2}{9}(1-\sqrt{y})\frac{1}{2\sqrt{y}}$$

$$= \frac{1}{9\sqrt{y}}(1+\sqrt{y}+1-\sqrt{y})$$

$$f_Y(y) = \frac{2}{9\sqrt{y}}$$

In the interval (1, 2), the transformation is a monotonic increasing function.

$$f_Y(y) = f_X(x)\frac{dx}{dy}$$

$$f_Y(y) = \frac{2}{9}(1+\sqrt{y})\frac{d\sqrt{y}}{dy}$$

$$= \frac{2}{9}(1+\sqrt{y})\frac{1}{2\sqrt{y}}$$

$$= \frac{1}{9\sqrt{y}}(1+\sqrt{y})$$

$$f_Y(y) = \frac{2}{9}\left(1+\frac{1}{\sqrt{y}}\right)$$

Hence the PDF is

$$f_Y(y) = \begin{cases} \frac{2}{9\sqrt{y}} & 0 < y < 1 \\ \frac{2}{9}\left(1+\frac{1}{\sqrt{y}}\right) & 1 < y < 4. \end{cases}$$

Example 3.4: Find the mean and variance of binomial distribution.

Solution: The expectation of discrete random variable is given by

$$E[X] = \sum_i x_i\, P(x_i) = \sum_{i=0}^{n} x_i\, P(x_i)$$

$$E[X] = \sum_{r=0}^{n} r\, {}^nC_r\, p^r q^{n-r}$$

$$= \sum_{r=0}^{n} r \frac{n!}{(n-r)!r!} p^r\, q^{n-r}$$

$$= pn \sum_{r=1}^{n} \frac{(n-1)!}{(n-r)!(r-1)!} p^{r-1}\, q^{n-1-(r-1)}$$

$$= pn \sum_{r=1}^{n} \frac{(n-1)!}{(n-r)!(r-1)!} p^{r-1}\, q^{(n-1)-(r-1)}$$

$$= pn(p+q)^{n-1} \qquad \because \sum_{r=0}^{n} \frac{(n)!}{(n-r)!(r)!} x^r y^{n-r} = (x+y)^n$$

$$E[X] = pn \qquad \because p + q = 1$$

$$E[X^2] = \sum_i x_i^2 P(x_i) = \sum_{i=0}^{n} x_i^2 P(x_i)$$

$$E[X] = \sum_{r=0}^{n} r^2\, {}^nC_r\, p^r q^{n-r}$$

$$= \sum_{r=0}^{n} \{r(r-1)+r\} \frac{n!}{(n-r)!r!} p^r q^{n-r}$$

$$= \sum_{r=0}^{n} r(r-1) \frac{n!}{(n-r)!r!} p^r q^{n-r} + \sum_{r=0}^{n} r \frac{n!}{(n-r)!r!} p^r q^{n-r}$$

$$= p^2 n(n-1) \sum_{r=2}^{n} \frac{(n-2)!}{(n-r)!(r-2)!} p^{r-2} q^{n-r} + np$$

$$= p^2 n(n-1)(p+q)^{n-2} + np$$

$$E[X^2] = p^2 n(n-1) + np$$

$$\sigma^2 = E[X^2] - \{E[X]\}^2 = p^2 n(n-1) + np - (np)$$

$$\sigma^2 = np(1-p) = npq$$

Example 3.5: Find the mean and variance of Poisson distribution.
Solution: The expectation of discrete random variable is given by

$$E[X] = \sum_{i} x_i P(x_i) = \sum_{i=0}^{\infty} x_i P(x_i)$$

$$E[X] = \sum_{r=0}^{\infty} r \frac{e^{-\lambda}\lambda^r}{r!}$$

$$= e^{-\lambda}\lambda \sum_{r=1}^{\infty} \frac{\lambda^{r-1}}{(r-1)!} = e^{-\lambda}\lambda e^{\lambda} = \lambda \qquad \because \sum_{r=1}^{\infty} \frac{\lambda^{r-1}}{(r-1)!} = e^{\lambda}$$

$$E[X] = \lambda$$

$$E[X] = \sum_{r=0}^{\infty} r \frac{e^{-\lambda}\lambda^r}{r!}$$

$$= e^{-\lambda}\lambda \sum_{r=1}^{\infty} \frac{\lambda^{r-1}}{(r-1)!} = e^{-\lambda}\lambda e^{\lambda} = \lambda \qquad \because \sum_{r=1}^{\infty} \frac{\lambda^{r-1}}{(r-1)!} = e^{\lambda}$$

$$E[X] = \lambda$$

$$E[X^2] = \sum_{r=0}^{\infty} r^2 \frac{e^{-\lambda}\lambda^r}{r!}$$

$$= \sum_{r=0}^{\infty} \{r(r-1) + r\} \frac{e^{-\lambda}\lambda^r}{r!}$$

$$= \sum_{r=0}^{\infty} r(r-1) \frac{e^{-\lambda}\lambda^r}{r!} + \sum_{r=0}^{\infty} r \frac{e^{-\lambda}\lambda^r}{r!}$$

$$= \lambda^2 e^{-\lambda} \sum_{r=2}^{\infty} \frac{\lambda^{r-2}}{(r-2)!} + \lambda$$

$$= \lambda^2 e^{-\lambda} e^{\lambda} + \lambda$$

$$E[X^2] = \lambda^2 + \lambda$$

$$\sigma^2 = E[X^2] - (E[X])^2 = \lambda^2 + \lambda - (\lambda)^2$$

$$\sigma^2 = \lambda.$$

Example 3.6: If the density function of continuous random variable X is $f_X(x) = \frac{1}{2} e^{-|x|}$ for $-\infty < x < \infty$, find moment generating function of X and also its mean and variance.

Solution: $f_X(x) = \frac{1}{2} e^{-|x|}$ for $-\infty < x < \infty$.

$$M_X(v) = E[e^{vx}] = \int_{-\infty}^{\infty} \frac{1}{2} e^{-|x|} e^{vx} dx$$

$$M_X(v) = \int_{-\infty}^{0} \frac{1}{2} e^{x} e^{vx} dx + \int_{0}^{\infty} \frac{1}{2} e^{-x} e^{vx} dx$$

$$M_X(v) = \int_{-\infty}^{0} \frac{1}{2} e^{x(1+v)} dx + \int_{0}^{\infty} \frac{1}{2} e^{-x(1-v)} dx$$

$$= \frac{1}{2} \frac{e^{x(1+v)}}{(1+v)}\bigg|_{-\infty}^{0} + \frac{1}{2} \frac{e^{-x(1-v)}}{-(1-v)}\bigg|_{0}^{\infty}$$

$$= \frac{1}{2}\left(\frac{1}{1+v} + \frac{1}{1-v}\right)$$

$$M_X(v) = \frac{1}{1-v^2}$$

$$E[X] = \frac{d}{dv}\left[\frac{1}{1-v^2}\right]_{v=0} = -\frac{2v}{(1-v^2)^2}\bigg|_{v=0} = 0$$

$$E[X^2] = \frac{d^2}{dv^2}\left[\frac{1}{1-v^2}\right]_{v=0} = \frac{d^2}{dv^2}\left[(1-v^2)^{-1}\right]_{v=0} = \frac{d}{dv}\left[-1(1-v^2)^{-2}(-2v)\right]\bigg|_{v=0}$$

$$= \frac{d}{dv}\left[(2v)(1-v^2)^{-2}\right]\bigg|_{v=0} = 2(1-v^2)^{-2} + (-2)(1-v^2)^{-3}(-2v)\bigg|_{v=0}$$

$$E[X^2] = 2$$

Variance $(X) = E[X^2] - (E[X])^2 = 2$

Example 3.7: Find the density function corresponding to the characteristic function defined as

$$\phi(f) = \begin{cases} 1-|t| & \text{for } |t| \le 1 \\ 0 & \text{for } |t| \ge 1 \end{cases}$$

Solution: We know that

$$\phi(\omega) = \int_{-\infty}^{\infty} f_X(x) e^{j\omega x} dx$$

$$f_X(x) = \frac{1}{2\pi} \int_{-\infty}^{\infty} \phi(f) e^{-j\omega x} dx$$

$$f_X(x) = \frac{1}{2\pi} \int_{-\infty}^{\infty} \phi(t) e^{-jxt} dt$$

$$= \frac{1}{2\pi} \int_{-1}^{1} (1-|t|) e^{-jxt} dt$$

$$= \frac{1}{2\pi}\left\{\int_{-1}^{0}(1+t)e^{-jxt}dt + \int_{0}^{1}(1-t)e^{-jxt}dt\right\}$$

$$= \frac{1}{2\pi}\left[\left\{(1+t)\frac{e^{-jxt}}{-jx} - \int_{-1}^{0}\frac{e^{-jxt}}{-jx}dx\right\}_{-1}^{0} + \left\{(1-t)\frac{e^{-jxt}}{-jx} + \int_{0}^{1}\frac{e^{-jxt}}{-jx}dt\right\}_{0}^{1}\right]$$

$$= \frac{1}{2\pi}\left[\left\{(1+t)\frac{e^{-jxt}}{-jx} - \frac{e^{-jxt}}{(-jx)^2}\right\}_{-1}^{0} + \left\{(1-t)\frac{e^{-jxt}}{-jx} + \frac{e^{-jxt}}{(-jx)^2}\right\}_{0}^{1}\right]$$

$$= \frac{1}{2\pi}\left[\left\{(1+t)\frac{e^{-jxt}}{-jx} + \frac{e^{-jxt}}{x^2}\right\}_{-1}^{0} + \left\{(1-t)\frac{e^{-jxt}}{-jx} - \frac{e^{-jxt}}{x^2}\right\}_{0}^{1}\right]$$

$$= \frac{1}{2\pi}\left[\left\{\frac{1}{-jx} + \frac{1}{x^2} - \frac{e^{jx}}{x^2}\right\} + \left\{\frac{1}{jx} - \frac{e^{-jx}}{x^2} + \frac{1}{x^2}\right\}\right]$$

$$= \frac{1}{2\pi}\left\{\frac{2}{x^2} - \left(\frac{e^{jx}}{x^2} + \frac{e^{-jx}}{x^2}\right)\right\}$$

$$= \frac{1}{\pi}\left\{\frac{1}{x^2} - \frac{1}{x^2}\left(\frac{e^{jx}+e^{-jx}}{2}\right)\right\}$$

$$= \frac{1}{\pi}\left(\frac{1-\cos x}{x^2}\right).$$

Example 3.8: A random variable X has mean $\mu = 12$, variance $\sigma^2 = 9$ and an unknown PDF. Find $P(6 < x < 18)$.

Solution: Since the probability distribution of X is unknown, we cannot find the value of the required probability but we can find lower bound for the probability using Chebyshev's inequality.

According to Chebyshev's inequality,

$$P(|x-\mu| \geq \varepsilon) \leq \frac{\sigma^2}{\varepsilon^2}, \varepsilon > 0$$

$$P(\mu-\varepsilon \leq x \leq \mu+\varepsilon) \geq 1 - \frac{\sigma^2}{\varepsilon^2}$$

$$P(\mu-\varepsilon \leq x \leq \mu+\varepsilon) \geq 1 - \frac{\sigma^2}{\varepsilon^2}$$

$\mu + \varepsilon = 18$ and $\mu - \varepsilon = 6$

By solving, $\varepsilon = 6$

$$P(6 < x < 18) \geq 1 - \frac{9}{36} = \frac{27}{36} = \frac{3}{4}.$$

Example 3.9: A discrete random variable X takes the values $-1, 0, 1$ with probabilities $\frac{1}{8}, \frac{3}{4}, \frac{1}{8}$ respectively. Evaluate $P(|x-\mu| \geq 2\sigma)$ and compare it with upper bound given by Chebyshev's inequality.

Solution: $$E[X] = -1\left(\frac{1}{8}\right)+0\left(\frac{3}{4}\right)+1\left(\frac{1}{8}\right)=0$$

$$E[X^2] = 1\left(\frac{1}{8}\right)+0\left(\frac{3}{4}\right)+1\left(\frac{1}{8}\right)=\frac{1}{4}$$

$$\sigma^2 = E[X^2]-(E[X])^2=\frac{1}{4}$$

$$P\{|x-\mu| \geq 2\sigma\} = P\{|x| \geq 1\}$$

$$= P\{x = -1 \text{ or } x = 1\}$$

$$= \frac{1}{8}+\frac{1}{8}=\frac{1}{4}$$

$$P(|x-\mu| > 2\sigma) \leq \frac{1}{4}.$$

Example 3.10: Find the characteristic function of the random variable with PDF

$$f_X(x) = \begin{cases} \frac{x}{2} & \text{for } 0 \leq x \leq 2 \\ 0 & \text{otherwise} \end{cases}$$

Solution: $$\phi(\omega) = E\left[e^{j\omega x}\right]= \int_{-\infty}^{\infty} f_X(x)e^{j\omega x}dx$$

$$= \int_0^2 \frac{x}{2}e^{j\omega x}dx$$

$$= \frac{1}{2}\left[\frac{xe^{j\omega x}}{j\omega}-\frac{e^{j\omega x}}{(j\omega)^2}\right]_0^2$$

$$= \frac{1}{2}\left[\frac{2e^{2j\omega}}{j\omega}-\frac{e^{j2\omega}}{(j\omega)^2}-0+\frac{1}{(j\omega)^2}\right]$$

$$= \frac{1}{2}\left[\frac{2e^{2j\omega}}{j\omega}+\frac{e^{j2\omega}}{(\omega)^2}-\frac{1}{(\omega)^2}\right]$$

$$= \frac{1}{2}\left[\frac{-2je^{2j\omega}+e^{j2\omega}-1}{(\omega)^2}\right]$$

$$\phi_X(\omega) = \frac{1}{2\omega^2}\left(e^{j2\omega}-2je^{2j\omega}-1\right)$$

Example 3.11: The density function of a random variable X is given by

$$f_X(x) = \begin{cases} 5e^{-5x} & 0 \le x < \infty \\ 0 & \text{elsewhere} \end{cases}$$

Find: i. $E(X)$ ii. $E\left[(X-1)^2\right]$ iii. $E(3X-1)$.

Solution: i. $E[X] = \int_0^\infty x5e^{-5x}dx$

$$= \left[x5\frac{e^{-5x}}{-5} - \int_0^\infty \frac{5e^{-5x}}{-5}dx\right]_0^\infty$$

$$= \left[-xe^{-5x} - \frac{e^{-5x}}{5}\right]_0^\infty$$

$$= \frac{1}{5}$$

ii. $$E\left[(X-1)^2\right] = \int_0^\infty \left(x^2 - 2x + 1\right) 5e^{-5x}dx$$

$$= \int_0^\infty x^2 5e^{-5x}dx - 2\int_0^\infty x5e^{-5x}dx + \int_0^\infty 5e^{-5x}dx$$

$$\int_0^\infty x^2 5e^{-5x}dx = \left[5x^2\frac{e^{-5x}}{-5} - 2\int_0^\infty x5\frac{e^{-5x}}{-5}dx\right]_0^\infty$$

$$= \left[\frac{2}{5}\int_0^\infty 5xe^{-5x}dx\right]_0^\infty$$

$$= \frac{2}{5}\cdot\frac{1}{5} = \frac{2}{25}$$

$$E\left[(X-1)^2\right] = \int_0^\infty x^2 5e^{-5x}dx - 2\int_0^\infty x5e^{-5x}dx + \int_0^\infty 5e^{-5x}dx = \frac{2}{25} - \frac{2}{5} + \left.\frac{5e^{-5x}}{-5}\right|_0^\infty$$

$$= \frac{2}{25} - \frac{2}{5} + 1 = \frac{17}{25}$$

iii. $$E[3X-1] = 3E[X] - 1 = 3\cdot\frac{1}{5} - 1 = \frac{-2}{5}.$$

Example 3.12: For the discrete probability distribution, find the mean and variance of X.

x_i	1	2	3	4	5	6
$P(x_i)$	1/6	1/6	1/6	1/6	1/6	1/6

Solution: $E[X] = \sum_{i=1}^{N} x_i P(x_i)$

$$E[X] = \sum_{i=1}^{6} x_i P(x_i) = x_1P(x_1) + x_2P(x_2) + x_3P(x_3) + x_4P(x_4) + x_5P(x_5) + x_6P(x_6)$$

$$= 1\cdot\frac{1}{6} + 2\cdot\frac{1}{6} + 3\cdot\frac{1}{6} + 4\cdot\frac{1}{6} + 5\cdot\frac{1}{6} + 6\cdot\frac{1}{6} = \frac{21}{6}$$

$$E[X] = \frac{7}{2}$$

$$E[X^2] = \sum_{i=1}^{N} x_i^2 P(x_i)$$

$$E[X^2] = \sum_{i=1}^{N} x_i^2 P(x_i) = x_1^2P(x_1) + x_2^2P(x_2) + x_3^2P(x_3) + x_4^2P(x_4) + x_5^2P(x_5) + x_6^2P(x_6)$$

$$= 1\cdot\frac{1}{6} + 4\cdot\frac{1}{6} + 9\cdot\frac{1}{6} + 16.\frac{1}{6} + 25\cdot\frac{1}{6} + 36\cdot\frac{1}{6} = \frac{91}{6}$$

$$\sigma^2 = E[X^2] - (E[X])^2 = \frac{91}{6} - \frac{49}{4} = \frac{35}{12}.$$

Example 3.13: If the mean and variance of the binomial distribution are 6 and 1.5 respectively, find $E[X - P(X > 3)]$.

Solution: $P(X > 3) = 1 - P(X \le 3)$

$$np = E[X] = 6$$

$$npq = \sigma^2 = 1.5, \qquad \frac{npq}{np} = q = \frac{1.5}{6} = 0.25$$

$$p = (1 - 0.25) = 0.75, \qquad n = \frac{6}{p} = \frac{6}{0.75} = 8$$

$$P(X \le 3) = \sum_{k=0}^{3} {}^nC_k p^k q^{n-k} = \sum_{k=0}^{3} {}^8C_k (0.75)^k (0.25)^{8-k}$$

$$= {}^8C_0(0.75)^0(0.25)^{8-0} + {}^8C_1(0.75)^1(0.25)^{8-1} + {}^8C_2(0.75)^2(0.25)^{8-2} + {}^8C_3(0.75)^3(0.25)^{8-3}$$

$$= (0.25)^8 + 8(0.75)(0.25)^7 + 28(0.75)^2(0.25)^6 + 56(0.75)^3(0.25)^5 = 0.02767$$

$$P(X > 3) = 1 - P(X \le 3) = 1 - 0.02767 = 0.9723$$

$$E[X - P(X > 3)] = E[X] - 0.9723 = 5.0277.$$

Example 3.14: Consider the random variable X with PDF $f_X(x)$ given by

$$f_X(x) = \begin{cases} \dfrac{1}{b-a} & a \le x \le b \\ 0 & \text{otherwise} \end{cases}$$

Show that the characteristic function is $\phi_X(\omega) = \dfrac{1}{j\omega(b-a)}\left(e^{j\omega b} - e^{j\omega a}\right)$.

Solution: $\phi_X(\omega) = \displaystyle\int_a^b \frac{1}{b-a} e^{j\omega x} dx$

$$= \frac{1}{b-a} \frac{e^{j\omega x}}{j\omega}\Bigg|_a^b = \frac{1}{j\omega(b-a)}\left(e^{j\omega b} - e^{j\omega a}\right).$$

EXERCISE

1. Find the mean and variance of Gaussian distribution function.
2. The mean of binomial distribution is 3 and the variance is 9/4. Find: (i) The value of n (ii) $P(X \ge 7)$ (iii) $P(1 \le X \le 7)$.
3. A random variable has a PDF $f_X(x) = \begin{cases} \dfrac{5}{4}(1 - x^4) & 0 < x \le 1 \\ 0 & \text{else} \end{cases}$

   Find: (i) $E[X]$ (ii) $E[4X + 2]$ (iii) $E[X^2]$.
4. A Gaussian random variable X having a mean value of zero and variance one is transformed to another random variable Y by a square law transformation. Find the density function of Y.
5. Let X be an exponential random variable with $f_X(x) = 2e^{-2x} u(x)$ and let the transformation be $Y = X^3$. Find $f_Y(y)$.
6. Suppose the average life span of a person is 75 years. Find the probability that a human lives for 110 years.
7. For a Gaussian random variable, derive expressions for the skewness and the kurtosis in terms of the mean (μ) and variance (σ^2).
8. Let X be a random variable with $E[X] = 1$ and $\text{Var}(X) = 4$. Find the following:

   (a) $E[2X - 4]$ (b) $E[X]^2$ (c) $E\left[(2X - 4)^2\right]$.
9. In a laboratory assignment, if the equipment is working, the density function of the observed outcome X is

   $$f_X(x) = \begin{cases} 2(1-x) & 0 < x < 1 \\ 0 & \text{else} \end{cases}$$

   Find the variance and standard deviation of X.

4

Multiple Random Variables

4.0 INTRODUCTION

Sometimes we encounter problems that deal with two or more random variables defined on the same sample space. In this chapter, we consider these multivariate systems. We first consider bivariate random variables and later consider systems with more than two random variables. One of our main goals here is to develop a vector/matrix notation that will allow us to represent potentially large sequences of random variables with a compact notation. Many of the concepts developed in Chapter 2 can be extended to multiple dimensions in a very straightforward manner.

4.1 VECTOR RANDOM VARIABLES

In many experiments, the observations are expressible not as a single quantity, but as a family of quantities. For example, to record the height and weight of each person in a community or the number of people and the total income in a family, we need two random variables.

Definition: Let S be the sample space associated with a random experiment E. Let X and Y be two random variables, $X = X(s)$ and $Y = Y(s)$ be two functions each assigning a real number to each outcomes $s \in S$. Then (X, Y) is called a two dimensional random variable.

Note: i. If the possible values of (X, Y) are finite or countable infinite, (X, Y) is called a two dimensional discrete random variable.

ii. If (X, Y) can assume all values in a specified region R in the xy-plane, (X, Y) is called a two dimensional continuous random variable.

In more general case, where N random variables are defined on a sample space S, we consider them to be components of an N-dimensional random vector or N-dimensional random variable. The sample space S_j is now N-dimensional.

Random vector: A random vector is a vector whose elements are random variables. Let X be a n-dimensional random vector whose elements are $\{X_1, X_2,..., X_n\}$.

It can expressed as

$$X = \begin{pmatrix} X_1 \\ X_2 \\ \vdots \\ X_n \end{pmatrix} \qquad ..(4.1)$$

It is a $(n \times 1)$ random vector.

Each element of X, i.e. $X_j, j = 1,..., n$, is a random variable with its own mean, variance, and probability distribution.

4.2 JOINT DISTRIBUTION FUNCTION

If (X, Y) is a two-dimensional random variable (discrete or continuous), the joint distribution function is denoted by $F_{XY}(x, y)$. It can be defined as

$$F_{XY}(x, y) = P\{X \le x \text{ and } Y \le y\} \quad ...(4.2)$$

It should be clear that

$$P\{X \le x \text{ and } Y \le y\} = P(A \cap B)$$

where A is an event defined as $A = \{X \le x\}$ and another event defined as $B = \{Y \le y\}$.

In the discrete case, let X have N possible values x_n and Y have M possible values y_m, then

$$F_{XY}(x, y) = \sum_{n=1}^{N}\sum_{m=1}^{M} P(x_n, y_m)u(x - x_n)u(y - y_m) \quad ...(4.3)$$

$P(x_n, y_m)$ is the probability of the joint event $\{X = x_n, Y = y_m\}$ and $u(\cdot)$ is the unit step function.

In the continuous case,

$$F_{XY}(x, y) = \int_{-\infty}^{y}\int_{-\infty}^{x} f_{XY}(x,y)\,dx\,dy \quad ...(4.4)$$

For N random variables X_n, $n = 1, 2,..., N$

$$F_{X_1,X_2,...,X_N}(x_1, x_2,..., x_N) = P\{X_1 \le x_1, X_2 \le x_2,..., X_N \le x_N\} \quad ...(4.5)$$

4.2.1 Properties

i. $F_{XY}(-\infty, -\infty) = 0, \quad F_{XY}(-\infty, y) = 0, \quad F_{XY}(x, -\infty) = 0$ ...(4.6a)

$F_{XY}(x, y) = P\{X \le x, Y \le y\}$

Let $x = -\infty$, $y = -\infty$

$$F_{XY}(-\infty, -\infty) = P\{X \le -\infty, Y \le -\infty\} = 0 \qquad \because P(X \le -\infty) = 0$$

Let $x = -\infty$, $y = y$

$$F_{XY}(-\infty, y) = P\{X \le -\infty, Y \le y\} = P\{X \le -\infty \cap Y \le y\} = P(\phi) = 0$$

Let $x = x$, $y = -\infty$

$$F_{XY}(x, -\infty) = P\{X \le x, Y \le -\infty\} = P\{X \le x \cap Y \le -\infty\} = P(\phi) = 0$$

ii. $F_{XY}(+\infty, +\infty) = 1$ ...(4.6b)

$$F_{XY}(x, y) = P\{X \le x, Y \le y\}$$

Let $x = +\infty$, $y = +\infty$

$$F_{XY}(+\infty, +\infty) = P\{X \le +\infty, Y \le +\infty\} = 1 \qquad \because P(X \le +\infty) = 1$$

iii. $0 \le F_{XY}(x, y) \le 1$ ...(4.6c)

From the properties (i) and (ii), the minimum value of joint distribution function is 0 and maximum value is 1.

iv. $F_{XY}(x_2,y_2)+F_{XY}(x_1,y_1)-F_{XY}(x_1,y_2)-F_{XY}(x_2,y_1)=P\{x_1 \le X \le x_2,\ y_1 \le Y \le y_2\}$ ...(4.6d)

From Fig. 4.1,

$$(x_2, y_2) = (x_1,y_2)\cup(x_2,y_1)\cup\{x_1 \le X \le x_2,\ y_1 \le Y \le y_2\}-(x_1,y_1)$$

Apply probability on both sides,

$$P(x_2, y_2) = P(x_1,y_2)+P(x_2,y_1)+P\{x_1 \le X \le x_2,\ y_1 \le Y \le y_2\}-P(x_1,y_1)$$

$$F_{XY}(x_2, y_2) = F_{XY}(x_1,y_2)+F_{XY}(x_2,y_1)+P\{x_1 \le X \le x_2,\ y_1 \le Y \le y_2\}-F_{XY}(x_1,y_1)$$

$$P\{x_1 \le X \le x_2,\ y_1 \le Y \le y_2\} = F_{XY}(x_2,y_2)+F_{XY}(x_1,y_1)-F_{XY}(x_1,y_2)-F_{XY}(x_2,y_1)$$

v. $F_{XY}(x, y)$ is a nondecreasing function of both x and y.

vi. $F_{XY}(x, \infty) = F_X(x)$ and $F_{XY}(\infty, y) = F_Y(y)$ ...(4.6e)

$$F_{XY}(x, y) = P\{X \le x,\ Y \le y\}$$

Let $x = x,\ \ y = +\infty$

$$F_{XY}(x, +\infty) = P\{X \le x, Y \le +\infty\} = P(X \le x) = F_X(x)$$

$$\because\ P\{X \le x, Y \le +\infty\} = P\{X \le x \cap Y \le +\infty\} = P\{X \le x\}$$

Similarly, let $x = +\infty,\ \ y = y$

$$F_{XY}(+\infty, y) = P\{X \le +\infty, Y \le y\} = P(Y \le y) = F_Y(y)$$

$$\because\ P\{X \le +\infty, Y \le y\} = P\{X \le +\infty \cap Y \le y\} = P\{X \le y\}$$

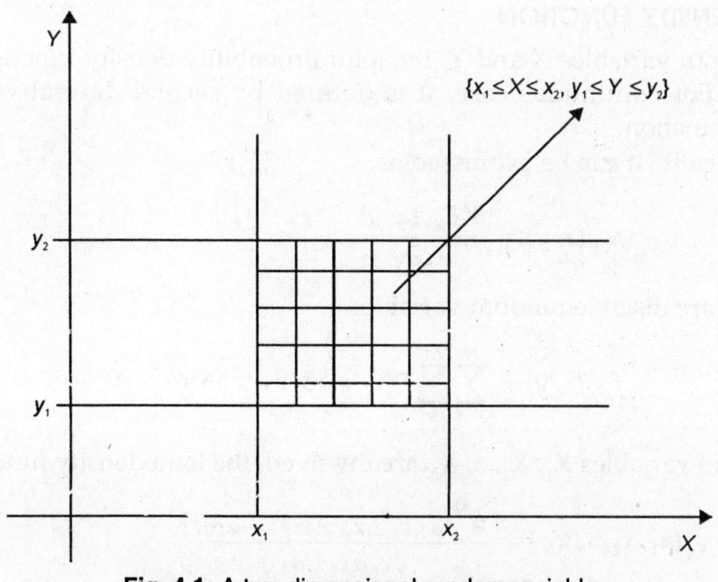

Fig. 4.1: A two dimensional random variable

Example 4.2.1: The joint CDF of two discrete random variables X and Y is given as follows:

$$P(x, y) = \begin{cases} \frac{1}{8} & x = 1, y = 1 \\ \frac{5}{8} & x = 1, \; y = 2 \\ \frac{1}{8} & x = 2, \; y = 1 \\ \frac{1}{8} & x = 2, \; y = 2 \end{cases}$$

Determine the joint distribution function of X and Y.

Solution: $F_{XY}(x, y) = P(X \le x, Y \le y)$

$$F_{XY}(1, 1) = P(X \le 1, Y \le 1) = P(X = 1, Y = 1) + P(X < 1, Y < 1) = \frac{1}{8}$$

$$F_{XY}(1, 2) = P(X \le 1, Y \le 2) = P(X = 1, Y = 1) + P(X = 1, Y = 2) = \frac{1}{8} + \frac{5}{8} = \frac{3}{4}$$

$$F_{XY}(2, 1) = P(X \le 2, Y \le 1) = P(X = 1, Y = 1) + P(X = 2, Y = 1) = \frac{1}{8} + \frac{1}{4} = \frac{3}{8}$$

$$F_{XY}(2, 2) = P(X \le 2, Y \le 2)$$

$$= P(X = 1, Y = 1) + P(X = 2, Y = 1) + P(X = 2, Y = 1) + P(X = 2, Y = 1)$$

$$= \frac{1}{8} + \frac{5}{8} + \frac{1}{8} + \frac{1}{8} = 1.$$

4.3 JOINT DENSITY FUNCTION

For two random variables X and Y, the joint probability density function is denoted by $f_{XY}(x, y)$. For continuous case, it is defined by second derivative of the joint distribution function.

Mathematically, it can be expressed as,

$$f_{XY}(x, y) = \frac{\partial^2 F_{XY}(x, y)}{\partial x \, \partial y} \qquad \text{...(4.7)}$$

If X and Y are discrete random variables,

$$f_{XY}(x, y) = \sum_{i=1}^{N} \sum_{j=1}^{M} P(x_i, y_j) \delta(x - x_i) \delta(y - y_j) \qquad \text{...(4.8)}$$

If N random variables $X_1, X_2, ..., X_N$ are involved, the joint density function becomes

$$f_{X_1, X_2, ..., X_N}(x_1, x_2, ..., x_N) = \frac{\partial^N F_{X_1, X_2, ..., X_N}(x_1, x_2, ..., x_N)}{\partial x_1 \, \partial x_2 ... \partial x_N} \qquad \text{...(4.9)}$$

4.3.1 Properties

i. $f_{XY}(x, y) \geq 0$ ...(4.10a)

ii. $\int_{-\infty}^{\infty}\int_{-\infty}^{\infty} f_{XY}(x,y)\,dx\,dy = 1$ ...(4.10b)

From Eq. (4.4),

$$F_{XY}(x, y) = \int_{-\infty}^{y}\int_{-\infty}^{x} f_{XY}(x,y)\,dx\,dy$$

Let $x = \infty$ and $y = \infty$

$$F_{XY}(\infty, \infty) = \int_{-\infty}^{\infty}\int_{-\infty}^{\infty} f_{XY}(x,y)\,dx\,dy$$

$$1 = \int_{-\infty}^{\infty}\int_{-\infty}^{\infty} f_{XY}(x,y)\,dx\,dy \qquad \because F_{XY}(\infty,\infty) = 1$$

Hence $\int_{-\infty}^{\infty}\int_{-\infty}^{\infty} f_{XY}(x,y)\,dx\,dy = 1$

iii. $$F_{XY}(x, y) = \int_{-\infty}^{y}\int_{-\infty}^{x} f_{XY}(\xi_1,\xi_2)\,d\xi_1\,d\xi_2$$ ...(4.10c)

$$F_{XY}(x, y) = \int_{-\infty}^{y}\left(\int_{-\infty}^{x} f_{XY}(\xi_1,\xi_2)\,d\xi_1\right)d\xi_2$$

$$= \int_{-\infty}^{y} f_{XY}(x,\xi_2)\,d\xi_2$$

$$= F_{XY}(x, y)$$

iv. $$F_X(x) = \int_{-\infty}^{\infty}\int_{-\infty}^{x} f_{XY}(\xi_1,\xi_2)\,d\xi_1\,d\xi_2 \text{ and } F_Y(y) = \int_{-\infty}^{y}\int_{-\infty}^{\infty} f_{XY}(\xi_1,\xi_2)\,d\xi_1\,d\xi_2$$ ...(4.10d)

$$F_{XY}(x, y) = \int_{-\infty}^{x}\left(\int_{-\infty}^{\infty} f_{XY}(\xi_1,\xi_2)\,d\xi_2\right)d\xi_1$$

$$= \int_{-\infty}^{x} f_X(\xi_1)\,d\xi_1 = F_X(x)$$

Similarly,

$$F_Y(y) = \int_{-\infty}^{\infty}\int_{-\infty}^{y} f_{XY}(\xi_1,\xi_2)\,d\xi_1\,d\xi_2$$

$$F_{XY}(x, y) = \int_{-\infty}^{y}\left(\int_{-\infty}^{\infty} f_{XY}(\xi_1,\xi_2)d\xi_1\right)d\xi_2$$

$$= \int_{-\infty}^{y} f_Y(\xi_2)d\xi_2$$

$$= F_Y(y)$$

Hence $F_X(x) = \int_{-\infty}^{\infty}\int_{-\infty}^{x} f_{XY}(\xi_1,\xi_2)d\xi_1\, d\xi_2$ and $F_Y(y) = \int_{-\infty}^{y}\int_{-\infty}^{\infty} f_{XY}(\xi_1,\xi_2)d\xi_1\, d\xi_2$

v. $P(x_1 < X \le x_2, y_1 < Y \le y_2) = \int_{y_1}^{y_2}\int_{x_1}^{x_2} f_{XY}(x,y)dx\,dy$ ...(4.10e)

From the property of distribution function, we know that

$$P\{x_1 \le X \le x_2, y_1 \le Y \le y_2\} = F_{XY}(x_2,y_2) + F_{XY}(x_1,y_1) - F_{XY}(x_1,y_2) - F_{XY}(x_2,y_1)$$

$$= \int_{x_1}^{x_2}\int_{y_1}^{y_2} \frac{\partial^2 F_{XY}(x,y)}{\partial x \partial y} dx\,dy$$

$$= \int_{x_1}^{x_2}\int_{y_1}^{y_2} f_{XY}(x,y)dx\,dy$$

vi. $F_X(x) = \int_{-\infty}^{\infty} f_{XY}(x,y)dy$ and $f_Y(y) = \int_{-\infty}^{\infty} f_{XY}(x,y)dx$ ...(4.10f)

$$F_{XY}(x, y) = \int_{-\infty}^{y}\int_{-\infty}^{x} f_{XY}(x,y)dx\,dy$$

Let $y = \infty$,

$$F_{XY}(x, y) = \int_{-\infty}^{\infty}\int_{-\infty}^{x} f_{XY}(x,y)dx\,dy$$

But we know that

$$F_X(x) = \int_{-\infty}^{\infty}\int_{-\infty}^{x} f_{XY}(x,y)dx\,dy$$

$$\frac{dF_X(x)}{dx} = \frac{d\left(\int_{-\infty}^{x}\int_{-\infty}^{\infty} f_{XY}(x,y)dx\,dy\right)}{dx}$$

$$F_X(x) = \int_{-\infty}^{\infty}\left[f_{XY}(x,y)\frac{dx}{dx} - f_X(-\infty,y)\frac{d(-\infty)}{dx} + \int_{-\infty}^{x}\frac{\partial f_{XY}(x,y)}{\partial x}dx\right]dy$$

$$f_X(x) = \int_{-\infty}^{\infty} f_{XY}(x,y)dy$$

Similarly,

$$f_Y(y) = \int_{-\infty}^{\infty} f_{XY}(x,y)\,dx.$$

Example 4.3.1: Two random variables X and Y have a joint probability density function

$$f_{XY}(x, y) = \begin{cases} ke^{-(x+y)} & 0 \le x < \infty, 0 \le y < \infty \\ 0 & \text{otherwise} \end{cases}$$

Find: (a) k (b) $f_X(x)$ (c) $P(0 \le x \le 2, 2 \le y \le 3)$.

Solution: (a) We know that

$$\int_{-\infty}^{\infty}\int_{-\infty}^{\infty} f_{XY}(x,y)\,dx\,dy = 1$$

$$\int_0^{\infty}\int_0^{\infty} ke^{-(x+y)}\,dx\,dy = 1$$

$$k\left(\int_0^{\infty} e^{-x}dx\right)\left(\int_0^{\infty} e^{-y}dy\right) = 1$$

$$k\left(\frac{e^{-x}}{-1}\right)_0^{\infty}\left(\frac{e^{-y}}{-1}\right)_0^{\infty} = 1$$

$$k = 1$$

(b) $f_X(x) = \int_{-\infty}^{\infty} f_{XY}(x,y)\,dy$

$$= \int_0^{\infty} e^{-(x+y)}dy$$

$$= e^{-x}\int_0^{\infty} e^{-(y)}dy$$

$$= e^{-x}\left(\frac{e^{-y}}{-1}\right)_0^{\infty} = e^{-x}$$

(c) $P(0 \le x \le 2, 2 \le y \le 3) = \int_2^3\int_0^2 e^{-(x+y)}dx\,dy$

$$= \int_2^3\int_0^2 e^{-x}e^{-y}dx\,dy = \int_0^2 e^{-x}dx\int_2^3 e^{-y}\,dy$$

$$= \left(\frac{e^{-x}}{-1}\right)_0^2\left(\frac{e^{-y}}{-1}\right)_2^3 = (-e^{-2}+1)(e^{-2}-e^{-3})$$

$$= (1-e^{-2})(e^{-2}-e^{-3}).$$

4.4 CONDITIONAL DISTRIBUTION AND DENSITY

4.4.1 Point Conditioning

We know that the conditional distribution function of a random variable X, given some event B, is defined as

$$F_X(x/B) = P(X \le x/B) = \frac{P(X \le x \cap B)}{P(B)}$$

and density function is defined as

$$f_X(x/B) = \frac{dF_X(x/B)}{dx}$$

The above two functions are extended to include a second variable through suitable definitions of event B.

In practical problems, we are interested in the distribution function of one random variable conditioned by a second random variable Y, having some specific value y. This is called point conditioning and we can handle such problem by defining event B by

$$B = \{y < Y < y + h\} \qquad \text{...(4.11)}$$

where h is a small quantity.

$$F_X(x/y < Y < y+h) = \frac{\int_{y}^{y+h}\int_{-\infty}^{x} f_{XY}(t,u)\,dt\,du}{\int_{y}^{y+h} f_Y(u)\,du} \qquad \text{...(4.12)}$$

As $h \to 0$, the denominator becomes 0. However, we can still show that the conditional density $f_X(x/Y = y)$ may exist. If h is very small, Eq. (4.12) can be written as

$$F_X(x/y < Y < y+h) = \frac{\int_{-\infty}^{x} f_{XY}(t,y)\,dt\,h}{f_Y(y)h} \qquad \text{...(4.13)}$$

As $h \to 0$

$$F_X(x/Y = y) = \frac{\int_{-\infty}^{x} f_{XY}(t,y)\,dt}{f_Y(y)} \qquad \text{...(4.14)}$$

$$f_X(x/Y = y) = \frac{dF_X(x/Y=y)}{dx} = \frac{f_{XY}(x,y)}{f_Y(y)} \qquad \text{...(4.15)}$$

Similarly,

$$f_X(y/X = x) = \frac{f_{XY}(x,y)}{f_X(x)} \qquad \text{...(4.16)}$$

If X and Y both are discrete random variables with values x_i, $i = 1, 2, ..., N$ and y_j, $j = 1, 2, ..., M$ respectively, while the probabilities of these values are denoted by $P(x_i)$ and $P(y_j)$ respectively, the probability of joint occurrence of x_i and y_j is denoted by $P(x_i, y_j)$. Thus

$$f_Y(y) = \sum_{j=1}^{M} P(y_j)\delta(y - y_j) \quad ...(4.17)$$

$$f_{XY}(x, y) = \sum_{i=1}^{N}\sum_{j=1}^{M} P(x_i, y_j)\delta(y - y_j)\delta(x - x_i) \quad ...(4.18)$$

Let us assume the specific value of y of interest be y_k.

Substitute Eqs. (4.17) and (4.18) into Eq. (4.12) and apply $h \to 0$, we can get the distribution function as

$$F_X(x/y = y_k) = \sum_{i=1}^{N} \frac{P(x_i, y_k)}{P(y_k)} u(x - x_i) \quad ...(4.19)$$

Similarly, the density function is

$$f_X(x/y = y_k) = \sum_{i=1}^{N} \frac{P(x_i, y_k)}{P(y_k)} \delta(x - x_i) \quad ...(4.20)$$

4.4.2 Interval Conditioning

Consider an event B defined as

$$B = \{y_1 < Y < y_2) \quad ...(4.21)$$

where y_1 and y_2 are real numbers and $P(B) \neq 0$.

$$F_X(x/y_1 < Y < y_2) = \frac{F_{XY}(x, y_2) - F_{XY}(x, y_1)}{F_X(y_2) - F_Y(y_1)} \quad ...(4.22)$$

$$= \frac{\int_{y_1}^{y_2}\int_{-\infty}^{x} f_{XY}(t, y)\,dt\,dy}{\int_{y_1}^{y_2}\int_{-\infty}^{\infty} f_{XY}(x, y)\,dx\,dy} \quad ...(4.23)$$

$$f_X(x/y_1 < Y < y_2) = \frac{\int_{y_1}^{y_2} f_{XY}(x, y)\,dy}{\int_{y_1}^{y_2} f_Y(y)\,dy} \quad ...(4.24)$$

4.5 INDEPENDENCE OF TWO RANDOM VARIABLES

Two random variables X and Y are said to be independent if any event A_1 defined in terms of X is independent of any event A_2 defined in terms of Y, i.e.

$$P(X \text{ in } A_1, Y \text{ in } A_2) = P(X \text{ in } A_1)\, P(Y \text{ in } A_2) \quad ...(4.25)$$

Suppose that X and Y are a pair of discrete random variables and if we let $A_1 = \{X = x_j\}$ and $A_2 = \{Y = y_k\}$, then the independence of X and Y implies that

$$P(x_j, y_k) = P(X = x_j, Y = y_k) = P(X = x_j)\, P(Y = y_k)$$

$$P(x_j, y_k) = P(x_j)\, P(y_k) \quad \text{for all } x_j \text{ and } y_k \qquad ...(4.26)$$

In general, it can be shown that the random variables X and Y are independent if and only if their joint CDF is equal to the product of their marginal CDF

$$F_{XY}(x, y) = F_X(x)\, F_Y(y) \text{ for all values of } x \text{ and } y \qquad ...(4.27)$$

Similarly, if X and Y are jointly continuous, then X and Y are independent if and only if their joint PDF is equal to the product of their marginal PDF

$$f_{XY}(x, y) = f_X(x)\, f_Y(y) \text{ for all values of } x \text{ and } y \qquad ...(4.28)$$

If X and Y are independent variables, then the conditional distribution function is

$$F_X(x/Y \le y) = \frac{P(X \le x, Y \le y)}{P(Y \le y)} = \frac{P(X \le x)\; P(Y \le y)}{P(Y \le y)} = P(X \le x)$$

$$F_X(x/Y \le y) = P(X \le x) = F_X(x) \qquad ...(4.29)$$

Similarly, $\quad F_Y(y/X \le x) = F_Y(y) \qquad ...(4.30)$

In the same way,

$$f_X(x/Y \le y) = f_X(x) \qquad ...(4.31)$$

$$f_Y(y/X \le x) = f_Y(y) \qquad ...(4.32)$$

Example 4.5.1: X and Y are two continuous random variables whose joint PDF is given by

$$f_{XY}(x, y) = \begin{cases} e^{-(x+y)} & 0 \le x < \infty, 0 \le y < \infty \\ 0 & \text{otherwise} \end{cases}$$. Are X and Y independent?

Solution: From Eq. (4.28), the condition for independence is

$$f_{XY}(x, y) = f_X(x) f_Y(y)$$

Marginal densities are given by

$$f_X(x) = \int_{-\infty}^{\infty} f_{XY}(x,y)dy \quad \text{and} \quad f_Y(y) = \int_{-\infty}^{\infty} f_{XY}(x,y)dx$$

$$f_X(x) = \int_{-\infty}^{\infty} f_{XY}(x,y)dy$$

$$f_X(x) = \int_0^{\infty} e^{-(x+y)}dy = e^{-x}\int_0^{\infty} e^{-y}dy = e^{-x}\left.\frac{e^{-y}}{-1}\right|_0^{\infty} = e^{-x}$$

$$f_Y(y) = \int_{-\infty}^{\infty} f_{XY}(x,y)dx$$

$$f_Y(y) = \int_0^{\infty} e^{-(x+y)}dx = e^{-y}\int_0^{\infty} e^{-x}dx = e^{-y}\left.\frac{e^{-x}}{-1}\right|_0^{\infty} = e^{-y}$$

$$f_X(x)\, f_Y(y) = e^{-x} e^{-y} = e^{-(x+y)}$$

$$f_{XY}(x, y) = f_X(x)\, f_Y(y)$$

Hence X and Y are statistically independent of each other.

4.6 DISTRIBUTION AND DENSITY OF SUM OF TWO OR MORE RANDOM VARIABLES

4.6.1 Sum of Two Random Variables

Let us consider a simple electrical circuit. Assume random variable X to represent a random voltage signal and random variable Y to represent a random noise signal at some instance of time. The sum Z would represent a signal-plus-noise voltage available to some receiver. In this topic, we discuss how to find the PDF of sum of two random variables, if known their joint PDF.

Let Z be a random variable equal to the sum of two random variables X and Y, i.e.

$$Z = X + Y \qquad ...(4.33)$$

In particular, we find the PDF of Z using the two-step procedure in which we first find the CDF $F_Z(z)$ by integrating the joint PDF $f_{XY}(x, y)$ over the region $X + Y \leq Z$ as shown in Fig. 4.2 and next by derivating CDF we can find the PDF.

The cumulative distribution function of Z is defined as

$$F_Z(z) = P(Z \leq z) = P(X + Y \leq z) \qquad ...(4.34)$$

Now from the probability corresponding to an elemental area $dx\,dy$ in the xy-plane located at the point (x, y) is $F_{XY}(x, y)dx\,dy$. If we sum all such probabilities over the region where $X + Y \leq Z$, we will obtain $F_Z(z)$. Thus,

$$F_Z(z) = P(X+Y \leq z) = \int_{-\infty}^{\infty}\left(\int_{-\infty}^{z-x} f_{XY}(x,y)dy\right)dx \qquad ...(4.35)$$

Taking the derivative of the CDF to find the PDF, we have

$$f_Z(z) = \frac{dF_Z(z)}{dz} = \int_{-\infty}^{\infty}\left[\frac{d}{dz}\left(\int_{-\infty}^{z-x} f_{XY}(x,y)dy\right)\right]dx$$

$$f_Z(z) = \int_{-\infty}^{\infty} f_{XY}(x,z-x)dx = \int_{-\infty}^{\infty} f_{XY}(z-y,y)dy \qquad ...(4.36)$$

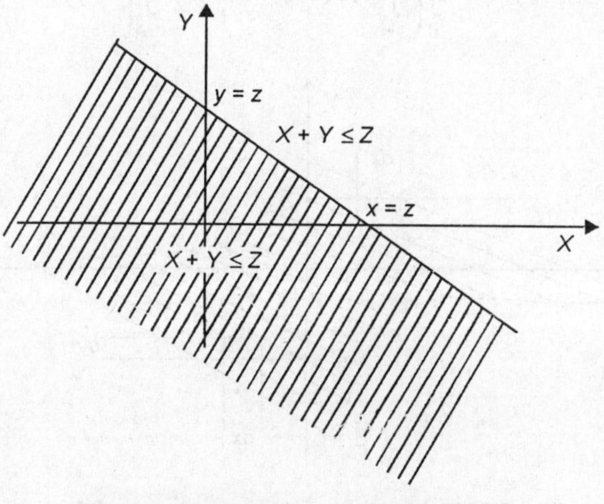

Fig. 4.2: Sum of two random variables

If X and Y are independent,

$$F_Z(z) = \int_{-\infty}^{\infty} f_X(x)\left(\int_{-\infty}^{z-x} f_Y(y)dy\right)dx \quad ...(4.37)$$

$$f_Z(z) = \int_{-\infty}^{\infty} f_X(x) f_Y(z-x)dx = \int_{-\infty}^{\infty} f_X(z-y) f_Y(y)dy \quad ...(4.38)$$

The above expression is recognized as a convolution integral. Consequently, we have shown that the density function of sum of two statistically independent random variables is the convolution of their individual density functions.

4.6.2 Sum of *N* Random Variables

For sum of N random variables, $X_1, X_2,..., X_N$, the density function is defined as

$$f_Z(z) = f_{Z=X_1+X_2+...+X_N}(x_1, x_2, ..., x_N) = f_{X_1}(x_1)\times f_{X_2}(x_2)\times ...\times f_{X_N}(N) \quad ...(4.39)$$

When X and Y are independent discrete random variables,

$$P_Z(z) = \sum_{k=-\infty}^{\infty} P_X(k)P_Y(z-k) = \sum_{k=-\infty}^{\infty} P_X(z-k)P_Y(k) \quad ...(4.40)$$

4.6.3 Multiplication of Two Random Variables

Let us consider two independent random variables X and Y and a random variable Z defined as

$$Z = XY \quad (4.41)$$

The PDF of Z can be obtained as follows.

From Fig. 4.3,

$$F_Z(z) = P(XY \le z) = \iint_{XY\le z} f_{XY}(x,y)\,dx\,dy$$

$$F_Z(z) = \int_0^{\infty}\left(\int_{-\infty}^{z/y} f_{XY}(x,y)dx\right)dy + \int_{-\infty}^{0}\left(\int_{z/y}^{\infty} f_{XY}(x,y)dx\right)dy \quad ...(4.42)$$

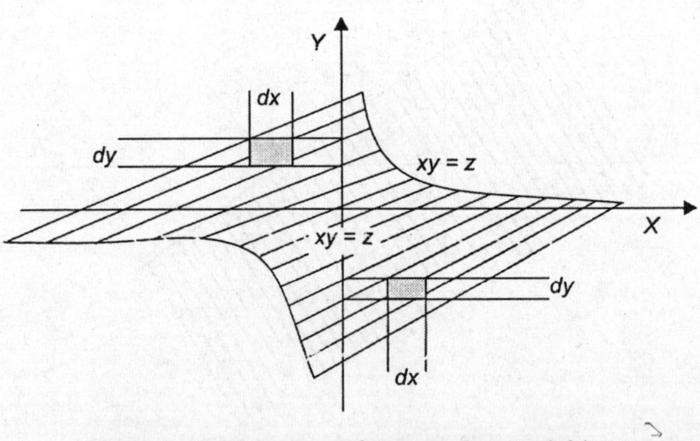

Fig. 4.3: Multiplication of two random variables

Because y can be negative, which would change the direction of the inequality when dividing, the outer integral must be split over two regions: $y \in (-\infty, 0]$ and $y \in [0, \infty)$. The corresponding inequalities that determine the limits of the inner integral are $x \geq z/y$ and $x \leq z/y$, respectively. By using Leibniz's rule, differentiating both terms with respect to z gives

$$f_Z(z) = \int_0^{\infty} \frac{1}{y} f_{XY}\left(\frac{z}{y}, y\right) dy + \int_{-\infty}^{0} \frac{-1}{y} f_{XY}\left(\frac{z}{y}, y\right) dy$$

$$f_Z(z) = \int_{-\infty}^{\infty} \left|\frac{1}{y}\right| f_{XY}\left(\frac{z}{y}, y\right) dy \qquad ...(4.43)$$

Similarly for discrete random variables,

$$P(z) = P(XY = z) = \sum_{y} \sum_{z/y} P(x)P(y) \qquad ...(4.44)$$

4.6.4 Division of Two Random Variables

Let us consider two independent random variables X and Y and a random variable Z defined as

$$Z = X/Y \qquad ...(4.45)$$

The PDF of Z can be obtained, in the similar manner to the PDF of multiplication of two random variables, as follows.

From Fig. 4.4,

$$F_Z(z) = P(X/Y \leq z) = \iint_{X/Y \leq z} f_{XY}(x, y)\, dx\, dy$$

$$F_Z(z) = \int_0^{\infty} \left(\int_{-\infty}^{zy} f_{XY}(x, y)\, dx \right) dy + \int_{-\infty}^{0} \left(\int_{zy}^{\infty} f_{XY}(x, y)\, dx \right) dy \qquad ...(4.46)$$

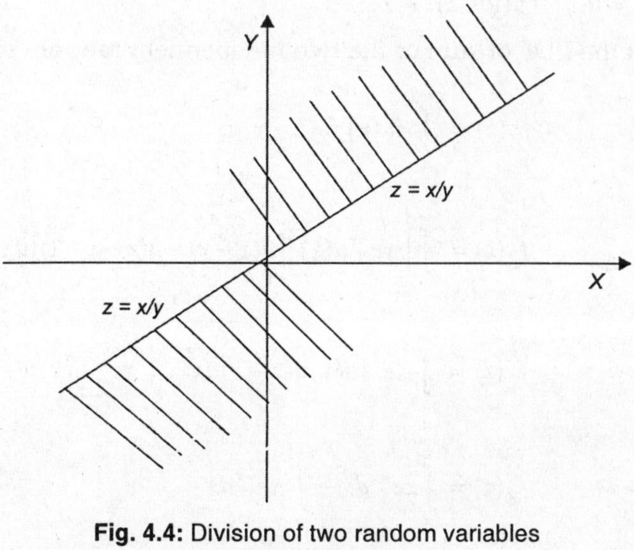

Fig. 4.4: Division of two random variables

By using Leibniz's rule, differentiating both terms with respect to z gives

$$f_Z(z) = \int_0^{\infty} y f_{XY}\left(\frac{z}{y}, y\right) dy + \int_{-\infty}^{0} -y f_{XY}\left(\frac{z}{y}, y\right) dy$$

$$f_Z(z) = \int_{-\infty}^{\infty} |y| f_{XY}\left(\frac{z}{y}, y\right) dy \qquad \text{...(4.47)}$$

Similarly for discrete random variables,

$$P(z) = P(X/Y = z) = \sum_{y}\sum_{zy} P(x)P(y). \qquad \text{...(4.48)}$$

Example 4.6.1: Two independent random variables X and Y have the probability density functions as follows:

$$f_X(x) = \begin{cases} xe^{-x} & x \geq 0 \\ 0 & \text{otherwise} \end{cases}$$

and $$f_Y(y) = \begin{cases} 1 & 0 \leq y \leq 1 \\ 0 & \text{otherwise} \end{cases}$$

Calculate the probability distribution and density functions of the random variable Z = X + Y.

Solution: $$f_X(x) = \begin{cases} xe^{-x} & x \geq 0 \\ 0 & \text{otherwise} \end{cases}$$

$$f_X(x) = xe^{-x}u(x)$$

$$f_Y(y) = \begin{cases} 1 & 0 \leq y \leq 1 \\ 0 & \text{otherwise} \end{cases}$$

$$f_Y(y) = u(y) - u(y-1)$$

We know that the PDF of sum of the two independent random variables is

$$f_Z(z) = \int_{-\infty}^{\infty} f_X(x) f_Y(z-x) dx$$

$$f_Z(z) = \int_{-\infty}^{\infty} \left[xe^{-x}u(x)\right]\left[u(z-x) - u(z-x-1)\right] dx$$

$$f_Z(z) = \int_{-\infty}^{\infty} xe^{-x}u(x)u(z-x)dx - \int_{-\infty}^{\infty} xe^{-x}u(x)u(z-x-1)dx$$

$$f_Z(z) = \int_0^{z} xe^{-x}dx - \int_0^{z-1} xe^{-x}dx$$

Consider $$\int_0^z xe^{-x}dx = x\int_0^z e^{-x}dx - \int_0^z \frac{dx}{dx}\int_0^z e^{-x}dx$$

$$\int_0^z xe^{-x}dx = x\frac{e^{-x}}{-1}\bigg|_0^z - \int_0^z 1\frac{e^{-x}}{-1}$$

$$= -ze^{-z} - e^{-x}\Big|_0^z = -ze^{-z} - e^{-z} + 1 = 1 - e^{-z}(1+z) \quad \text{for } z > 0$$

Consider $$\int_0^{z-1} xe^{-x}dx = x\int_0^{z-1} e^{-x}dx - \int_0^{z-1} \frac{dx}{dx}\int_0^{z-1} e^{-x}dx$$

$$\int_0^{z-1} xe^{-x}dx = x\frac{e^{-x}}{-1}\bigg|_0^{z-1} - \int_0^{z-1} 1\frac{e^{-x}}{-1} = -(z-1)e^{-(z-1)} - e^{-x}\Big|_0^{z-1}$$

$$= -(z-1)e^{-(z-1)} - e^{-(z-1)} + 1 = 1 - e^{-(z-1)}(1+z-1) = 1 - ze^{-(z-1)} \quad \text{for } z-1>0$$

$$f_Z(z) = \begin{cases} 1-e^{-z}(1+z) & \text{for } 0 \le z \le 1 \\ 1-ze^{-z} & \text{for } z > 1 \\ 0 & \text{otherwise.} \end{cases}$$

4.7 CENTRAL LIMIT THEOREM

The central limit theorem explains why the Gaussian random variable appears in so many diverse applications. In nature, many macroscopic phenomena result from the addition of numerous independent, microscopic processes; this gives rise to the Gaussian random variable. Many random phenomena that arise in physical situations result from the combined actions of many individual events. This is true for such things as thermal agitation of electrons in a conductor, shot noise from electrons or holes in a vacuum tube or transistor, atmospheric noise, turbulence in a medium, ocean waves, and many other physical sources of random disturbances. Hence, regardless of the probability density functions of the individual components (and these density functions are usually not even known), one would expect to find that the observed disturbance has a Gaussian density function. The central limit theorem provides a theoretical justification for assuming this, and, in almost all cases, experimental measurements bear out the soundness of this assumption.

Statement: Let $\{X_1, X_2, \dots, X_n\}$ be independent and identically distributed (iid) random variables with finite mean μ and finite variance σ^2, and let S_n be the sum of the first n random variables in the sequence

$$S_n = X_1 + X_2 + \cdots + X_n \qquad \text{...(4.49)}$$

then $$\lim_{n\to\infty} P\left(a \le \frac{S_n - n\mu}{\sigma\sqrt{n}} \le b\right) = \frac{1}{2\pi}\int_a^b e^{-v^2/2}dv$$

where the random variable $\frac{S_n - n\mu}{\sigma\sqrt{n}}$, which is standardized variable corresponding to S_n, is normal or Gaussian.

This theorem indicates that the probability density of a sum of n independent random variables tends to approach a Gaussian density as n increases. The mean and variance of Gaussian density are respectively the sum of the means and the sum of the variances of n independent random variables.

Proof: The theorem's proof consists of showing that the moment generating function S'_n is of zero mean, unit variance Gaussian random variable, which is

$$M_X(v) = \exp\left(\frac{v^2}{2}\right)$$

We have $\quad S_n = X_1 + X_2 + \cdots + X_n$

Each X_n has a mean μ and finite variance σ^2.

$$E[S_n] = E[X_1 + X_2 + \cdots + X_n] = E[X_1] + E[X_2] + \cdots + E[X_n] = nE[X] = n\mu$$

$\because$ each random variable is iid, each of equal mean

$$\text{Var}[S_n] = \text{Var}[X_1 + X_2 + \cdots + X_n] = \text{Var}[X_1] + \text{Var}[X_2] + \cdots + \text{Var}[X_n] = n\text{Var}[X] = n\sigma^2$$

$\because$ each random variable is iid, each of equal variance

Consider the standard normal variable S'_n associated with S_n, i.e.

$$S'_n = \frac{S_n - n\mu}{\sigma\sqrt{n}}$$

The moment generating function for S'_n is

$$E\left[e^{vS'_n}\right] = E\left[e^{v\left(\frac{S_n - n\mu}{\sigma\sqrt{n}}\right)}\right] = E\left[e^{v\left(\frac{X_1 + X_2 + \cdots + X_n - (\mu + \mu + \cdots + \mu)}{\sigma\sqrt{n}}\right)}\right]$$

$$= E\left[e^{v\left(\frac{X_1 - \mu}{\sigma\sqrt{n}}\right)} e^{v\left(\frac{X_2 - \mu}{\sigma\sqrt{n}}\right)} e^{v\left(\frac{X_3 - \mu}{\sigma\sqrt{n}}\right)} \ldots e^{v\left(\frac{X_n - \mu}{\sigma\sqrt{n}}\right)}\right]$$

$$M_X(v) = \left[E\left\{e^{v\left(\frac{X - \mu}{\sigma\sqrt{n}}\right)}\right\}\right]^n \qquad \because \text{all } X'_n\text{s are iid}$$

$$E\left[e^{v\left(\frac{X - \mu}{\sigma\sqrt{n}}\right)}\right] = E\left[1 + \frac{v(X - \mu)}{\sigma\sqrt{n}} + \frac{v^2(X - \mu)^2}{2!(\sigma\sqrt{n})^2} + R(n)\right]$$

$$M_X(v) = \left[E\left\{e^{v\left(\frac{X - \mu}{\sigma\sqrt{n}}\right)}\right\}\right]^n = \left\{1 + \frac{E[v(X - \mu)]}{\sigma\sqrt{n}} + \frac{E\left[v^2(X - \mu)^2\right]}{2!(\sigma\sqrt{n})^2} + E[R(n)]\right\}^n$$

$$= \left\{1 + \frac{v \cdot 0}{\sigma\sqrt{n}} + \frac{v^2}{2!n} + E[R(n)]\right\}^n$$

$$\ln M_X(v) = \ln\left\{1+0+\frac{\sigma^2}{2!(\sigma\sqrt{n})^2}+E[R(n)]\right\}^n = n\ln\left\{1+\frac{v^2}{2!n}+E[R(n)]\right\}$$

$$\because \qquad \ln(1+z) = \left\{z+\frac{z^2}{2}+\frac{z^3}{3}+\cdots\right\}$$

$$\ln M_X(v) = n\left\{\frac{v^2}{2!n}+E[R(n)]+\frac{1}{2}\left(\frac{v^2}{2!n}+E[R(n)]\right)^2+\cdots\right\}$$

$$\ln M_X(v) = \left\{\frac{v^2}{2!}+nE[R(n)]+n\frac{1}{2}\left(\frac{v^2}{2!n}+E[R(n)]\right)^2+\cdots\right\}$$

$$\lim_{n\to\infty}\ln M_X(v) = \lim_{n\to\infty}\left\{\frac{v^2}{2!}+nE[R(n)]+n\frac{1}{2}\left(\frac{v^2}{2!n}+E[R(n)]\right)^2+\cdots\right\}$$

$$\lim_{n\to\infty}\ln M_X(v) = \frac{v^2}{2}$$

$$\ln\lim_{n\to\infty} M_X(v) = \frac{v^2}{2}$$

$$\lim_{n\to\infty} M_X(v) = e^{v^2/2}$$

This is same as the moment generating function of a standard normal random variable.

Example 4.7.1: Assume that the random variable S_n is the sum of 48 independent experiment values of random variable X whose PDF is given by

$$f_X(x) = \begin{cases}\frac{1}{3} & 0 \le x \le 4\\ 0 & \text{otherwise}\end{cases}$$

Find the probability that S_n lies in the range $108 \le S_n \le 126$.

Solution: The expected value and variance of X is given by

$$E[X] = \int_1^4 \frac{1}{3}x\,dx = \frac{1}{3}\frac{x^2}{2}\bigg|_1^4 = \frac{1}{6}(16-1) = \frac{15}{6} = 2.5$$

$$E[X^2] = \int_1^4 \frac{1}{3}x^2\,dx = \frac{1}{3}\frac{x^3}{3}\bigg|_1^4 = \frac{1}{9}(64-1) = \frac{63}{9} = 7$$

$$\sigma^2 = E[X^2]-(E[X])^2 = 7-\frac{25}{4} = \frac{3}{4}$$

Thus, the mean and variance of S_n are given by

$$E[S_n] = (48)(2.5) = 120$$

$$\sigma_{S_n}^2 = (48)\left(\frac{3}{4}\right) = 36$$

$$P(S_n \le s) = F_{S_n}(s) = \phi\left(\frac{S_n - E(S_n)}{\sigma_{S_n}}\right) = \phi\left(\frac{S_n - 120}{6}\right)$$

$$P(108 \le S_n \le 126) = F_{S_n}(126) - F_{S_n}(108) = \phi\left(\frac{126-120}{6}\right) - \phi\left(\frac{108-120}{6}\right)$$

$$= \phi\left(\frac{6}{6}\right) - \phi\left(\frac{-12}{6}\right) = \phi(1) - \phi(-2) = \phi(1) - [1 - \phi(2)]$$

$$= \phi(1) - 1 + \phi(2) = 0.8413 + 0.9772 - 1 = 0.8185.$$

EXAMPLES

Example 4.1: Two random variables X and Y have the following joint PDF:

$$f_{XY}(x, y) = \begin{cases} \frac{5}{16} e^{-x} e^{-y} & 0 \le y \le x < \infty \\ 0 & \text{otherwise} \end{cases}$$

Find: i. The marginal density of X and Y.

ii. If X and Y are statistically independent.

Solution: We know that the marginal density of X is given by

$$f_X(x) = \int_{-\infty}^{\infty} f_{XY}(x, y)\,dy$$

$$f_X(x) = \int_0^x \frac{5}{16} e^{-x} e^{-y} dy = \frac{5}{16} e^{-x} \int_0^x e^{-y} dy$$

$$= \frac{5}{16} e^{-x} \left.\frac{e^{-y}}{-1}\right|_0^x = \frac{5}{16} e^{-x}(1 - e^{-x})$$

$$f_Y(y) = \int_{-\infty}^{\infty} f_{XY}(x, y)\,dx$$

$$f_Y(y) = \int_y^{\infty} \frac{5}{16} e^{-x} e^{-y} dx = \frac{5}{16} e^{-y} \int_y^{\infty} e^{-x} dx$$

$$= \frac{5}{16} e^{-y} \left.\frac{e^{-x}}{-1}\right|_y^{\infty} = \frac{5}{16} e^{-y}(e^{-y}) = \frac{5}{16} e^{-2y}$$

$$f_X(x)\,f_Y(y) = \frac{5}{16}e^{-x}(1-e^{-x})\frac{5}{16}e^{-2y} \neq f_{XY}(x,y)$$

Hence X and Y are not statistically independent.

Example 4.2: The joint PDF of a two dimensional random variable (X, Y) is given by

$$f_{XY}(x,y) = \begin{cases} xy^2 + \dfrac{x^2}{8} & 0 \le x \le 2, 0 \le y \le 1 \\ 0 & \text{otherwise} \end{cases}$$

Compute: i. $P(X > 1)$ ii. $P\left(Y < \frac{1}{2}\right)$ iii. $P\left(X > 1 / Y < \frac{1}{2}\right)$

iv. $P\left(Y < \frac{1}{2} / X > 1\right)$ v. $P(X + Y \le 1)$

Solution: i. $P(X > 1) = \int_1^2\int_0^1\left(xy^2 + \frac{x^2}{8}\right)dx\,dy = \int_1^2\int_0^1 (xy^2)\,dx\,dy + \int_1^2\int_0^1\left(\frac{x^2}{8}\right)dx\,dy$

$$= \int_1^2 x\,dx\left(\int_0^1 y^2 dy\right) + \int_1^2 \frac{x^2}{8}dx\int_0^1 1\,dy = \left(\frac{x^2}{2}\Big|_1^2\right)\left(\frac{y^3}{3}\Big|_0^1\right) + \left(\frac{x^3}{24}\Big|_1^2\right)\left(y\big|_0^1\right)$$

$$= \left(\frac{4}{2} - \frac{1}{2}\right)\left(\frac{1}{3}\right) + \left(\frac{8}{24} - \frac{1}{24}\right)(1) = \left(\frac{3}{2}\right)\left(\frac{1}{3}\right) + \frac{7}{24} = \frac{19}{24}$$

ii. $P\left(Y < \frac{1}{2}\right) = \int_0^2\int_0^{1/2}\left(xy^2 + \frac{x^2}{8}\right)dx\,dy = \int_0^2\int_0^{1/2} (xy^2)\,dx\,dy + \int_0^2\int_0^{1/2}\left(\frac{x^2}{8}\right)dx\,dy$

$$= \int_0^2 x\,dx\left(\int_0^{1/2} y^2 dy\right) + \int_0^2 \frac{x^2}{8}dx\int_0^{1/2} 1\,dy = \left(\frac{x^2}{2}\Big|_0^2\right)\left(\frac{y^3}{3}\Big|_0^{1/2}\right) + \left(\frac{x^3}{24}\Big|_0^2\right)\left(y\big|_0^{1/2}\right)$$

$$= \left(\frac{4}{2}\right)\left(\frac{1}{24}\right) + \left(\frac{8}{24}\right)\left(\frac{1}{2}\right) = \left(\frac{1}{12}\right)\left(\frac{1}{6}\right) + \frac{3}{12} = \frac{1}{4}$$

iii. $$P\left(X > 1 / Y < \frac{1}{2}\right) = \frac{P\left(X > 1, Y < \frac{1}{2}\right)}{P\left(Y < \frac{1}{2}\right)}$$

$$P\left(X > 1, Y < \frac{1}{2}\right) = \int_1^2\int_0^{1/2}\left(xy^2 + \frac{x^2}{8}\right)dx\,dy = \int_1^2\int_0^{1/2} (xy^2)\,dx\,dy + \int_1^2\int_0^{1/2}\left(\frac{x^2}{8}\right)dx\,dy$$

$$= \int_1^2 x\,dx\left(\int_0^{1/2} y^2 dy\right) + \int_1^2 \frac{x^2}{8}dx\int_0^{1/2} 1\,dy = \left(\frac{x^2}{2}\Big|_1^2\right)\left(\frac{y^3}{3}\Big|_0^{1/2}\right) + \left(\frac{x^3}{24}\Big|_1^2\right)\left(y\big|_0^{1/2}\right)$$

$$= \left(\frac{4}{2}-\frac{1}{2}\right)\left(\frac{1}{24}\right)+\left(\frac{8}{24}-\frac{1}{24}\right)\left(\frac{1}{2}\right)=\left(\frac{3}{2}\right)\left(\frac{1}{24}\right)+\left(\frac{7}{24}\right)\left(\frac{1}{2}\right)=\frac{5}{24}$$

$$P\left(X>1/Y<\frac{1}{2}\right)=\frac{P\left(X>1,Y<\frac{1}{2}\right)}{P\left(Y<\frac{1}{2}\right)}=\frac{5/24}{1/4}=\frac{5}{6}$$

iv. $$P\left(Y<\frac{1}{2}/X>1\right)=\frac{P\left(Y<\frac{1}{2},X>1\right)}{P(X>1)}=\frac{5/24}{19/24}=\frac{5}{19}$$

v. $$P(X+Y\le 1)=\int_0^1\int_0^{1-y}\left(xy^2+\frac{x^2}{8}\right)dx\,dy=\int_0^{1-y}\int_0^1\left(xy^2\right)dx\,dy+\int_0^1\int_0^{1-y}\left(\frac{x^2}{8}\right)dx\,dy$$

$$=\int_0^1 y^2\left(\int_0^{1-y}xdx\right)dy+\int_0^1 1\left(\int_0^{1-y}\frac{x^2}{8}dx\right)dy=\int_0^1\left(\frac{x^2}{2}\Big|_0^{1-y}\right)y^2dy+\int_0^1\left(\frac{x^3}{24}\Big|_0^{1-y}\right)dy$$

$$=\int_0^1\frac{(1-y)^2}{2}y^2dy+\int_0^1\left(\frac{(1-y)^3}{24}\right)dy=\int_0^1\left(\frac{y^4-2y^3+y^2}{2}\right)dy+\int_0^1\left(\frac{(1-y)^3}{24}\right)dy$$

$$=\frac{y^5}{10}\Big|_0^1-\frac{y^4}{4}\Big|_0^1+\frac{y^3}{6}\Big|_0^1+\frac{(1-y)^4}{96}\Big|_0^1$$

$$=\frac{1}{10}-\frac{1}{4}+\frac{1}{6}-\frac{1}{96}=\frac{48-120+80-5}{480}=\frac{1}{160}.$$

Example 4.3: The location of hits of darts thrown at a round dart board of radius r is determined by a vector random variable with components X and Y. The joint density of X and Y is uniform and is given by

$$f_{XY}(x,y)=\begin{cases}\dfrac{1}{\pi r^2} & x^2+y^2\le r^2\\ 0 & \text{otherwise}\end{cases}$$

Find the densities of X and Y.

Solution: $$f_X(x)=\int_{-\infty}^{\infty}f_{XY}(x,y)dy$$

$$f_X(x)=\int_{-\sqrt{r^2-x^2}}^{\sqrt{r^2-x^2}}\frac{1}{\pi r^2}dy=\frac{1}{\pi r^2}\int_{-\sqrt{r^2-x^2}}^{\sqrt{r^2-x^2}}1\cdot dy$$

$$=\frac{1}{\pi r^2}2\sqrt{r^2-x^2}=\frac{2\sqrt{r^2-x^2}}{\pi r^2}\quad \text{for } -r<x<r$$

$$f_Y(y) = \int_{-\infty}^{\infty} f_{XY}(x,y)dx$$

$$f_Y(y) = \int_{-\sqrt{r^2-y^2}}^{\sqrt{r^2-y^2}} \frac{1}{\pi r^2} dx = \frac{1}{\pi r^2} \int_{-\sqrt{r^2-y^2}}^{\sqrt{r^2-y^2}} 1 \cdot dx$$

$$= \frac{1}{\pi r^2} 2\sqrt{r^2 - y^2} = \frac{2\sqrt{r^2 - y^2}}{\pi r^2} \quad \text{for } -r < y < r.$$

Example 4.4: Two discrete random variables X and Y have probability functions $P_X(k) = \frac{1}{k!}e^{-a}a^k$ and $P_Y(j) = \frac{1}{j!}e^{-b}b^j$. If Z = X + Y, find $P_Z(z)$.

Solution: $P_Z(z) = \sum_{k=0}^{n} P_X(x)P_Y(n-k) = \sum_{k=0}^{n} \frac{1}{k!}\frac{1}{(n-k)!}e^{-(a+b)}a^k b^{(n-k)}$

$$= \frac{1}{n!}e^{-(a+b)}\sum_{k=0}^{n} \frac{n!}{k!}\frac{1}{(n-k)!}a^k b^{(n-k)} = \frac{1}{n!}e^{-(a+b)}\sum_{k=0}^{n} {}^nC_k a^k b^{(n-k)}$$

$$= \frac{1}{n!}e^{-(a+b)}(a+b)^n \qquad \because \sum_{k=0}^{n} {}^nC_k a^k b^{n-k} = (a+b)^n$$

$$P_Z(z) = \frac{1}{n!}e^{-(a+b)}(a+b)^n \qquad \text{for } n \geq 0.$$

Example 4.5: Find the conditional density functions $f_X(x/y_1), f_X(x/y_2), f_X(y/x_1)$ and $f_X(y/x_2)$ for joint function defined by $P(x_1,y_1)=\frac{2}{15}$, $P(x_2,y_1)=\frac{3}{15}$, $P(x_2,y_2)=\frac{1}{15}$, $P(x_1,y_3)=\frac{4}{15}$ and $P(x_2,y_3)=\frac{5}{15}$.

Solution: $P(y_3) = \frac{5}{15}+\frac{4}{15}=\frac{9}{15}$, $P(y_1) = \frac{2}{15}+\frac{3}{15}=\frac{5}{15}$, $P(y_2) = \frac{1}{15}$

$$P(x_1) = \frac{2}{15}+\frac{4}{15}=\frac{6}{15}$$

and $$P(x_2) = \frac{3}{15}+\frac{1}{15}+\frac{5}{15}=\frac{9}{15}$$

$f_X(x/y_1)$, i.e. it combines $f_X(x_1/y_1)$ and $f_X(x_2/y_1)$:

$$f_X(x_1/y_1) = \frac{P(x_1,y_1)}{P(y_1)} = \frac{2/15}{5/15} = \frac{2}{5}$$

$$f_X(x_2/y_1) = \frac{P(x_2,y_1)}{P(y_1)} = \frac{3/15}{5/15} = \frac{3}{5}$$

$f_X(x/y_2)$:

$$f_X(x_1/y_2) = \frac{P(x_1,y_2)}{P(y_2)} = 0$$

$$f_X(x_2/y_2) = \frac{P(x_2,y_2)}{P(y_2)} = \frac{1/15}{1/15} = 1$$

$f_X(y/x_1)$:

$$f_X(y_1/x_1) = \frac{P(x_1,y_1)}{P(x_1)} = \frac{2/15}{6/15} = \frac{2}{6}$$

$$f_X(y_2/x_1) = \frac{P(x_1,y_2)}{P(x_1)} = 0$$

$$f_X(y_3/x_1) = \frac{P(x_1,y_3)}{P(x_1)} = \frac{4/15}{6/15} = \frac{4}{6}$$

$f_X(y/x_2)$:

$$f_X(y_1/x_2) = \frac{P(x_2,y_1)}{P(x_2)} = \frac{3/15}{9/15} = \frac{3}{9}$$

$$f_X(y_2/x_2) = \frac{P(x_2,y_2)}{P(x_2)} = \frac{1/15}{9/15} = \frac{1}{9}$$

$$f_X(y_3/x_2) = \frac{P(x_2,y_3)}{P(x_2)} = \frac{5/15}{9/15} = \frac{5}{9}.$$

Example 4.6: Given two random variables X and Y with the joint CDF $F_{XY}(x, y)$ and marginal CDFs $F_X(x)$ and $F_Y(y)$ respectively, compute the joint probability that X is greater than a and Y is greater than b.

Solution: We can obtain the desired probability as follows. From the De Morgan's second law, we know that $\overline{A \cap B} = \overline{A} \cup \overline{B}$. Thus,

$$\begin{aligned} P[X > a, Y > b] &= P[\{X > a\} \cap \{Y > b\}] = 1 - P\left[\overline{\{X > a\} \cap \{Y > b\}}\right] \\ &= 1 - P\left[\overline{\{X > a\}} \cup \overline{\{y > b\}}\right] = 1 - P[\{X \le a\} \cup \{Y \le b\}] \\ &= 1 - \{P[X \le a] + P[Y \le b] - P[X \le a \cdot Y \le b]\} \\ &= 1 - F_X(a) - F_Y(b) + F_{XY}(a,b). \end{aligned}$$

Example 4.7: A fair coin is tossed three times. Let X be a random variable that takes the value 0 if the first toss is a tail and the value 1 if the first toss is a head. Also, let Y be a random variable that defines the total number of heads in the three tosses.

(a) Determine the joint PMF of X and Y.

(b) Are X and Y independent?

Solution. Let H denote the event that a head appears on a toss and T the event that a tail appears on a toss. Table 4.1 shows the sample space and the values of the two random variables.

(a) Since X takes values 0 and 1, and Y takes values 0, 1, 2, and 3, the joint PMF of X and Y can then be constructed as follows:

Table 4.1: Sample space and values of random variables

Sample space	*Value of X*	*Value of Y*
HHH	1	3
HHT	1	2
HTH	1	2
THH	0	2
THT	0	1
TTH	0	1
HTT	0	1
TTT	0	0

$$P_{XY}(0,0) = P[X=0, Y=0] = P[\text{TTT}] = \frac{1}{8}$$

$$P_{XY}(0,1) = P[X=0, Y=1] = P[\{\text{THT}\} \cup \{\text{TTH}\}] = \frac{1}{4}$$

$$P_{XY}(0,2) = P[X=0, Y=2] = P[\text{THH}] = \frac{1}{8}$$

$$P_{XY}(0,3) = P[X=0, Y=3] = 0$$

$$P_{XY}(1,0) = P[X=1, Y=0] = 0$$

$$P_{XY}(1,1) = P[X=1, Y=1] = P[\text{HTT}] = \frac{1}{8}$$

$$P_{XY}(1,2) = P[X=1, Y=2] = P[\{\text{HHT}\} \cup \{\text{HTH}\}] = \frac{1}{4}$$

$$P_{XY}(1,3) = P[X=1, Y=3] = P[\text{HHH}] = \frac{1}{8}$$

(b) If X and Y are independent, then for all x and y we have that $P_{XY}(x, y) = P_X(x)\,P_Y(y)$. Thus, to show that X and Y are not independent, all we have to do is to find a pair of x and y at which the joint PMF does not satisfy the above condition. Consider the point $(x, y) = (0, 0)$.

$$P_X(0) = \sum_y P_{XY}(0,y) = P_{XY}(0,0) + P_{XY}(0,1) + P_{XY}(0,2) + P_{XY}(0,3) = \frac{1}{8} + \frac{1}{4} + \frac{1}{8} = \frac{1}{2}$$

$$P_Y(0) = \sum_x P_{XY}(x,0) = P_{XY}(0,0) + P_{XY}(1,0) + P_{XY}(2,10) + P_{XY}(3,0) = \frac{1}{8}$$

$$P_Y(0)\,P_X(0) = \left(\frac{1}{8}\right)\left(\frac{1}{2}\right) = \left(\frac{1}{16}\right)$$

But $P_{XY}(0,0) = \frac{1}{8} \neq P_X(0) P_Y(0)$. Hence, we conclude that X and Y are not statistically independent.

Example 4.8: Two random variables X and Y have the following joint PDF:

$$f_{XY}(x,y) = \begin{cases} xe^{-x(y+1)} & 0 \le x < \infty;\ 0 \le y < \infty \\ 0 & \text{otherwise} \end{cases}$$

Determine the conditional PDF of X given Y and the conditional PDF of Y given X.

Solution: To determine the conditional PDFs, we first evaluate the marginal PDFs, which are given by

$$f_X(x) = \int_{-\infty}^{\infty} f_{XY}(x,y)\,dy = \int_0^{\infty} xe^{-x(y+1)}dy = xe^{-x}\int_0^{\infty} e^{-xy}dy = xe^{-x}\left.\frac{e^{-xy}}{-x}\right|_0^{\infty} = e^{-x} \qquad 0 \le x < \infty$$

$$f_Y(y) = \int_{-\infty}^{\infty} f_{XY}(x,y)\,dy = \int_0^{\infty} xe^{-x(y+1)}dx = x\int_0^{\infty} e^{-x(y+1)}dx - \int_0^{\infty}\left(\frac{dx}{dx}\int_0^{\infty} e^{-x(y+1)}dx\right)dx$$

$$= x\left.\frac{e^{-x(y+1)}}{-(y+1)}\right|_0^{\infty} - \int_0^{\infty}\frac{e^{-x(y+1)}}{-(y+1)}\,dx = 0 + \frac{1}{(y+1)}\left.\frac{e^{-x(y+1)}}{-(y+1)}\right|_0^{\infty} = \frac{1}{(y+1)^2} \qquad 0 \le y < \infty$$

$$f_{Y/X}(y/x) = \frac{f_{XY}(x,y)}{f_X(x)} = \frac{xe^{-x(y+1)}}{e^{-x}} = xe^{-xy} \qquad 0 \le y < \infty$$

$$f_{X/Y}(x/y) = \frac{f_{XY}(x,y)}{f_Y(y)} = \frac{xe^{-x(y+1)}}{1/(y+1)^2} = x(y+1)^2 e^{-x(y+1)} \qquad 0 \le x < \infty$$

Example 4.9: Find the density of $W = X + Y$ where the densities of X and Y are assumed to be $f_X(x) = \frac{1}{a}[u(x) - u(x-a)]$ and $f_Y(y) = \frac{1}{b}[u(y) - u(y-b)]$ with $0 < a < b$.

Solution: The density functions are shown in Fig. 4.5.

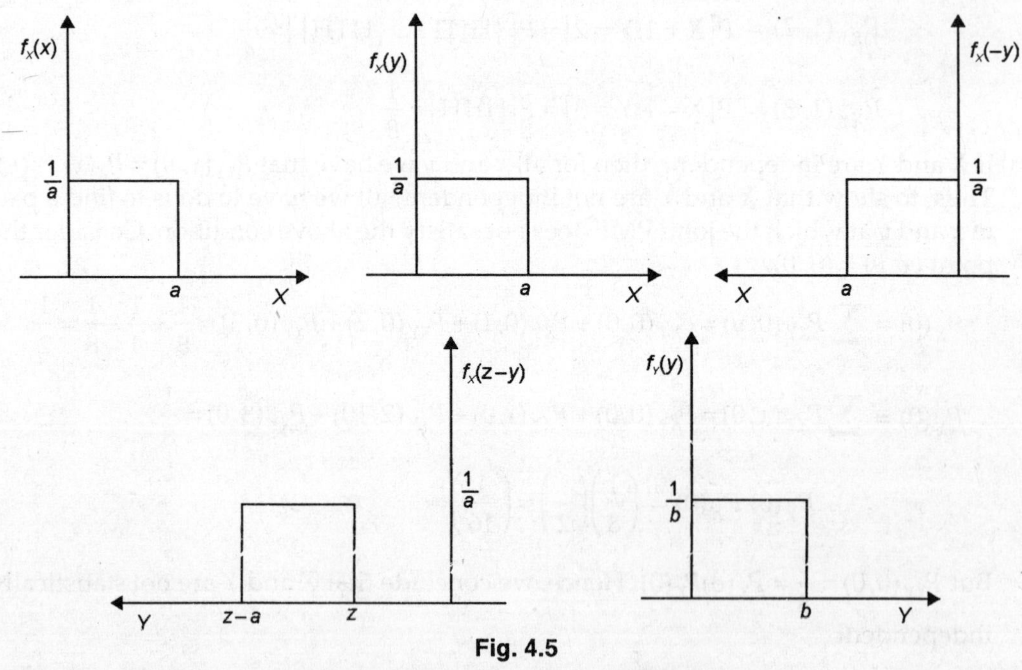

Fig. 4.5

Here X and Y are discrete, so that we cannot obtain the density of Z with single expression. It consists of five different regions, $Z < 0$, $0 \le Z < a$, $a \le Z < b$, $b \le Z < a + b$, $Z \ge a + b$.

***Region* 1** ($Z < 0$): For this region, the graph is shown in Fig. 4.6.

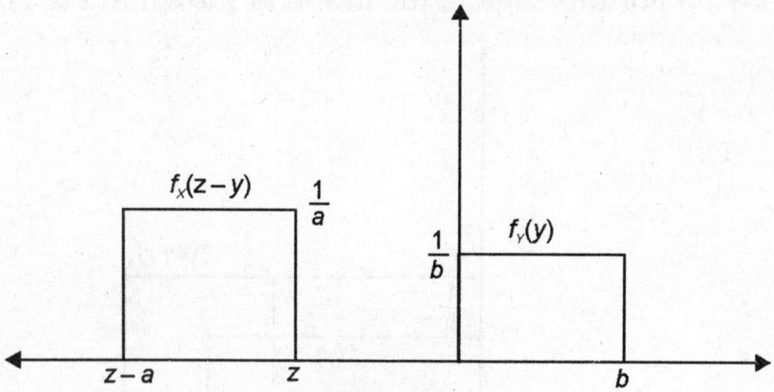

Fig. 4.6

In this case, there is no common area between $f_X(z-y)$ and $f_Y(y)$. Hence $f_Z(z) = 0$.

***Region* 2** ($0 \le Z < a$): For this region, the graph is shown in Fig. 4.7.

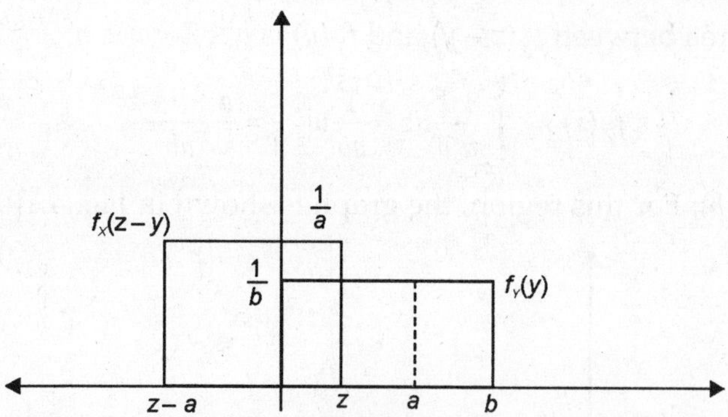

Fig. 4.7

The common area between $f_X(z-y)$ and $f_Y(y)$ is $0 \le Z < a$.

$$f_Z(z) = \int_0^z \frac{1}{a}\frac{1}{b} dz = \frac{1}{ab} y\Big|_0^z = \frac{z}{ab}$$

***Region* 3** ($a \le Z < b$): For this region, the graph is shown in Fig. 4.8.

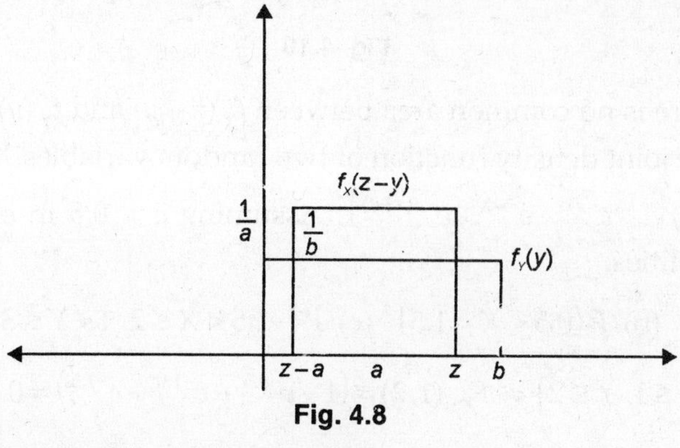

Fig. 4.8

The common area between $f_X(z-y)$ and $f_Y(y)$ is $a \le Z < b$.

$$f_Z(z) = \int_{z-a}^{z} \frac{1}{a}\frac{1}{b} dz = \frac{1}{ab} y\Big|_{z-a}^{z} = \frac{a}{ab} = \frac{1}{b}$$

Region 4 ($b \le Z < a + b$): For this region, the graph is shown in Fig. 4.9.

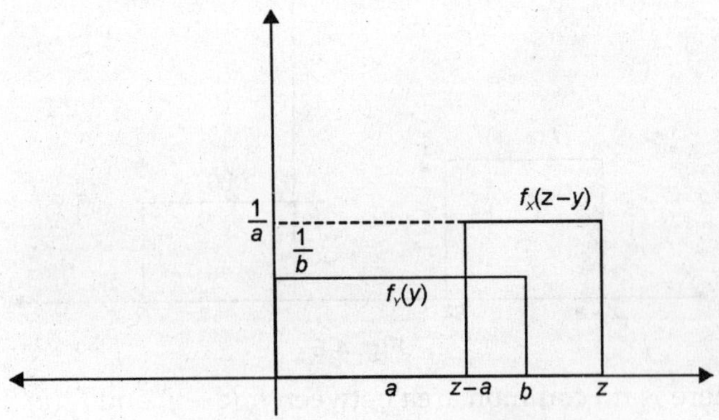

Fig. 4.9

The common area between $f_X(z-y)$ and $f_Y(y)$ is $b \le Z < a + b$.

$$f_Z(z) = \int_{z-a}^{b} \frac{1}{a}\frac{1}{b} dz = \frac{1}{ab} y\Big|_{z-a}^{b} = \frac{a+b-z}{ab}$$

Region 5 ($Z \ge a + b$): For this region, the graph is shown in Fig. 4.10.

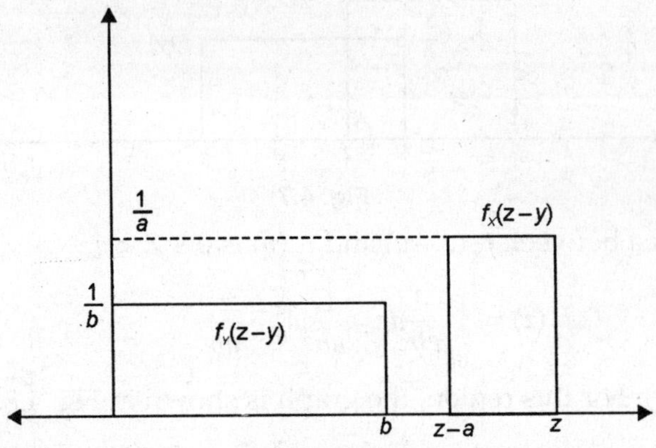

Fig. 4.10

In this case, there is no common area between $f_X(z-y)$ and $f_Y(y)$. Hence $f_Z(z) = 0$.

Example 4.10: The joint density function of two random variables X and Y is given by $f_{XY}(x,y) = u(x)u(y)\{1 - e^{-ax} - e^{-ay} + e^{-a(x+y)}\}$. Assuming $a = 0.5$ in each case, find the following probabilities:

(a) $P\{X \le 1, Y \le 2\}$ (b) $P\{0.5 < X < 1.5\}$ (c) $P\{-1.5 < X \le 2, 1 < Y \le 3\}$

Solution: (a) $P\{X \le 1, Y \le 2\} = F_{XY}(1,2) = (1 - e^{-0.5} - e^{-1.0} + e^{-1.5}) = 0.249$

(b) $P\{0.5 < X < 1.5\} = F_X(1.5) - F_X(0.5) = F_{XY}(1.5,\infty) - F_{XY}(0.5,\infty)$

$$= \{1 - e^{-1.5(0.5)}\} - \{1 - e^{-0.5(0.5)}\} = 0.306$$

(c) $P\{-1.5 < X \le 2, 1 < Y \le 3\} = F_{XY}(2,3) + F_{XY}(-1.5,1) - F_{XY}(-1.5,3) - F_{XY}(2,1)$

$$= \{1 - e^{-0.5(2)} - e^{-0.5(3)} + e^{-0.5(5)}\} + 0 - 0 - \{1 - e^{-0.5(2)} - e^{-0.5(1)} + e^{-0.5(3)}\}$$

$$= \{1 - 0.367 - 0.223 + 0.0820\} - \{1 - 0.367 - 0.606 + 0.223\} = 0.492 - 0.25 = 0.242$$

Example 4.11:- Given the function

$$f_{XY}(x,y) = \begin{cases} b(x+y)^2 & -2 < x < 2, -3 < y < 3 \\ 0 & \text{otherwise} \end{cases}$$

(a) Find the constant b such that the joint density function is a valid density function.
(b) Determine the marginal density functions of X and Y.

Solution: (a) For valid density function

$$\int_{-\infty}^{\infty}\int_{-\infty}^{\infty} f_{XY}(x,y)\,dx\,dy = 1$$

$$\int_{-3}^{3}\int_{-2}^{2} b(x+y)^2\,dx\,dy = \int_{-3}^{3} \frac{b(x+y)^3}{3}\Bigg|_{-2}^{2} = \int_{-3}^{3} \frac{b}{3}\left[(2+y)^3 - (-2+y)^3\right]dy$$

$$= \frac{b}{3}\left\{\frac{(2+y)^4}{4} - \frac{(-2+y)^4}{4}\right\}\Bigg|_{-3}^{3}$$

$\Rightarrow\ 104b = 1$

$\Rightarrow\quad b = \dfrac{1}{104}$

(b) $f_X(x) = \displaystyle\int_{-\infty}^{\infty} f_{XY}(x,y)\,dy = \int_{-3}^{3} b(x+y)^2\,dy = \frac{b(x+y)^3}{3}\Bigg|_{-3}^{3}$

$$= \frac{b}{3}\left[(3+x)^3 - (-3+x)^3\right] = \frac{1}{312}\left[(x+3)^3 - (x-3)^3\right] - 2 < x < 2$$

$$f_Y(y) = \int_{-\infty}^{\infty} f_{XY}(x,y)\,dx$$

$$f_Y(y) = \int_{-2}^{2} b(x+y)^2\,dx = \frac{b(x+y)^3}{3}\Bigg|_{-2}^{2}$$

$$= \frac{b}{3}\left[(2+y)^3 - (-2+y)^3\right] = \frac{1}{312}\left[(y+2)^3 - (y-2)^3\right] - 3 < y < 3$$

Example 4.12: Determine a constant b such that each of the following are valid density functions:

(a) $f_{XY}(x, y) = \begin{cases} 3xy & 0<x<1,\ 0<y<b \\ 0 & \text{otherwise} \end{cases}$

(b) $f_{XY}(x, y) = \begin{cases} bx(1-y) & 0<x<0.5,\ 0<y<1 \\ 0 & \text{otherwise} \end{cases}$

(c) $f_{XY}(x, y) = \begin{cases} b(x^2+4y^2) & 0<|x|<1,\ 0<y<2 \\ 0 & \text{otherwise} \end{cases}$

Solution: (a) $\int_0^1\int_0^b 3xy\,dx\,dy = \frac{3}{4}b^2 = 1 \Rightarrow b = \frac{2}{\sqrt{3}}$

(b) $\int_0^1\int_0^{0.5} bx(1-y)dx\,dy = \frac{b}{16} = 1 \Rightarrow b = 16$

(c) $\int_{-1}^1\int_0^2 b(x^2+4y^2)dx\,dy = \frac{68b}{3} = 1 \Rightarrow b = \frac{3}{68}$

Example 4.13: Given the function

$$f_{XY}(x, y) = \begin{cases} \frac{(x^2+y^2)}{8\pi} & x^2+y^2<b \\ 0 & \text{otherwise} \end{cases}$$

(a) Find a constant b such that this is a valid density function.

(b) Find $P\{0.5b < x^2+y^2 < 0.8b\}$.

Solution: Let $r = \sqrt{x^2+y^2}, \theta = \tan^{-1}\left(\frac{y}{x}\right)$ and $r dr d\theta = dx\,dy$

(a) $$\iint_{x^2+y^2<b} \frac{(x^2+y^2)}{8\pi}dx\,dy = \int_{\theta=0}^{2\pi}\int_{r=0}^{\sqrt{b}} \frac{r^2}{8\pi} r dr d\theta = \int_{\theta=0}^{2\pi}\int_{r=0}^{\sqrt{b}} \frac{r^3}{8\pi} dr d\theta = \frac{2\pi}{8\pi}\frac{r^4}{4}\Bigg|_0^{\sqrt{b}} = \frac{b^2}{16} = 1 \Rightarrow b = 4$$

(b) $P\{0.5b < x^2+y^2 \le 0.8b\} = P\{0.5b < r^2 \le 0.8b\} = P\{\sqrt{0.5b} < r \le \sqrt{0.8b}\}$

$$= \int_{\theta=0}^{2\pi}\int_{\sqrt{0.5b}}^{\sqrt{0.8b}} \frac{r^3}{8\pi} dr d\theta = \frac{2\pi}{8\pi}\frac{r^4}{4}\Bigg|_{\sqrt{0.5b}}^{\sqrt{0.8b}} = \frac{0.39b^2}{16} = 0.39.$$

Example 4.14: $X_1, X_2, \ldots, X_{36}$ are independent random variables each of which is uniformly distributed between 0 and 1. Let $S = X_1 + X_2 + \cdots + X_{36}$. What is the probability that $S > 22$?

Solution: The expected value and variance of X are given by

$$f_X(x) = \frac{1}{1-0} = 1 \qquad 0 \le x \le 1$$

$$E[X] = \int_0^1 x dx = \frac{x^2}{2}\bigg|_0^1 = \frac{1}{2} = 0.5$$

$$E[X^2] = \int_0^1 x^2 dx = \frac{x^3}{3}\bigg|_0^1 = \frac{1}{3}$$

$$\sigma^2 = E[X^2] - (E[X])^2 = \frac{1}{3} - \frac{1}{4} = \frac{1}{12}$$

Thus, the mean and variance of S_n are given by

$$E[S_n] = (36)(0.5) = 18$$

$$\sigma_{S_n}^2 = (36)\left(\frac{1}{12}\right) = 3$$

$$P(S_n \le s) = F_{S_n}(s) = \phi\left(\frac{S_n - E(S_n)}{\sigma_{S_n}}\right) = \phi\left(\frac{S_n - 18}{\sqrt{3}}\right)$$

$$P(S_n > 22) = 1 - P(S_n \le 22) = 1 - F_{S_n}(22)$$

$$= 1 - \phi\left(\frac{22-18}{\sqrt{3}}\right) = 1 - \phi(2.3094) = 1 - 0.9916 = 0.0084.$$

Example 4.15: A modem transmits one million bits. Each bit is 0 or 1 independently with equal probability. Estimate the probability of at least 502,000 ones.

Solution: Let X_i be the value of bit i (either 0 or 1). The number of ones in one million bits is $\sum_{i=1}^{10^5} X_i$. Because X_i is a Bernoulli (0.5) random variable, $E[X_i] = 0.5$ and $\text{Var}[X_i] = 0.25$ for all i. Note that $E[S_n] = 10^6 E[X_i] = 500{,}000$ and $\text{Var}[S_n] = 10^6\,\text{Var}[X_i] = 250{,}000$. Therefore, $\sigma_{S_n} = 500$. By the central limit theorem approximation,

$$P(S_n \le s) = F_{S_n}(s) = \phi\left(\frac{S_n - E(S_n)}{\sigma_{S_n}}\right) = \phi\left(\frac{S_n - 500000}{500}\right)$$

$$P(S_n > 502{,}000) = 1 - P(S_n \le 502{,}000) = 1 - F_{S_n}(502{,}000)$$

$$= 1 - \phi\left(\frac{502{,}000 - 500{,}000}{500}\right) = 1 - \phi(4) = 3.17 \times 10^{-5}.$$

Example 4.16: The joint density function of X and Y is given by

$$f_{XY}(x, y) = \begin{cases} \frac{15}{2}x(2-x-y) & 0<x<1, 0<y<1 \\ 0 & \text{otherwise} \end{cases}$$

Find the conditional density function X given $Y = y$, where $0 < y < 1$.

Solution: $f_X(x/y) = \dfrac{f_{XY}(x,y)}{\int_{-\infty}^{\infty} f_{XY}(x,y)dx} = \dfrac{f_{XY}(x,y)}{f_Y(y)}$

$$f_X(x/y) = \frac{\frac{15}{2}x(2-x-y)}{\int_0^1 \frac{15}{2}x(2-x-y)dx} = \frac{x(2-x-y)}{\left(2\frac{x^2}{2}-\frac{x^3}{3}-\frac{x^2}{2}y\right)\Big|_0^1}$$

$$= \frac{x(2-x-y)}{\frac{2}{3}-\frac{y}{2}} = \frac{6x(2-x-y)}{4-3y}$$

EXERCISE

1. If 30 fair dice are rolled, what is the probability that the sum obtained is between 95 and 125?
2. The random variable X has the following PDF:
$$f_X(x) = \begin{cases} \frac{1}{3} & -1 < x < 2 \\ 0 & \text{otherwise} \end{cases}$$
Find PDF of Y which is defined as $Y = 2X + 3$.
3. A random variable X has the PDF
$$f_X(x) = \begin{cases} 2x & 0 \le x \le 1 \\ 0 & \text{otherwise} \end{cases}$$
Independently of X, a random variable Y is uniformly distributed between -2 and 2. Let the variable $Z = X + Y$. Determine the PDF of Z.
4. The joint PDF of continuous random variables X and Y is given by
$$f_{XY}(x, y) = \begin{cases} 2e^{-(x+y)} & 0 < x < \infty,\ 0 < y < \infty \\ 0 & \text{otherwise} \end{cases}$$
If we define the random variable $Z = X/Y$, find the PDF of Z.
5. Let X and Y be two independent random variables that are uniformly distributed between -1 and $+1$. If we define $W = XY$, find the PDF of W.
6. A fair coin is tossed three times. Let the random variable X denote the number of heads in the first two tosses, and let the random variable Y denote the number of heads in the third toss. Determine the joint PMF $P_{XY}(x, y)$ of X and Y.
7. The joint PDF of two continuous random variables X and Y is given by
$$f_{XY}(x, y) = \begin{cases} \frac{kx}{5} & 0 < y \le x < 1 \\ 0 & \text{otherwise} \end{cases}$$
   i. Determine the value of the constant k.
   ii. Find the marginal PDFs of X and Y.
   iii. Find $P\left\{0 < X < \frac{1}{4}, 0 < Y < \frac{1}{6}\right\}$.

8. The joint PDF of two continuous random variables X and Y is given by

$$f_{XY}(x, y) = \begin{cases} 6e^{-(x+2y)} & x \geq 0, y \geq 0 \\ 0 & \text{otherwise} \end{cases}$$

Find the conditional density of (i) X given Y (ii) Y given X.

9. The joint PDF of two continuous random variables X and Y is given by

$$f_{XY}(x, y) = \begin{cases} 1 - e^{-ax}e^{-by} + e^{-(ax+by)} & x \geq 0, y \geq 0 \\ 0 & \text{otherwise} \end{cases}$$

   i. Find the marginal PDFs of X and Y.
   ii. Find if X and Y are statistically independent or not.

10. The joint PDF of two continuous random variables X and Y is given by

$$f_{XY}(x, y) = \begin{cases} a(2x + y^2) & 0 \leq x \leq 2,\ 2 \leq y \leq 4 \\ 0 & \text{otherwise} \end{cases}$$

   i. Find the value of a.
   ii. Find $P\{X \leq 1, Y > 3\}$.

11. Discrete random variables X and Y have a joint distribution function
$F_{XY}(x, y) = 0.1u(x + 4)\, u(y - 1) + 0.15u(x + 3)u(y + 5) + 0.17u(x + 1)u(y - 3)$
$+ 0.05u(x)u(y - 1) + 0.18u(x - 2)u(y + 2) + 0.23u(x - 3)u(y - 4) + 0.12u(x - 4)\, u(y + 3)$
   i. Sketch $F_{XY}(x, y)$.
   ii. Find marginal distribution functions of X and Y.
   iii. Find $P\{-1 < X \leq 4, -3 < Y \leq 3\}$.
   iv. Find $P\{X < 1, Y \leq 2\}$.

5 Operations on Multiple Random Variables

5.0 INTRODUCTION

In previous chapters, we have discussed the concepts of distribution functions, density functions of single random variable as well as multiple random variables and also discussed expectations, variance and higher order moments of single random variable. This chapter extends the concepts of expectations, variance and higher order moments of single random variable to multiple random variables, so that we can make probability statements about collections and sequences of random variables. The most important instrument in this chapter is joint moments, and we look at the important ideas of covariance and correlation. Finally, we consider conditional expectation in this new setting.

5.1 EXPECTATION AND VARIANCE

Let $g(X, Y)$ be some arbitrary two-dimensional function. The expected value of $g(X, Y)$, where X and Y are random variables, is

$$E[g(X,Y)] = \begin{cases} \sum_j \sum_i g(x_i, y_j) P(x_i, y_j) & X \text{ and } Y \text{ are discrete} \\ \int_{-\infty}^{\infty}\int_{-\infty}^{\infty} g(x,y) f_{XY}(x,y)\, dx\, dy & X \text{ and } Y \text{ are continuous} \end{cases} \quad \text{...(5.1)}$$

For N random variables $X_1, X_2, \ldots, X_N$, some function of these N random variables is represented by $g(X_1, X_2, \ldots, X_N)$. The expectation is defined as

$$E[g(X_1, X_2, \ldots, X_N] = \int_{-\infty}^{\infty}\int_{-\infty}^{\infty}\cdots\int_{-\infty}^{\infty} g(x_1, x_2, \ldots, x_N) f_{X_1, X_2, \ldots, X_N}(x_1, x_2, \ldots, x_N)\, dx_1,\, dx_2, \ldots, dx_N \quad \text{...(5.2)}$$

Example 5.1.1: Show that $E[aX + bY] = aE[X] + bE[Y]$, where a and b are constants

Solution: $E[aX + bY] = \int_{-\infty}^{\infty}\int_{-\infty}^{\infty} (aX + bY) f_{XY}(x,y)\, dx\, dy \qquad \because\ E[X] = \int_{-\infty}^{\infty} x f_X(x)\, dx$

$$E[aX + bY] = a\int_{-\infty}^{\infty}\int_{-\infty}^{\infty} X f_{XY}(x,y)\,dx\,dy + b\int_{-\infty}^{\infty}\int_{-\infty}^{\infty} Y f_{XY}(x,y)\,dx\,dy$$

$$= aE[X] + bE[Y]$$

Example 5.1.2: Show that the mean value of a weighted sum of random variables equals the weighted sum of mean values.

Solution: Let $g(x_1, x_2, ..., x_N) = a_1x_1 + a_2x_2 + \cdots + a_Nx_N = \sum_{i=1}^{N} a_i x_i$

$$E[a_1x_1 + a_2x_2 + \cdots + a_Nx_N] = E\left[\sum_{i=1}^{N} a_i x_i\right]$$

$$\Rightarrow \quad a_1E[x_1] + a_2E[x_2] + \cdots + a_NE[x_N] = \sum_{i=1}^{N} a_i E[x_i]$$

$$E\left[\sum_{i=1}^{N} a_i x_i\right] = \sum_{i=1}^{N} a_i E[x_i]$$

Hence, the mean value of a weighted sum of random variables equals the weighted sum of mean values.

5.2 JOINT MOMENTS

Although expectation is an important 'summary' number for the behavior of a random variable, it is far from adequate in describing the complete behavior of the random variable. Even though for two random variables their expectation could be the same but their standard deviation could be different. Under certain conditions, it is possible to completely describe the behavior of the random variable, i.e. reconstruction of its PDF from knowledge of all the moments.

5.2.1 Joint Moments About the Origin

The joint moments about the origin are denoted by m_{nk}

$$m_{nk} = \begin{cases} \sum_{j=1}^{L}\sum_{i=1}^{N} x_i^n y_j^k & X \text{ and } Y \text{ are discrete} \\ \int_{-\infty}^{\infty} x^n y^k f_{XY}(x,y)\,dx\,dy & X \text{ and } Y \text{ are continuous} \end{cases} \quad ...(5.3)$$

Special Cases

The sum $n + k$ is the order of moments.

Suppose if $k = 0$, the joint moment is

$$m_{n0} = E(X^n) \qquad n^{th} \text{ moment of random variable } X \quad ...(5.4)$$

Similarly if $n = 0$,

$$m_{0k} = E(Y^k) \qquad n^{th} \text{ moment of random variable } Y \quad ...(5.5)$$

The moments m_{11}, m_{02}, m_{20} are called second order moments.

If $n = 1$ and $m = 1$,

$$m_{11} = E(XY) = \int_{-\infty}^{\infty} xy\, f_{XY}(x,y)\,dx\,dy \qquad ...(5.6)$$

Equation (5.6) is called as the correlation of X and Y, which is denoted by R_{XY}.

If correlation can be written in the form

$$R_{XY} = E[X]E[Y] \qquad ...(5.7)$$

then X and Y are said to be uncorrelated.

If $R_{XY} = 0$, then X and Y are said to be orthogonal.

For N random variables $X_1, X_2,..., X_N$, the $n_1 + n_2 +\cdots+n_N$ order joint moments are defined as

$$E\left[X_1^{n_1}, X_2^{n_2}, ..., X_N^{n_N}\right] = \int_{-\infty}^{\infty}\int_{-\infty}^{\infty}...\int_{-\infty}^{\infty}\left(X_1^{n_1}, X_2^{n_2}, ..., X_N^{n_N}\right) f_{X_1,X_2,...,X_N}(x_1,x_2,...,x_N)\,dx_1, dx_2,..., dx_N \qquad ...(5.8)$$

5.2.2. Joint Moments About the Mean

The joint central moments of random variables X and Y are denoted by μ_{nk}. It is defined as

$$\mu_{nk} = E\left[(X-\bar{X})^n (Y-\bar{Y})^k\right]$$

$$\mu_{nk} = \begin{cases} \sum\limits_j \sum\limits_i (x_i-\bar{X})^n (y_j-\bar{Y})^k P(x_i,y_j) & X \text{ and } Y \text{ are discrete} \\ \int_{-\infty}^{\infty}\int_{-\infty}^{\infty} (x-\bar{X})^n (y-\bar{Y})^k f_{XY}(x,y)\,dx\,dy & X \text{ and } Y \text{ are continuous} \end{cases} \qquad ...(5.9)$$

Special Cases

For $n = 2$ and $k = 0$,

$$\mu_{20} = E\left[(X-\bar{X})^2\right] = \int_{-\infty}^{\infty} (x-\bar{X})^2 f_X(x)\,dx = \sigma_X^2 \qquad ...(5.10)$$

This is the variance of random variable X.

For $n = 0$ and $k = 2$,

$$\mu_{02} = E\left[(Y-\bar{Y})^2\right] = \int_{-\infty}^{\infty} (y-\bar{Y})^2 f_Y(y)\,dy = \sigma_Y^2 \qquad ...(5.11)$$

This is the variance of random variable Y.

For $n = 1$ and $k = 1$,

$$\mu_{11} = E\left[(X-\bar{X})(Y-\bar{Y})\right] = \int_{-\infty}^{\infty}\int_{-\infty}^{\infty} (x-\bar{X})(y-\bar{Y}) f_{XY}(x,y)\,dx\,dy \qquad ...(5.12)$$

The second order joint moment μ_{11} is very important. It is called the *covariance* of X and Y and is represented by C_{XY}. Hence

$$C_{XY} = E\left[(X-\overline{X})(Y-\overline{Y})\right] = \int_{-\infty}^{\infty}\int_{-\infty}^{\infty}(x-\overline{X})(y-\overline{Y})\, f_{XY}(x,y)\,dx\,dy$$

$$= \int_{-\infty}^{\infty}\int_{-\infty}^{\infty}\left(xy-\overline{X}y-x\overline{Y}+\overline{XY}\right) f_{XY}(x,y)\,dx\,dy$$

$$= \int_{-\infty}^{\infty}\int_{-\infty}^{\infty} xyf_{XY}(x,y)\,dx\,dy - \overline{X}\int_{-\infty}^{\infty}\int_{-\infty}^{\infty} y\, f_{XY}(x,y)\,dx\,dy - \overline{Y}\int_{-\infty}^{\infty}\int_{-\infty}^{\infty} xf_{XY}(x,y)\,dx\,dy$$

$$+\overline{XY}\int_{-\infty}^{\infty}\int_{-\infty}^{\infty} f_{XY}(x,y)\,dx\,dy$$

$$= R_{XY} - \overline{X}\,\overline{Y} - \overline{X}\,\overline{Y} + \overline{X}\,\overline{Y}$$

$$C_{XY} = R_{XY} - \overline{X}\,\overline{Y} \qquad \text{...(5.13)}$$

When $\quad C_{XY} = 0,$

$R_{XY} = \overline{X}\,\overline{Y} = E[X]E[Y]$, i.e. if X and Y are independent or uncorrelated then $C_{XY} = 0$.

If X and Y are two orthogonal random variables, then

$$C_{XY} = \overline{X}\,\overline{Y} = E[X]E[Y] \qquad \because R_{XY} = 0 \qquad \text{...(5.14)}$$

The normalized second order moment is also called as correlation coefficient of X and Y, which is given by

$$\rho = \frac{\mu_{11}}{\sqrt{\mu_{20}\mu_{02}}} = \frac{C_{XY}}{\sigma_X\sigma_Y} \qquad \text{...(5.15)}$$

$$\rho = E\left[\frac{(X-\overline{X})}{\sigma_X}\,\frac{(Y-\overline{Y})}{\sigma_Y}\right] \qquad \text{...(5.16)}$$

Example 5.2.1: Show that the correlation coefficient satisfies the expression

$$|\rho| = \frac{|\mu_{11}|}{\sqrt{\mu_{20}\mu_{02}}} \le 1$$

Solution: Consider a non-negative expression

$$\rho = E\left[\left\{\lambda(X-\overline{X})-(Y-\overline{Y})\right\}^2\right] = E\left[\lambda^2(X-\overline{X})^2 - 2\lambda(X-\overline{X})(Y-\overline{Y}) + (Y-\overline{Y})^2\right]$$

$$= \lambda^2 E\left[(X-\overline{X})^2\right] - 2\lambda E\left[(X-\overline{X})(Y-\overline{Y})\right] + E\left[(Y-\overline{Y})^2\right] \ge 0$$

$$\Rightarrow \qquad \lambda^2\mu_{20} - 2\lambda\mu_{11} + \mu_{02} \ge 0$$

The above equation have at least one real root. Hence, its determinant must satisfy

$$b^2 - 4ac \le 0$$

$$(2\lambda\,\mu_{11})^2 - 4\lambda^2\,\mu_{20}\,\mu_{02} \le 0$$

$$4\lambda^2\mu_{11}^2 - 4\lambda^2\mu_{20}\,\mu_{02} \le 0$$

$$\left(\frac{\mu_{11}}{\mu_{20}}\right)^2 - \left(\frac{\mu_{02}}{\mu_{20}}\right) \leq 0$$

$$\mu_{11}^2 \leq \mu_{02}\mu_{20}$$

$$\mu_{11} \leq \sqrt{\mu_{02}\mu_{20}}$$

$$\frac{|\mu_{11}|}{\sqrt{\mu_{02}\mu_{20}}} \leq 1$$

$$\rho = \frac{|\mu_{11}|}{\sqrt{\mu_{02}\mu_{20}}}.$$

5.3 FUNCTIONS THAT GENERATE JOINT MOMENTS

There are two functions that allow moments to be calculated for random variables X and Y. They are:

i. Joint moment generating function
ii. Joint characteristic function.

5.3.1 Joint Moment Generating Function

The joint moment generating function is defined as

$$M_{XY}(u, v) = E[e^{ux+vy}] = \int_{-\infty}^{\infty}\int_{-\infty}^{\infty} e^{ux+vy} f_{XY}(x,y)\,dx\,dy \qquad ...(5.17)$$

where u and v are real numbers, $-\infty < u < \infty$ and $-\infty < v < \infty$.

By substituting $u = 0$,

$$M_Y(v) = M_{XY}(0, v) \qquad ...(5.18)$$

Equation (5.18) is called marginal moment generating function of random variable Y.

Similarly by substituting $v = 0$,

$$M_X(u) = M_{XY}(u, 0) \qquad ...(5.19)$$

Equation (5.19) is called marginal moment generating function of random variable X.

The joint moments m_{nk} found by using joint moment generating function is

$$m_{nk} = \left.\frac{d^{n+k}[M_{XY}(u,v)]}{du^n dv^n}\right|_{u=0,v=0} \qquad ...(5.20)$$

For N random variables, the joint moment generating function is

$$M_{X_1X_2...X_N}(u_1,u_2,...,u_N) = E[e^{u_1x_1+u_2x_2+\cdots+u_Nx_N}] \qquad ...(5.21)$$

Joint moments are

$$m_{n_1n_2....n_N} = \left.\frac{d^{n_1+n_2+...+n_N}\left(M_{X_1X_2...X_N}(u_1,u_2,...,u_N)\right)}{du_1^{n_1}du_2^{n_2}...du_N^{n_N}}\right|_{\text{for all } u_i} \qquad ...(5.22)$$

5.3.2 Joint Characteristic Function

The joint characteristic function is defined as

$$\phi_{XY}(\omega_1, \omega_2) = E\left[e^{j\omega_1 x + j\omega_2 y}\right] = \int_{-\infty}^{\infty}\int_{-\infty}^{\infty} e^{j\omega_1 x + j\omega_2 y} f_{XY}(x,y)\,dx\,dy \qquad ...(5.23)$$

where ω_1 and ω_2 are real numbers, $-\infty < \omega_1 < \infty$ and $-\infty < \omega_2 < \infty$.

From the above, it is clear that Eq. (5.23) is similar to the Fourier transform equation. By using inverse Fourier transform we can also find the joint density function $f_{XY}(x, y)$.

$$f_{XY}(x, y) = \left(\frac{1}{2\pi}\right)^2 \int_{-\infty}^{\infty}\int_{-\infty}^{\infty} e^{-j\omega_1 x - j\omega_2 y} \phi_{XY}(\omega_1, \omega_2)\,d\omega_1 d\omega_2 \qquad ...(5.24)$$

By letting $\omega_1 = 0$,

$$\phi_Y(\omega_2) = \phi_{XY}(0, \omega_2) \qquad ...(5.25)$$

Equation (5.25) is called marginal characteristic function of random variable Y.

Similarly by letting $\omega_2 = 0$,

$$\phi_X(\omega_1) = \phi_{XY}(\omega_1, 0) \qquad ...(5.26)$$

Equation (5.26) is called marginal characteristic function of random variable X.

The joint moments m_{nk} can be found by using joint characteristic function as

$$m_{nk} = \left.\frac{\partial^{n+k}\left[\phi_{XY}(\omega_1, \omega_2)\right]}{\partial\omega_1^n \partial\omega_2^n}\right|_{\omega_1=0,\omega_2=0} \qquad ...(5.27)$$

For N random variables, the joint characteristic function is

$$\phi_{X_1X_2\ldots X_N}(\omega_1, \omega_2, \ldots, \omega_N) = E\left[e^{j\omega_1 x_1 + j\omega_2 x_2 + \cdots + j\omega_N x_N}\right] \qquad ...(5.28)$$

The joint moments are

$$m_{n_1 n_2 \ldots n_N} = \left.\frac{\partial^{n_1+n_2+\ldots+n_N}\left[\phi_{X_1X_2\ldots X_N}(\omega_1, \omega_2, \ldots, \omega_N)\right]}{\partial\omega_1^{n_1}\partial\omega_2^{n_2}\ldots\partial\omega_N^{n_N}}\right|_{\text{for all } \omega_i} \qquad ...(5.29)$$

Note: It is very useful to obtain the PDF for sum of N statistically independent random variables.

5.3.3 Joint Gaussian Random Variables

The jointly Gaussian random variables appear in number of applications in electrical engineering. They are frequently used to model signals in signal processing applications, and they are the most important model used in communication systems that involve dealing with signals in the presence of noise. They also play a central role in many statistical methods.

Two random variables X and Y are said to be jointly gaussian if their joint density function is of the form

$$f_{XY}(x,y) = \frac{1}{2\pi\sigma_X\sigma_Y\sqrt{1-\rho_{XY}^2}} \exp\left[\frac{-1}{2\left(1-\rho_{XY}^2\right)}\left\{\left(\frac{x-\mu_X}{\sigma_X}\right)^2 - 2\rho_{XY}\frac{(x-\mu_X)(y-\mu_Y)}{\sigma_X\sigma_Y} + \left(\frac{y-\mu_Y}{\sigma_Y}\right)^2\right\}\right]$$

for $-\infty < x < \infty$, $-\infty < y < \infty$ ...(5.30)

where μ_X = expectation of X
μ_Y = expectation of Y
σ_X^2 = variance of X
σ_Y^2 = variance of Y
and ρ_{XY} = correlation coefficient.

Sometimes it is also called as Bivariate gaussian density

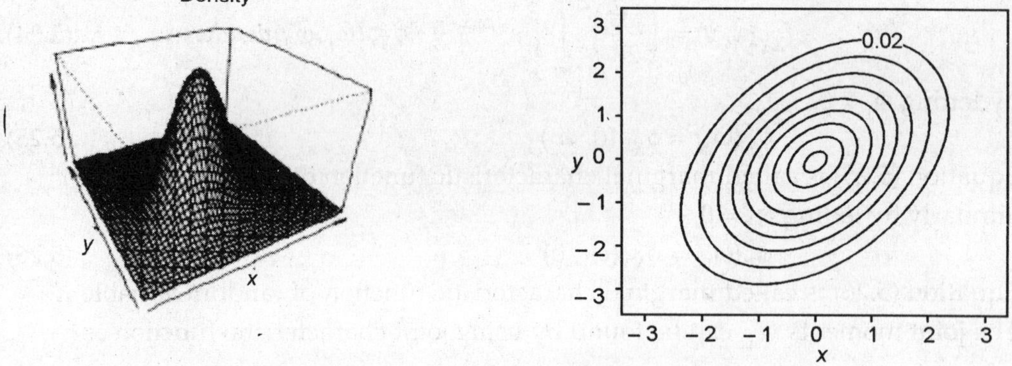

Fig. 5.1: Bivariate standard normal density (ρ = 0.5) and its contour

Its maximum value is at the point (μ_X, μ_Y). The maximum value is

$$f_{XY}(x, y) = f_{XY}(\mu_X, \mu_Y) = \frac{1}{2\pi\sigma_X\sigma_Y\sqrt{1-\rho_{XY}^2}} \quad \text{...(5.31)}$$

If random variables X and Y are uncorrelated,

$$f_X(x) = \frac{1}{\sqrt{2\pi\sigma_X^2}}\exp\left\{\frac{-1}{2}\left(\frac{x-\mu_X}{\sigma_X}\right)^2\right\} \quad \text{...(5.32)}$$

$$f_Y(y) = \frac{1}{\sqrt{2\pi\sigma_Y^2}}\exp\left\{\frac{-1}{2}\left(\frac{y-\mu_Y}{\sigma_Y}\right)^2\right\} \quad \text{...(5.33)}$$

For N random variables:

The N random variables $X_1, X_2, ..., X_N$ are said to be jointly gaussian if their joint density function is in the form

$$f_{X_1,X_2,...,X_N}(x_1, x_2, ..., x_N) = \frac{\left|[C_X]^{-1}\right|^{1/2}}{(2\pi)^{N/2}}\exp\left\{\frac{-1}{2}\left(\frac{[x-\mu_X]^T[C_X]^{-1}[x-\mu_X]}{\sigma_X}\right)^2\right\} \quad \text{...(5.34)}$$

where $[x-\mu_X] = \begin{bmatrix} x_1-\mu_{X_1} \\ x_2-\mu_{X_2} \\ \vdots \\ x_N-\mu_{X_N} \end{bmatrix}$ and $C_X = \begin{bmatrix} C_{11} & C_{12} & \cdots & C_{1N} \\ C_{21} & C_{22} & \cdots & C_{2N} \\ \vdots & \vdots & & \vdots \\ C_{N1} & C_{N2} & \cdots & C_{NN} \end{bmatrix}$

Matrix C_X is called covariance matrix.

We know that $C_{XY} = E[(x-\mu_X)(y-\mu_Y)]$

Similarly $C_{X_1X_2} = E[(x_1-\mu_1)(x_2-\mu_2)]$ and $C_{X_iX_j} = E\left[(x_i-\mu_i)(x_j-\mu_j)\right]$

In short cut form $C_{ij} = C_{X_iX_j}$

$$C_{ij} = \begin{cases} \sigma_{X_i} & \text{for } i = j \\ E\left[(x_i-\mu_i)(x_j-\mu_j)\right] & \text{for } i \neq j \end{cases} \qquad ...(5.35)$$

Theorem 5.1: Random variables X and Y have a joint gaussian PDF with parameters μ_1, μ_2, σ_1, σ_2 and ρ if

$$f_{XY}(x, y) = \frac{1}{2\pi\sigma_1\sigma_2\sqrt{1-\rho^2}} \exp\left[\frac{-1}{2(1-\rho^2)}\left\{\left(\frac{x-\mu_1}{\sigma_1}\right)^2 - 2\rho\frac{(x-\mu_1)(y-\mu_2)}{\sigma_1\sigma_2} + \left(\frac{y-\mu_2}{\sigma_2}\right)^2\right\}\right]$$

where μ_1 and μ_2 can be any real numbers, $\sigma_1 > 0$, $\sigma_2 > 0$ and $-1 < \rho < 1$.

From Eq. (5.34), for $N = 2$

$$f_{X_1,X_2}(x_1, x_2) = \frac{\left|[C_X]^{-1}\right|^{1/2}}{(2\pi)} \exp\left\{-\left(\frac{[x-\mu_X]^T [C_X]^{-1} [x-\mu_X]}{2}\right)^2\right\}$$

and $$C_X = \begin{bmatrix} C_{11} & C_{12} \\ C_{21} & C_{22} \end{bmatrix} \quad \text{for } N = 2$$

According to Eq. (5.35), $C_{11} = \sigma_1^2$, $C_{12} = \rho\sigma_1\sigma_2$, $C_{21} = \rho\sigma_1\sigma_2$ and $C_{22} = \sigma_2^2$.

Hence $$C_X = \begin{bmatrix} \sigma_1^2 & \rho\sigma_1\sigma_2 \\ \rho\sigma_1\sigma_2 & \sigma_2^2 \end{bmatrix}$$

$$[C_X]^{-1} = \frac{1}{\sigma_1^2\sigma_2^2(1-\rho^2)}\begin{bmatrix} \sigma_2^2 & -\rho\sigma_1\sigma_2 \\ -\rho\sigma_1\sigma_2 & \sigma_1^2 \end{bmatrix} = \frac{1}{(1-\rho^2)}\begin{bmatrix} \frac{1}{\sigma_1^2} & \frac{-\rho}{\sigma_1\sigma_2} \\ \frac{-\rho}{\sigma_1\sigma_2} & \frac{1}{\sigma_2^2} \end{bmatrix}$$

$$[C_X] = \sigma_1^2\sigma_2^2(1-\rho^2)$$

$$\left|[C_X]\right|^{-1/2} = \frac{1}{\sigma_1\sigma_2\sqrt{(1-\rho^2)}}$$

$$[x-\mu_X] = \begin{bmatrix} x_1-\mu_1 \\ x_2-\mu_2 \end{bmatrix}$$

$$[x-\mu_X]^T[C_X]^{-1} = [x_1-\mu_1 \quad x_2-\mu_2]\frac{1}{(1-\rho^2)}\begin{bmatrix}\frac{1}{\sigma_1^2} & \frac{-\rho}{\sigma_1\sigma_2}\\ \frac{-\rho}{\sigma_1\sigma_2} & \frac{1}{\sigma_2^2}\end{bmatrix}$$

$$= \frac{1}{(1-\rho^2)}\left[\frac{x_1-\mu_1}{\sigma_1^2}-\frac{\rho(x_2-\mu_2)}{\sigma_1\sigma_2} \quad \frac{x_2-\mu_2}{\sigma_2^2}-\frac{\rho(x_1-\mu_1)}{\sigma_1\sigma_2}\right]$$

$$[x-\mu_X]^T[C_X]^{-1}[x-\mu_X] = \frac{1}{(1-\rho^2)}\left[\frac{x_1-\mu_1}{\sigma_1^2}-\frac{\rho(x_2-\mu_2)}{\sigma_1\sigma_2} \quad \frac{x_2-\mu_2}{\sigma_2^2}-\frac{\rho(x_1-\mu_1)}{\sigma_1\sigma_2}\right]\begin{bmatrix}x_1-\mu_1\\ x_2-\mu_2\end{bmatrix}$$

$$= \frac{1}{(1-\rho^2)}\left[\frac{(x_1-\mu_1)^2}{\sigma_1^2}-\frac{\rho(x_2-\mu_2)(x_1-\mu_1)}{\sigma_1\sigma_2}+\frac{(x_2-\mu_2)^2}{\sigma_2^2}-\frac{\rho(x_1-\mu_1)(x_2-\mu_2)}{\sigma_1\sigma_2}\right]$$

$$= \frac{1}{(1-\rho^2)}\left[\frac{(x_1-\mu_1)^2}{\sigma_1^2}-\frac{2\rho(x_2-\mu_2)(x_1-\mu_1)}{\sigma_1\sigma_2}+\frac{(x_2-\mu_2)^2}{\sigma_2^2}\right]$$

$$f_{X_1,X_2}(x_1,x_2) = \frac{\left|[C_X]^{-1}\right|^{1/2}}{(2\pi)}\exp\left\{-\left(\frac{[x-\mu_X]^T[C_X]^{-1}[x-\mu_X]}{2}\right)^2\right\}$$

By substituting all the above values, $X_1 = X$ and $X_2 = Y$

$$f_{XY}(x,y) = \frac{1}{2\pi\sigma_1\sigma_2\sqrt{1-\rho^2}}\exp\left\{\frac{-1}{2(1-\rho^2)}\left[\left(\frac{x-\mu_1}{\sigma_1}\right)^2-2\rho\frac{(x-\mu_1)(y-\mu_2)}{\sigma_1\sigma_2}+\left(\frac{y-\mu_2}{\sigma_2}\right)^2\right]\right\}$$

Theorem 5.2: The angle required to convert correlated variables X and Y to statistically independent random variables is $\theta = \frac{1}{2}\tan^{-1}\left[\frac{2\rho\sigma_X\sigma_Y}{\sigma_X^2-\sigma_Y^2}\right]$.

Let the two random variables be X and Y, which are going to convert to statistically independent random variables.

The new variables are

$$Y_1 = X\cos\theta + Y\sin\theta$$
$$Y_2 = -X\sin\theta + Y\cos\theta$$

The mean of Y_1 is denoted by $\bar{Y}_1$ and the mean of Y_2 by $\bar{Y}_2$.

$$\bar{Y}_1 = \bar{X}\cos\theta + \bar{Y}\sin\theta$$

$$\bar{Y}_2 = -\bar{X}\sin\theta + \bar{Y}\cos\theta$$

$$C_{Y_1Y_2} = E\left[(Y_1-\bar{Y}_1)(Y_2-\bar{Y}_2)\right]$$

$$= E\left[(X\cos\theta + Y\sin\theta - \bar{X}\cos\theta - \bar{Y}\sin\theta)(-X\sin\theta + Y\cos\theta + \bar{X}\sin\theta - \bar{Y}\cos\theta)\right]$$

$$= (\sigma_Y^2 - \sigma_X^2)\cos\theta\sin\theta + C_{XY}[\cos^2\theta - \sin^2\theta]$$

$$C_{Y_1Y_2} = (\sigma_Y^2 - \sigma_X^2)\frac{1}{2}\sin 2\theta + C_{XY}\cos 2\theta$$

If we require Y_1 and Y_2 being uncorrelated, we must have $C_{Y_1Y_2} = 0$

$$0 = (\sigma_Y^2 - \sigma_X^2)\frac{1}{2}\sin 2\theta + C_{XY}\cos 2\theta$$

But $$C_{XY} = \rho\sigma_X\sigma_Y$$

$$(\sigma_Y^2 - \sigma_X^2)\frac{1}{2}\sin 2\theta = -\rho\sigma_X\sigma_Y\cos 2\theta$$

$$\frac{1}{2}\tan 2\theta = \frac{-\rho\sigma_X\sigma_Y}{(\sigma_Y^2 - \sigma_X^2)}$$

$$\theta = \frac{1}{2}\tan^{-1}\left(\frac{2\rho\sigma_X\sigma_Y}{(\sigma_X^2 - \sigma_Y^2)}\right)$$

Some important properties of Gaussian random variables are:

1. By using first and second order moments only, we can define Gaussian random variables completely.
2. The linear combination of two independent random variables having a normal distribution also have a normal distribution.
3. Uncorrelated random variables are also statistically independent.
4. Any *K*-dimensional marginal density function obtained from *N*-dimensional density function by integrating *N-K* random variables will be Gaussian.

Example 5.3.1: Statistically independent random variables X and Y have moments $m_{10} = 2$, $m_{20} = 14$, $m_{02} = 12$, $m_{11} = -6$. Find the moment m_{22}.

Solution: $m_{10} = E[X] = 2$, $m_{20} = E[X^2] = 14$, $m_{02} = E[Y^2] = 12$ and $m_{11} = E[XY] = -6$

For statistically independent random variables, $m_{11} = E[XY] = E[X]E[Y] = -6$

$$E[Y] = -\frac{6}{2} = -3$$

$$m_{22} = E\left[(x-\bar{X})^2(y-\bar{Y})^2\right] = E\left[x^2 - 2x\bar{X} + \bar{X}^2\right]E\left[y^2 - 2y\bar{Y} + \bar{Y}^2\right]$$

$$= \left(E[x^2] - 2\bar{X}E[x] + \bar{X}^2\right)\left(E[y^2] - 2\bar{Y}E[y] + \bar{Y}^2\right) = (14-4)(12-9) = 30$$

Example 5.3.2: The density function of two random variables X and Y is

$$f_{XY}(x, y) = \begin{cases} \frac{1}{24} & 0 < x < 6, 0 < y < 4 \\ 0 & \text{otherwise} \end{cases}$$

Find the expectation of the function $g(x, y) = (xy)^2$.

Solution: $E[g(x,y)] = \int_0^4\int_0^6 \frac{1}{24}(xy)^2\,dx\,dy = \frac{1}{24}\frac{x^3}{3}\Big|_0^6 \frac{y^3}{3}\Big|_0^4 = 64$

5.4 TRANSFORMATIONS OF MULTIPLE RANDOM VARIABLES

5.4.1 Single Function

Consider a transformation of one random variable into one function of several random variables.

Let us consider

$$Y = g(X_1, X_2, ..., X_N) \qquad ...(5.36)$$

In order to find $f_Y(y)$, first we define distribution function of Y and then the probability density function of Y.

The distribution function of Y is

$$F_Y(y) = P(Y \leq y) = P\{g(X_1, X_2, ..., X_N) \leq y\} \qquad ...(5.37)$$

This probability is associated with all points in the hyper space $(X_1, X_2, ..., X_N)$ that maps such that $g(X_1, X_2, ..., X_N) \leq y$ for any y.

$$F_Y(y) = P\{g(X_1, X_2, ..., X_N) \leq y\} = \iint ... \int f_{X_1, X_2, ..., X_N}(x_1, x_2, ..., x_N)\{g(x_1, x_2, ..., x_N) \leq y\} \qquad ...(5.38)$$

The density follows differentiation

$$f_Y(y) = \frac{dF_Y(y)}{dy} = \frac{d}{dy}\iint ... \int f_{X_1, X_2, ..., X_N}(x_1, x_2, ..., x_N)\{g(x_1, x_2, ..., x_N) \leq y\} \qquad ...(5.39)$$

5.4.2 Multiple Functions

Let X and Y be two random variables with joint probability density function $f_{XY}(x, y)$ and let two new random variables be defined as $Z = \varphi_1(X, Y)$ and $W = \varphi_2(X, Y)$ with the inverse relations $X = \varphi_1^{-1}(Z, W)$ and $Y = \varphi_2^{-1}(Z, W)$. Let $g(z, w)$ be the joint probability density function of Z and W and consider the case where if X and Y both increase, Z and W also increase. The probability that x and y lie in a particular region is equal to the probability that z and w lie in a corresponding region. This can be stated mathematically as follows.

$$P\{z_1 < Z < z_2, w_1 < W < w_2\} = P\{x_1 < X < x_2, y_1 < Y < y_2\} \qquad ...(5.40)$$

or equivalently,

$$\int_{w_1}^{w_2}\int_{z_1}^{z_2} g_{ZW}(z, w)\,dz\,dw = \int_{y_1}^{y_2}\int_{x_1}^{x_2} f_{XY}(x, y)\,dx\,dy \qquad ...(5.41)$$

We know that

$$x = \varphi_1^{-1}(z, w) \qquad ...(5.42)$$

$$y = \varphi_2^{-1}(z, w) \qquad ...(5.43)$$

$$dx = \frac{\partial \varphi_1^{-1}(z, w)}{\partial z}dz + \frac{\partial \varphi_1^{-1}(z, w)}{\partial w}dw$$

$$dy = \frac{\partial \varphi_2^{-1}(z, w)}{\partial z}dz + \frac{\partial \varphi_2^{-1}(z, w)}{\partial w}d\omega$$

From the above, the coefficients of dz and dw are the coefficients of jacobian matrix, i.e.

$$[J] = \begin{bmatrix} \dfrac{\partial \varphi_1^{-1}(z,w)}{\partial z} & \dfrac{\partial \varphi_1^{-1}(z,w)}{\partial w} \\ \dfrac{\partial \varphi_2^{-1}(z,w)}{\partial z} & \dfrac{\partial \varphi_2^{-1}(z,w)}{\partial w} \end{bmatrix}$$

The magnitude of jacobian matrix is $|J|$.

Hence $$dxdy = |J|\, dzdw \qquad ...(5.44)$$

Substitute Eqs. (5.42), (5.43) and (5.44) in Eq. (5.41),

$$\int_{w_1}^{w_2}\int_{z_1}^{z_2} g_{ZW}(z,w)dzdw = \int_{w_1}^{w_2}\int_{z_1}^{z_2} f_{XY}\left[x = \varphi_1^{-1}(z,w), y = \varphi_2^{-1}(z,w)\right]|J|dzdw \qquad ...(5.45)$$

By comparing LHS and RHS,

$$g_{ZW}(z, w) = f_{XY}(x, y)\,|J| \qquad ...(5.46)$$

Similarly, for N functions

$$Y_i = \varphi_i(X_1, X_2,...,X_N) \quad \text{for } i = 1, 2,...,N \qquad ...(5.47)$$

$$\int_{R_Y}\!\!\cdots\!\int f_{Y_1,Y_2,...,Y_N}(y_1,y_2,...,y_N)dy_1,dy_2,...,dy_N$$

$$= \int_{R_Y}\!\!\cdots\!\int f_{X_1,X_2,...,X_N}(x_1,x_2,...,x_N)|J|dy_1,dy_2,...,dy_N$$

$$= \int_{R_Y}\!\!\cdots\!\int f_{X_1,X_2,...,X_N}\begin{bmatrix}(x_1 = \varphi_1^{-1}(y_1,y_2,...,y_N), x_2 = \varphi_2^{-1}(y_1,y_2,...,y_N),..., \\ x_N = \varphi_N^{-1}(y_1,y_2,...,y_N)\end{bmatrix}|J|dy_1,dy_2,...,dy_N$$

where

$$[J] = \begin{bmatrix} \dfrac{\partial \varphi_1^{-1}(y_1,y_2,...,y_N)}{\partial y_1} & \dfrac{\partial \varphi_1^{-1}(y_1,y_2,...,y_N)}{\partial y_2} & \cdots & \dfrac{\partial \varphi_1^{-1}(y_1,y_2,...,y_N)}{\partial y_N} \\ \dfrac{\partial \varphi_2^{-1}(y_1,y_2,...,y_N)}{\partial y_1} & \dfrac{\partial \varphi_2^{-1}(y_1,y_2,...,y_N)}{\partial y_2} & \cdots & \dfrac{\partial \varphi_2^{-1}(y_1,y_2,...,y_N)}{\partial y_N} \\ \vdots & \vdots & & \vdots \\ \dfrac{\partial \varphi_N^{-1}(y_1,y_2,...,y_N)}{\partial y_1} & \dfrac{\partial \varphi_N^{-1}(y_1,y_2,...,y_N)}{\partial y_2} & \cdots & \dfrac{\partial \varphi_N^{-1}(y_1,y_2,...,y_N)}{\partial y_N} \end{bmatrix} \qquad ...(5.48)$$

By comparing LHS and RHS,

$$f_{Y_1,Y_2,...,Y_N}(y_1,y_2,...,y_N) = f_{X_1,X_2,...,X_N}(x_1,x_2,...,x_N)|J| \qquad ...(4.49)$$

5.4.3 Linear Transformations of Gaussian Random Variables

Let us consider a set of Gaussian random variables $X_1, X_2, \ldots, X_N$.

The new random variables $Y_1, Y_2, \ldots, Y_N$ are defined as

$$Y_1 = a_{11}X_1 + a_{12}X_2 + \cdots + a_{1N}X_N$$
$$Y_2 = a_{21}X_1 + a_{22}X_2 + \cdots + a_{2N}X_N$$
$$\vdots$$
$$Y_N = a_{N1}X_1 + a_{N2}X_2 + \cdots + a_{NN}X_N$$

where the coefficients a_{ij}, i and $j = 1, 2, \ldots, N$ are real numbers.

Now we define the following matrices,

$$[T] = \begin{bmatrix} a_{11} & a_{12} & \cdots & a_{1N} \\ a_{21} & a_{22} & \cdots & a_{2N} \\ \vdots & \vdots & \vdots & \vdots \\ a_{N1} & a_{N2} & \cdots & a_{NN} \end{bmatrix}$$

$$[Y] = \begin{bmatrix} Y_1 \\ Y_2 \\ \vdots \\ Y_N \end{bmatrix}, \quad [\bar{Y}] = \begin{bmatrix} \bar{Y}_1 \\ \bar{Y}_2 \\ \vdots \\ \bar{Y}_N \end{bmatrix}$$

$$[X] = \begin{bmatrix} X_1 \\ X_2 \\ \vdots \\ X_N \end{bmatrix}, \quad [\bar{X}] = \begin{bmatrix} \bar{X}_1 \\ \bar{X}_2 \\ \vdots \\ \bar{X}_N \end{bmatrix}$$

From the above, it is clear that

$$[Y] = [T][X]; \quad [Y - \bar{Y}] = [T][X - \bar{X}];$$
$$[X] = [T]^{-1}[Y] \text{ and } [X - \bar{X}] = [T]^{-1}[Y - \bar{Y}]$$
$$X_i = T_i^{-1}(Y_1, Y_2, \ldots, Y_N) = b_{i1}Y_1 + b_{i2}Y_2 + \cdots + b_{iN}Y_N$$

$$\frac{dX_i}{dY_j} = \frac{dT_i^{-1}}{dY_j} = b_{ij} \quad \text{...(5.50)}$$

$$X_i - \bar{X}_i = b_{i1}(Y_1 - \bar{Y}_1) + b_{i2}(Y_2 - \bar{Y}_2) + \cdots + b_{iN}(Y_N - \bar{Y}_N)$$

where b_{ij} represents the ij^{th} element of $[T]^{-1}$.

The density function of the new variables $Y_1, Y_2, \ldots, Y_N$ can be obtained in two steps. The first step is to determine $|J|$.

From Eq. (5.50) and Eq. (5.48), we can find that J equals to the determinant of matrix $[T]^{-1}$, i.e.

$$J = \left|[T]^{-1}\right|$$
$$|J| = \left\|[T]^{-1}\right\| \quad \text{...(5.51)}$$

The second step is to find the covariance.

$$C_{X_i,X_j} = E\left[(X_i - \overline{X}_i)(X_j - \overline{X}_j)\right]$$

$$C_{X_i,X_j} = \sum_{n=1}^{N} b_{in} \sum_{m=1}^{M} b_{jm} E\left[(Y_i - \overline{Y}_i)(Y_j - \overline{Y}_j)\right] \quad ...(5.52)$$

$\because C_{X_i,X_j}$ is the ij^{th} element in the covariance matrix $[C_X]$ and C_{Y_n,Y_m} is the nm^{th} element of the covariance of the new variables Y_i, which we denote by $[C_Y]$.

Equation (5.52) can be written as

$$[C_X] = [T]^{-1}\,[C_Y]\,([T]^t)^{-1}$$

where $[T]^t$ is the transpose of $[T]$

$$[T]\,[C_X] = [C_Y]\,([T]^t)^{-1}$$

$$[T][C_X]\,[T]^t = [C_Y]$$

$$\overline{[C_Y]} = \overline{[T]}\,\overline{[C_X]}\overline{[T]^t}$$

$$|[C_Y]| = |[C_X]|\left|\overline{|[T]|}\right|^2$$

$$|[C_X]^{-1}| = |[C_Y]^{-1}|\left|\overline{|[T]|}\right|^2$$

$$f_{X_1,X_2,...,X_N}(x_1 = T_1^{-1}, x_2 = T_2^{-1}, ..., x_N = T_N^{-1})$$

$$= \frac{\left(|[C_Y]^{-1}|\left|\overline{|[T]|}\right|^2\right)^{1/2}}{(2\pi)^{N/2}} \exp\left\{\frac{-1}{2}[x-\overline{x}]^t\,[T]^t[C_Y]^{-1}[T][x-\overline{x}]\right\}$$

$$f_{Y_1,Y_2,...,Y_N}(y_1, y_2, ..., y_N) = \frac{\left(|[C_Y]^{-1}|\right)^{1/2}}{(2\pi)^{N/2}} \exp\left\{\frac{-1}{2}[y-\overline{y}]^t[C_Y]^{-1}[y-\overline{y}]\right\} \quad ...(5.53)$$

Example 5.4.1: Two random variables X and Y are defined by $\overline{X} = 0, \overline{Y} = -1$, $\overline{X^2} = 2, \overline{Y^2} = 4$ and $R_{XY} = -2$, two new random variables W and V defined as $W = 2X + Y$, $V = -X - 3Y$. Find $\overline{W}, \overline{V}, \overline{W^2}, \overline{V^2}, R_{WV}, \sigma_X^2, \sigma_Y^2, \sigma_W^2$ and σ_V^2.

Solution: $C_{XY} = R_{XY} - E[X]E[Y] = -2 - 0 = -2$

$$E[W] = E[2X + Y] = 2\overline{X} + \overline{Y} = -1$$

$$E[V] = E[-X - 3Y] = -\overline{X} - 3\overline{Y} = 3$$

$$R_{WV} = E[(2X + Y)(-X - 3Y)]$$

$$= E[-2X^2 - 7XY - 3Y^2] = -(2)(2) - (7)(-2) - (3)(4) = -2$$

$$\sigma_X^2 = E[X^2] - (E[X])^2 = 2$$

$$\sigma_Y^2 = E[Y^2] - (E[Y])^2 = 4 - 1 = 3$$

$$\sigma_W^2 = E[W^2]-(E[W])^2$$

$$E[W^2] = E[(2X+Y)^2] = 4\overline{X^2}+4\overline{XY}+\overline{Y^2} = (4)(2)+(4)(-2)+4=4$$

$$E[V^2] = E[(-X-3Y)^2] = \overline{X^2}+6\overline{XY}+9\overline{Y^2} = 2+(4)(9)+(6)(-2)=26$$

$$\sigma_W^2 = E[W^2]-(E[W])^2 = 4-1=3$$

$$\sigma_V^2 = E[V^2]-(E[V])^2 = 26-9=17.$$

EXAMPLES

Example 5.1: Find the expected value of $g(X_1,X_2,X_3,X_4)=X_1^{n_1}X_2^{n_2}X_3^{n_3}X_4^{n_4}$, where n_1, n_2, n_3 and n_4 are integers ≥ 0 and

$$f_{X_1,X_2,X_3,X_4}(x_1,x_2,x_3,x_4) = \begin{cases} \dfrac{1}{abcd} & 0<x_1<a, 0<x_2<b, 0<x_3<c, 0<x_4<d \\ 0 & \text{elsewhere} \end{cases}$$

Solution: $E[g(X_1,X_2,X_3,X_4)] = \int_0^d\int_0^c\int_0^b\int_0^a x_1^{n_1}x_2^{n_2}x_3^{n_3}x_4^{n_4}\dfrac{1}{abcd}dx_1\,dx_2\,dx_3\,dx_4$

$$E[g(X_1,X_2,X_3,X_4)] = \left(\int_0^a\left(\int_0^b\left(\int_0^c x_3^{n_3}dx_3\left(\int_0^d x_4^{n_4}dx_4\right)\right)x_2^{n_2}dx_2\right)x_1^{n_1}dx_1\right)\frac{1}{abcd}$$

$$= \frac{1}{abcd}\left(\frac{d^{n_4+1}}{n_4+1}+\frac{c^{n_3+1}}{n_3+1}+\frac{b^{n_2+1}}{n_2+1}+\frac{a^{n_1+1}}{n_1+1}\right) = \frac{a^{n_1}b^{n_2}c^{n_3}d^{n_4}}{(n_4+1)(n_3+1)(n_2+1)(n_1+1)}.$$

Example 5.2: Two random variables have a uniform density on a circular region defined by

$$f_{XY}(x,y) = \begin{cases} \dfrac{1}{\pi r^2} & x^2+y^2 \leq r^2 \\ 0 & \text{else} \end{cases}$$

Find the mean of $g(X, Y) = X^2 + Y^2$.

Solution: Let $x = r\cos\theta$, $y = r\sin\theta$ and $dxdy = rdrd\theta$

$$E[g(X,Y)] = \int_{-\infty}^{\infty}\int_{-\infty}^{\infty} g(X,Y)f_{X,Y}(x,y)dx\,dy$$

$$= \int_0^{2\pi}\int_0^r r^2\frac{1}{\pi r^2}rdrd\theta = \frac{1}{\pi}\int_0^{\cdot}\frac{r^2}{2}d\theta = \frac{r^2}{2\pi}2\pi = r^2$$

Example 5.3: Two random variables X and Y have variances $\sigma_X^2 = 9$, and $\sigma_Y^2 = 4$ respectively. It is known that a coordinate rotation of $-\frac{\pi}{8}$ results in new random variables, Y_1 and Y_2, which are uncorrelated. Find the correlation coefficient.

Solution: $\theta = \frac{1}{2}\tan^{-1}\left(\frac{2\rho\sigma_X\sigma_Y}{\sigma_X^2 - \sigma_Y^2}\right)$

$$\frac{-\pi}{8} = \frac{1}{2}\tan^{-1}\left(\frac{2\rho(3)(2)}{9-4}\right)$$

$$\frac{-\pi}{4} = \tan^{-1}\left(\frac{12\rho}{5}\right)$$

$$\frac{12\rho}{5} = \tan\left(\frac{-\pi}{4}\right)$$

$$\rho = \frac{-5}{12}.$$

Example 5.4: Three uncorrelated random variables X_1, X_2 and X_3 have means 1, –3, 1.5 respectively and second moments $E[X_1^2] = 2.5$, $E[X_2^2] = 11$, $E[X_3^2] = 3.5$. Let $Y = X_1 - 2X_2 + 3X_3$ is new random variable. Find: (a) the mean value of Y (b) the variance of Y.

Solution: $E[Y] = E[X_1 - 2X_2 + 3X_3] = E[X_1] - 2E[X_2] + 3E[X_3] = 1 + 6 + 4.5 = 11.5$

$$\sigma_Y^2 = \text{Var}(Y) = \text{Var}(X_1 - 2X_2 + 3X_3)$$

$$\sigma_Y^2 = \text{Var}(X_1) - 4\text{Var}(X_2) + 9\text{Var}(X_3) \qquad \because \text{Var}(kx) = k^2\text{Var}(x)$$

$$\text{Var}(X_1) = E[X_1^2] - (E[X_1])^2 = 3.5 - 1 = 2.5$$

$$\text{Var}(X_2) = E[X_2^2] - (E[X_2])^2 = 11 - 9 = 2$$

$$\text{Var}(X_3) = E[X_3^2] - (E[X_3])^2 = 3.5 - 2.25 = 1.25$$

$$\sigma_Y^2 = \text{Var}(X_1) + 4\text{Var}(X_2) + 9\text{Var}(X_3) = 2.5 + 8 + 11.25 = 21.75$$

Example 5.5: Obtain the relationship between marginal and joint characteristic function.

Solution: Let us consider the joint characteristic function

$$\phi_{XY}(\omega_1, \omega_2) = E[e^{j\omega_1 X + j\omega_2 Y}]$$

Substitute $\omega_1 = \omega_1$ and $\omega_2 = 0$

$$\phi_{XY}(\omega_1, 0) = E[e^{j\omega_1 X + 0}] = E[e^{j\omega_1 X}] = \phi_X(\omega_1)$$

Similarly, substitute $\omega_1 = 0$ and $\omega_2 = \omega_2$

$$\phi_{XY}(0, \omega_2) = E[e^{j\omega_2 X + 0}] = E[e^{j\omega_2 X}] = \phi_Y(\omega_2).$$

Example 5.6: Let $U = g(X, Y) = X + Y$ and $W = h(X, Y) = X - Y$. Find $f_{UW}(u, w)$.

Solution: $u = x + y$ and $w = x - y$

$$x = \frac{u+w}{2} \text{ and } y = \frac{u-w}{2}$$

$$J(x, y) = \begin{bmatrix} \dfrac{\partial u}{\partial x} & \dfrac{\partial u}{\partial y} \\ \dfrac{\partial w}{\partial x} & \dfrac{\partial w}{\partial y} \end{bmatrix} = \begin{bmatrix} 1 & 1 \\ 1 & -1 \end{bmatrix}$$

$$f_{UW}(u, w) = \frac{f_{XY}(x,y)}{|J(x,y)|} = \frac{1}{|-2|} f_{XY}\left(\frac{u+w}{2}, \frac{u-w}{2}\right) = \frac{1}{2} f_{XY}\left(\frac{u+w}{2}, \frac{u-w}{2}\right).$$

Example 5.7: Two Gaussian random variables X_1 and X_2 have zero mean and variances $\sigma_{X_1}^2 = 4$ and $\sigma_{X_2}^2 = 9$ respectively, and their covariance $C_{X_1X_2} = 2$. If X_1 and X_2 are linearly transformed to new variables Y_1 and Y_2 according to

$$Y_1 = X_1 + 2X_2$$
$$Y_2 = 3X_1 + 4X_2$$

then find the mean, variance and covariance of Y_1 and Y_2.

Solution: $[T] = \begin{bmatrix} 1 & 2 \\ 3 & 4 \end{bmatrix}$ and $[C_X] = \begin{bmatrix} 4 & 2 \\ 2 & 9 \end{bmatrix}$

$\because$ X_1 and X_2 are zero mean gaussian random variables, Y_1 and Y_2 will also be zero mean Gaussian random variables.

$$[C_Y] = [T][C_X][T]^t = \begin{bmatrix} 1 & 2 \\ 3 & 4 \end{bmatrix}\begin{bmatrix} 4 & 2 \\ 2 & 9 \end{bmatrix}\begin{bmatrix} 1 & 3 \\ 2 & 4 \end{bmatrix} = \begin{bmatrix} 8 & 20 \\ 20 & 42 \end{bmatrix}\begin{bmatrix} 1 & 3 \\ 2 & 4 \end{bmatrix} = \begin{bmatrix} 48 & 104 \\ 104 & 128 \end{bmatrix}$$

$$\sigma_{Y_1}^2 = 48, \sigma_{Y_2}^2 = 128 \text{ and } C_{Y_1Y_2} = 104.$$

Example 5.8: Gaussian random variables X_1 and X_2 for which $\overline{X_1} = 2$, $\sigma_{X_1}^2 = 9$, $\overline{X_2} = -1$, $\sigma_{X_2}^2 = 4$ and $C_{X_1X_2} = -3$ are transformed to new random variables Y_1 and Y_2 according to $Y_1 = -X_1 + X_2$, $Y_2 = -2X_1 - 3X_2$. Find: (a) $\overline{X_1^2}$ (b) $\overline{X_2^2}$ (c) $\rho_{X_1X_2}$ (d) $\sigma_{X_1}^2$ (e) $\sigma_{X_2}^2$.

Solution: $[T] = \begin{bmatrix} -1 & 1 \\ -2 & -3 \end{bmatrix}$ and $[C_X] = \begin{bmatrix} 9 & -3 \\ -3 & 4 \end{bmatrix}$

$$\overline{X_1^2} = \sigma_{X_1}^2 + (E[X_1])^2 = 9 + 4 = 13$$

$$\overline{X_2^2} = \sigma_{X_2}^2 + (E[X_2])^2 = 4 + 1 = 5$$

$$\rho_{X_1X_2} = \frac{C_{X_1X_2}}{\sqrt{\sigma_{X_1}^2 \sigma_{X_2}^2}} = \frac{-3}{6} = \frac{-1}{2}$$

$$[C_Y] = [T][C_X][T]^t = \begin{bmatrix} -1 & 1 \\ -2 & -3 \end{bmatrix}\begin{bmatrix} 9 & -3 \\ -3 & 4 \end{bmatrix}\begin{bmatrix} -1 & -2 \\ 1 & -3 \end{bmatrix}$$

$$= \begin{bmatrix} -12 & 7 \\ -9 & -6 \end{bmatrix}\begin{bmatrix} -1 & -2 \\ 1 & -3 \end{bmatrix} = \begin{bmatrix} 19 & 3 \\ 3 & 36 \end{bmatrix}$$

$$\sigma_{Y_1}^2 = 19, \ \sigma_{Y_2}^2 = 36 \text{ and } C_{Y_1Y_2} = 3.$$

Example 5.9: The density function of two random variables X and Y is $f_{XY}(x, y) = 16e^{-4(x+y)}u(x)u(y)$. Find the expectation of the function

$$g(x, y) = \begin{cases} 5 & 0 < x \le \frac{1}{2},\ 0 < y \le \frac{1}{2} \\ -1 & \frac{1}{2} < x,\ \frac{1}{2} < y \\ 0 & \text{otherwise} \end{cases}$$

Solution: $E[g(x,\ y)] = \int_{-\infty}^{\infty}\int_{-\infty}^{\infty} g(x,y) f_{XY}(x,y)\,dx\,dy$

$$= \int_0^{1/2}\int_0^{1/2} (5)16e^{-4(x+y)}dx\,dy + \int_{1/2}^{\infty}\int_{1/2}^{\infty}(-1)16e^{-4(x+y)}dx\,dy$$

$$= 80\int_0^{1/2}\int_0^{1/2} e^{-4x}e^{-4y}dx\,dy - 16\int_{1/2}^{\infty}\int_{1/2}^{\infty} e^{-4x}e^{-4y}dx\,dy$$

$$= 80\left(\frac{e^{-4x}}{-4}\bigg|_0^{1/2}\frac{e^{-4y}}{-4}\bigg|_0^{1/2}\right) - 16\left(\frac{e^{-4x}}{-4}\bigg|_{1/2}^{\infty}\frac{e^{-4y}}{-4}\bigg|_{1/2}^{\infty}\right)$$

$$= \frac{80}{16}\left[(e^{-2}-1)(e^{-2}-1)\right] - \frac{16}{16}\left[(-e^{-2})(-e^{-2})\right]$$

$$= 5e^{-4} - 10e^{-2} + 5 - e^{-4} = 4e^{-4} - 10e^{-2} + 5 = 3.7199.$$

Example 5.10: State that the characteristic function of N independent random variables X_i having the characteristic function $\phi_{X_i}(\omega_i)$ is $\phi_{X_1,X_2,\ldots,X_N}(\omega_1,\omega_2,\ldots,\omega_N) = \prod_{i=1}^{N}\phi_{X_i}(\omega_i)$.

Solution: $\phi_{X_1,X_2,\ldots,X_N}(\omega_1,\omega_2,\ldots,\omega_N) = E[e^{j\omega_1x_1+j\omega_2x_2+\cdots+j\omega_Nx_N}]$

$$\phi_{X_1,X_2,\ldots,X_N}(\omega_1,\omega_2,\ldots,\omega_N) = \int_{-\infty}^{\infty}\int_{-\infty}^{\infty}\cdots\int_{-\infty}^{\infty} e^{j\omega_1x_1+j\omega_2x_2+\cdots+j\omega_Nx_N} f_{X_1,X_2,\ldots,X_N}(x_1,x_2,\ldots,x_N)dx_1dx_2\ldots dx_N$$

$\because X_1, X_2, \ldots, X_N$ are independent $f_{X_1,X_2,\ldots,X_N}(x_1,x_2,\ldots,x_N) = f_{X_1}(x_1)f_{X_2}(x_2)\ldots f_{X_N}(x_N)$

$$\phi_{X_1,X_2,\ldots,X_N}(\omega_1,\omega_2,\ldots,\omega_N) = \int_{-\infty}^{\infty}\int_{-\infty}^{\infty}\cdots\int_{-\infty}^{\infty} e^{j\omega_1x_1} f_{X_1}(x_1)e^{j\omega_2x_2} f_{X_2}(x_2)\ldots e^{j\omega_Nx_N} f_{X_N}(x_N)dx_1dx_2\ldots dx_N$$

$$= \int_{-\infty}^{\infty}\prod_{i=1}^{N} e^{j\omega_ix_i} f_{X_i}(x_i)dx_i = \prod_{i=1}^{N}\int_{-\infty}^{\infty} e^{j\omega_ix_i} f_{X_i}(x_i)dx_i$$

$$\phi_{X_1,X_2,\ldots,X_N}(\omega_1,\omega_2,\ldots,\omega_N) = \prod_{i=1}^{N}\phi_{X_i}(\omega_i).$$

Example 5.11: Two random variables X and Y have the joint characteristic function $\phi_{X,Y}(\omega_1,\omega_2)=[(1-2j\omega_1)(1-2j\omega_2)]^{-N/2}$ where $N>0$ is an integer.

Find: (a) The correlation and moments m_{20} and m_{02}.

(b) Expectations of X and Y.

(c) Correlation coefficient.

Solution: $m_{nk} = (-j)^{n+k}\left.\dfrac{\partial^{n+k}\phi_{X,Y}(\omega_1,\omega_2)}{\partial\omega_1^n\,\partial\omega_2^k}\right|_{\omega_1=0,\omega_2=0}$

Correlation $R_{XY} = E[XY] = m_{11}$

$$m_{11} = (-j)^2\left.\frac{\partial^2\phi_{X,Y}(\omega_1,\omega_2)}{\partial\omega_1\,\partial\omega_2}\right|_{\omega_1=0,\omega_2=0}$$

$$\frac{\partial\phi_{X,Y}(\omega_1,\omega_2)}{\partial\omega_2} = \frac{\partial}{\partial\omega_2}[(1-2j\omega_1)(1-2j\omega_2)]^{-N/2}$$

$$\frac{\partial\phi_{X,Y}(\omega_1,\omega_2)}{\partial\omega_2} = \frac{-N}{2}[(1-2j\omega_1)(1-2j\omega_2)]^{-N/2-1}(-2j)(1-2j\omega_1)$$

$$\frac{\partial^2\phi_{X,Y}(\omega_1,\omega_2)}{\partial\omega_1\partial\omega_2} = \frac{-N}{2}\left(\frac{-N}{2}-1\right)[(1-2j\omega_1)(1-2j\omega_2)]^{-N/2-2}(-2j)(1-2j\omega_1)(-2j)(1-2j\omega_2)$$

$$+\frac{-N}{2}[(1-2j\omega_1)(1-2j\omega_2)]^{-N/2-1}(-2j)(-2j)$$

$$\left.\frac{\partial^2\phi_{X,Y}(\omega_1,\omega_2)}{\partial\omega_1\partial\omega_2}\right|_{\omega_1=0,\omega_2=0} = \frac{N}{2}\left(\frac{N}{2}+1\right)(-2j)(-2j)+\frac{-N}{2}(-2j)(-2j)$$

$$\left.\frac{\partial^2\phi_{X,Y}(\omega_1,\omega_2)}{\partial\omega_1\partial\omega_2}\right|_{\omega_1=0,\,\omega_2=0} = -4\frac{N}{2}\left(\frac{N}{2}+1\right)+\frac{4N}{2}=-N^2$$

$$m_{11} = (-j)^2\left.\frac{\partial^2\phi_{X,Y}(\omega_1,\omega_2)}{\partial\omega_1\partial\omega_2}\right|_{\omega_1=0,\omega_2=0} = -N^2(-j)^2 = N^2$$

$$m_{10} = E[X] = (-j)^1\left.\frac{\partial\phi_{X,Y}(\omega_1,\omega_2)}{\partial\omega_1}\right|_{\omega_1=0,\omega_2=0}$$

$$= (-j)\frac{-N}{2}\left.[(1-2j\omega_1)(1-2j\omega_2)]^{-N/2-1}(-2j)(1-2j\omega_2)\right|_{\omega_1=0,\omega_2=0}$$

$$= (-j)\frac{-N}{2}(-2j) = N$$

Similarly $$m_{01} = E[Y] = (-j)^1 \frac{\partial \phi_{X,Y}(\omega_1,\omega_2)}{\partial \omega_2}\bigg|_{\omega_1=0,\omega_2=0} = N$$

$$m_{02} = (-j)^2 \frac{\partial^2 \phi_{X,Y}(\omega_1,\omega_2)}{\partial \omega_2^2}\bigg|_{\omega_1=0,\omega_2=0}$$

$$\frac{\partial \phi_{X,Y}(\omega_1,\omega_2)}{\partial \omega_2} = \frac{\partial}{\partial \omega_2}[(1-2j\omega_1)(1-2j\omega_2)]^{-N/2}$$

$$\frac{\partial \phi_{X,Y}(\omega_1,\omega_2)}{\partial \omega_2} = \frac{-N}{2}[(1-2j\omega_1)(1-2j\omega_2)]^{-N/2-1}(-2j)(1-2j\omega_1)$$

$$\frac{\partial^2 \phi_{X,Y}(\omega_1,\omega_2)}{\partial \omega_2^2} = \frac{-N}{2}\left(\frac{-N}{2}-1\right)[(1-2j\omega_1)(1-2j\omega_2)]^{-N/2-2} \times$$

$$(-2j)(1-2j\omega_1)(-2j)(1-2j\omega_1)$$

$$\frac{\partial^2 \phi_{X,Y}(\omega_1,\omega_2)}{\partial \omega_2^2}\bigg|_{\omega_1=0,\ \omega_2=0} = \frac{N}{2}\left(\frac{N}{2}+1\right)(-2j)(-2j) = -4\frac{N}{2}\left(\frac{N}{2}+1\right)$$

$$m_{02} = (-j)^2 \frac{\partial^2 \phi_{X,Y}(\omega_1,\omega_2)}{\partial \omega_2^2}\bigg|_{\omega_1=0,\omega_2=0} = -N(N+2)(-j)^2 = N(N+2)$$

Similarly, $$m_{20} = (-j)^2 \frac{\partial^2 \phi_{X,Y}(\omega_1,\omega_2)}{\partial \omega_1^2}\bigg|_{\omega_1=0,\omega_2=0} = N(N+2)$$

$$\sigma_X^2 = E[X^2] - (E[X])^2 = N(N+2) - N^2 = 2N$$

and $$\sigma_Y^2 = E[Y^2] - (E[Y])^2 = N(N+2) - N^2 = 2N$$

$$C_{XY} = R_{XY} - E[X]E[Y] = N^2 - (-N)(-N) = 0$$

$$\rho = \frac{C_{XY}}{\sigma_X \sigma_Y} = \frac{0}{(2N)(2N)} = 0$$

Example 5.12: Find the coefficient of variation between X and Y from the following data:

X	1	2	3	4	5	6	7	8	9
Y	8	9	10	11	12	13	14	15	16

Solution: Correlation coefficient is $\rho = \frac{C_{XY}}{\sigma_X \sigma_Y}$

$$C_{XY} = E[XY] - E[X]E[Y]$$

$$E[X] = \frac{1+2+3+4+5+6+7+8+9}{9} = 5$$

$$E[Y] = \frac{8+9+10+11+12+13+14+15+16}{9} = 12$$

$$E[XY] = \frac{1*8+2*9+3*10+4*11+5*12+6*13+7*14+8*15+9*16}{9} = 66.667$$

$$E[X^2] = \frac{1+4+9+16+25+36+49+64+81}{9} = 31.667$$

$$E[Y^2] = \frac{64+81+100+121+144+169+196+225+256}{9} = 150.667$$

$$\text{Var}(X) = E[X^2]-(E[X])^2 = 31.667-25 = 6.667$$

$$\text{Var}(Y) = E[Y^2]-(E[Y])^2 = 150.667-144 = 6.667$$

$$C_{XY} = E[XY]-E[X]E[Y] = 66.667-60 = 6.667$$

$$\rho = \frac{C_{XY}}{\sigma_X\sigma_Y} = \frac{6.667}{\sqrt{(6.667)(6.667)}} = 1.$$

Example 5.13: Let X and Y be two random variables having $\overline{X} = 1, \overline{Y} = 2, \sigma_X^2 = 6, \sigma_Y^2 = 9$ and $\rho = \frac{2}{3}$. Find: (a) covariance of X and Y (b) correlation of X and Y (c) moments m_{02} and m_{20}.

Solution: Correlation of X and Y is

$$C_{XY} = \rho\sigma_X\sigma_Y = \frac{2}{3}*\sqrt{6}*3 = 2\sqrt{6}$$

$$R_{XY} = C_{XY}+E[X]E[Y] = 2\sqrt{6}+2 = 2(1+\sqrt{6})$$

$$m_{20} = \sigma_X^2+(E[X])^2 = 6+1 = 7$$

$$m_{02} = \sigma_Y^2+(E[Y])^2 = 9+4 = 13.$$

Example 5.14: Two discrete random variables X and Y have the joint density function $f_{XY}(x, y) = 0.4\delta(x+a)\,\delta(y-2) + 0.3\delta(x-a)\,\delta(y-2) + 0.18\delta(x-a)\,\delta(y-a) + 0.2\delta(x-1)\,\delta(y-1)$. Determine the value of a, if any, that minimizes the correlation between X and Y and also find the minimum correlation.

Solution: $R_{XY} = E[XY] = \int_{-\infty}^{\infty}\int_{-\infty}^{\infty} xyf_{XY}(x,y)dx\,dy$

$$\int_{-\infty}^{\infty}\int_{-\infty}^{\infty} xy\left[0.4\delta(x+a)\delta(y-2)+0.3\delta(x-a)\delta(y-2)+0.18\delta(x-a)\delta(y-a)+0.2\delta(x-1)\delta(y-1)\right]dxdy$$

$$\because \int_{-\infty}^{\infty}\int_{-\infty}^{\infty} Q(g_1, g_2)\delta(g_1-x)\delta(g_2-2)dg_1\,dg_2 = Q(x,y)$$

$$R_{XY} = 0.4(-a)(2)+0.3(a)(2)+0.18(a)(a)+0.2(1)(1) = 0.1a^2-0.2a+0.2$$

The value of a, when correlation is minimum, is

$$\frac{dR_{XY}}{da} = \frac{d(0.1a^2 - 0.2a + 0.2)}{da} = 0.2a - 0.2 = 0$$

$$\Rightarrow \qquad a = 1$$

The minimum correlation is

$$R_{XY} = (0.1a^2 - 0.2a + 0.2)\Big|_{a=1} = 0.1.$$

EXERCISE

1. Let the transformations be linear and given by
$$Y_1 = T_1(X_1, X_2) = aX_1 + bX_2$$
$$Y_2 = T_2(X_1, X_2) = cX_1 + dX_2$$
where a, b, c and d are real constants. Show that the joint density of new variables is
$$f_{Y_1Y_2}(y_1, y_2) = \frac{f_{X_1X_2}\left(\dfrac{dy_1 - by_2}{ad - bc}, \dfrac{-cy_1 + ay_2}{ad - bc}\right)}{|ad - bc|}$$
2. If the joint PDF of X and Y is given by
$$f_{XY}(x, y) = \begin{cases} x + y & 0 \le x \le 1, 0 \le y \le 1 \\ 0 & \text{otherwise} \end{cases}$$
show that $\rho_{XY} = \dfrac{-1}{11}$.
3. Two random variables X and Y have joint characteristic function
$$\phi_{XY}(\omega_1, \omega_2) = \exp\left(-2\omega_1^2 - 8\omega_2^2\right)$$
(a) Show that X and Y are zero mean random variables.
(b) Show that X and Y are uncorrelated.
4. Zero mean random variables X_1, X_2, X_3 having a covariance matrix
$$[C_X] = \begin{bmatrix} 4 & 2.05 & 1.05 \\ 2.05 & 4 & 1.05 \\ 1.05 & 2.05 & 4 \end{bmatrix}$$
are transformed to new variables $Y_1 = 5X_1 + 2X_2 - X_3$, $Y_2 = -X_1 + 3X_2 + X_3$, and $Y_3 = 2X_1 - X_2 + 2X_3$. Find the covariance matrix of Y.
5. Two random variables X and Y have the following joint probability density function:
$$f_{XY}(x, y) = \begin{cases} 2 - x - y & 0 \le x \le 1, 0 \le y \le 1 \\ 0 & \text{otherwise} \end{cases}$$
Find the Var(X) and Var(Y).
6. If X is a random variable for which $E(X) = 6$ and Var(X) = 36, for what possible values of a and b does $Y = aX + b$ have expectation 0 and variance 1?
7. Let X_i, $i = 1, 2, 3, 4$ be four zero mean Gaussian random variables. Use the joint characteristic function to show that $E[X_1 X_2 X_3 X_4] = E[X_1 X_2]\, E[X_3 X_4] + E[X_1 X_3]\, E[X_2 X_4] + E[X_2 X_3]\, E[X_1 X_4]$.

8. Two random variables X and Y have the following joint probability density function:

$$f_{XY}(x, y) = \begin{cases} \frac{1}{16} & |x|< 4, 2 < y < 4 \\ 0 & \text{otherwise} \end{cases}$$

Show that X and Y are independent and orthogonal.

9. Three random variables X_1, X_2, and X_3 represent samples of random noise voltage taken at three times. Their covariance matrix is defined by

$$[C_X] = \begin{bmatrix} 3.0 & 1.8 & 1.1 \\ 1.8 & 3.0 & 1.8 \\ 1.1 & 1.8 & 3.0 \end{bmatrix}$$

The transformation matrix is

$$[T] = \begin{bmatrix} 4 & -1 & -2 \\ 2 & 2 & 1 \\ -3 & -1 & 3 \end{bmatrix}$$

Convert the variables to new random variables Y_1, Y_2 and Y_3. Find the covariance matrix of the new random variables.

10. Show that the variance of a weighted sum of uncorrected random variables equals the weighted sum of the variances of the random variables.

11. Three statistically independent random variables X_1, X_2 and X_3 have mean values of 3, 6 and –2 respectively. Find the mean values of the following functions:
(a) $g(X_1, X_2, X_3) = X_1 + 3X_2 + 4X_3$
(b) $g(X_1, X_2, X_3) = X_1X_2X_3$
(c) $g(X_1, X_2, X_3) = -2X_1X_2 - 3X_1X_3 + 4X_2X_3$
(d) $g(X_1, X_2, X_3) = X_1 + X_2 + X_3$.

6

Random Processes: Temporal Characteristics

6.0 INTRODUCTION

In communication engineering, generally we come across two types of signals: Deterministic signal and random signal. Deterministic signal $x(t)$ is one for which the waveform shape with respect to time remains same at any time $t = t_1$, i.e. there is no uncertainty about the value of $x(t)$ at that time. For deterministic signals, the analysis such as frequency domain representation, power calculations, etc. can be carried out using conventional methods available in signals and systems theory, e.g. $A\cos \omega_0 t$. Random signal $y(t)$ is one for which the signal waveform shape with respect to time is unpredictable, i.e. the signal at $t = t_1$ may not be similar to that at $t = t_0$, i.e. the signal shape is random in nature. Conventional methods are not useful for analysis of random signals since the future values of these signals are unpredictable. The analysis of random signals is carried out by using probability theory.

Example: Thermal noise.

In communication system, we deal with time varying waveforms whose value at any time is random in nature. For example, the speech waveform recorded by a microphone, the signal received by communication receiver or the daily record of stock market data represents random variables that change with time. In describing random signals, we find that each sample point in our sample space is a function of time. This sample space or ensemble comprising time is called *random or stochastic process*.

6.1 RANDOM PROCESS

A random variable X is defined as a function of possible outcomes of an experiment. If we include time function with function of possible outcomes, the random variable becomes *random process*, i.e. the random process concept is based on enlarging the random variable concept to include time. Let S denote the possible outcome and t denotes time, then the random process is denote by $X(t, s)$. Random process is also known as *stochastic process*. The random process $X(t, s)$ represents a family or ensemble of time functions. In $X(t, s)$ both t and s are variables. If we consider random process for a fixed possible outcome $s = s_1$, then the random process $X(t, s_i) = x_i(t)$ is called the ith *sample function* of the random process $X(t, s)$. Generally sample functions are represented by lower case letters. Let us consider a random experiment with outcomes s in sample space S as shown in Fig. 6.1. Let every outcome s belongs to sample space S, i.e. $s \in S$. Let the random process be observed over an interval T as $X(t, s), -T \leq t \leq T$.

Now for a fixed sample point s_1, the random process represents a particular waveform as a function of time. This particular waveform is the sample function of the random process for that particular fixed sample point. Therefore, $X(t, s_1) = x_1(t)$ is the sample function for sample point s_1 as shown in Fig. 6.1. Similarly, for sample point s_2 we get sample function $x_2(t)$. In this way we get several sample functions $x_1(t)$, $x_2(t),\ldots,x_n(t)$ where n represents very large integer value. Therefore, random process $X(t, s)$ represents *ensemble* or family of all the sample functions.

Now let us consider all the sample functions at a fixed time $t = t_1$. The sample function $x_1(t_1)$ represents just a point on that sample function at $t = t_1$. Similarly, $x_2(t_1)$ represents its value at $t = t_1$. The ensemble of all the sample functions at $t = t_1$, i.e. $\{x_1(t_1), x_2(t_2),\ldots,x_n(t_n)\}$ represents the random process at specific time $t = t_1$. This ensemble is denoted by $X(t_1, s) = X_1$. The random process at a fixed time $t = t_1$ represents random variable X_1 as shown in Fig. 6.1. Similarly at time $t = t_2$, the random process $X(t_2, s)$ represents another random variable. In this way, we can get infinite number of random variables in any random process.

Therefore, the random process $X(t, s)$ becomes sample function $x_i(t)$ when s is fixed, $(s = s_i)$ and t is variable, and the random process becomes random variable X_j when t is fixed $(t = t_j)$ and sample point s is variable.

Definition: A family or ensemble of signals that correspond to every possible outcome of a certain signal measurement. Each signal in this collection is referred to as realization or sample function of the process.

6.1.1 Classification of Random Processes

It is convenient to classify random processes according to the characteristics of t and the random variable $X = X(t)$ at time t. We shall consider only four cases based on t and X having values in the ranges $-\infty \leq t \leq \infty$ and $-\infty \leq x \leq \infty$.

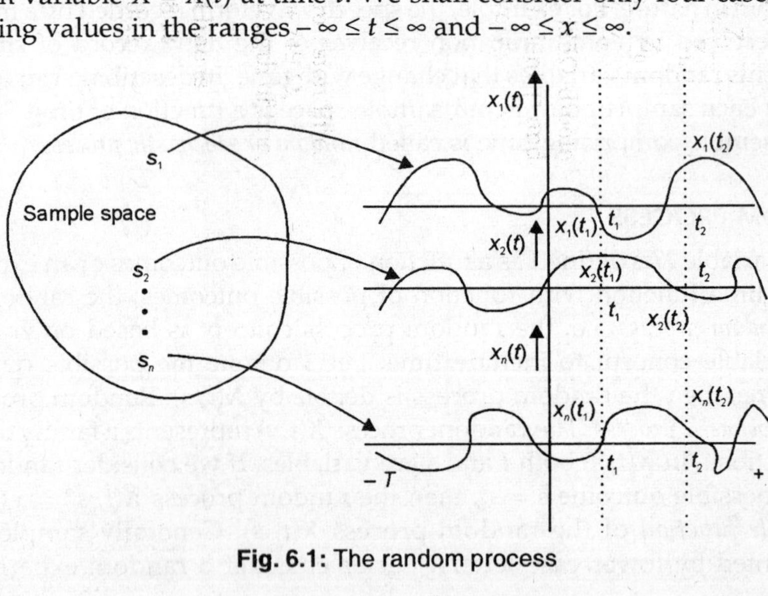

Fig. 6.1: The random process

If X is continuous and t can have continuous values, then $X(t)$ is called a continuous random process. Figure 6.2 is an illustration of this class of process. Thermal noise generated by any realizable network is a practical example of a waveform that is modeled as a sample function of a continuous random process.

A second class of random process, called discrete random process, corresponds to the random variable X having only discrete values while t is continuous. Figure 6.3 is an illustration of this class of random process.

A random process for which X is continuous but time has only discrete values is called a continuous random sequence. Since a continuous random sequence is defined at only discrete times, it is also called a discrete time random process and is

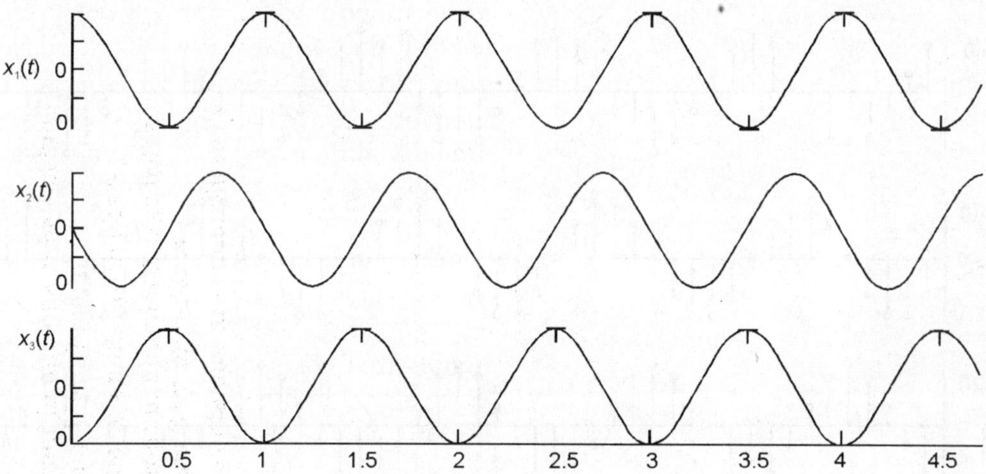

Fig. 6.2: Continuous random process

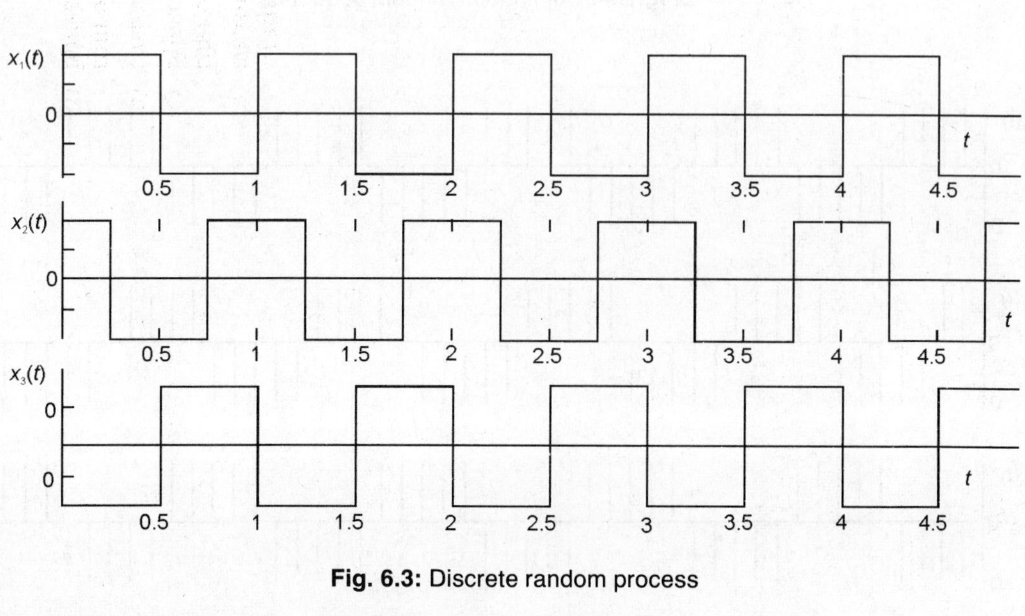

Fig. 6.3: Discrete random process

shown in Fig. 6.4. These types of processes are important in the analysis of various digital signal processing systems.

A fourth class of random process, called a discrete random sequence, corresponds to both time and the random variable being discrete and is shown in Fig. 6.5.

6.1.2 Deterministic and Nondeterministic Random Processes

If the future values of any sample function can be predicted from past values, the random process is called *deterministic random process*. If future values of any sample function cannot be predicted exactly from observed past values, the process is called *nondeterministic random process*.

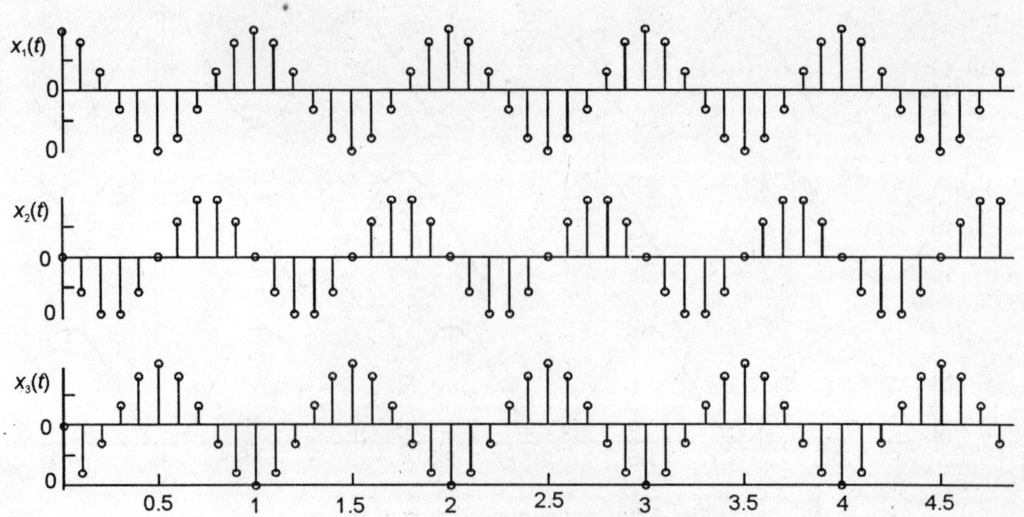

Fig. 6.4: Continuous random sequence

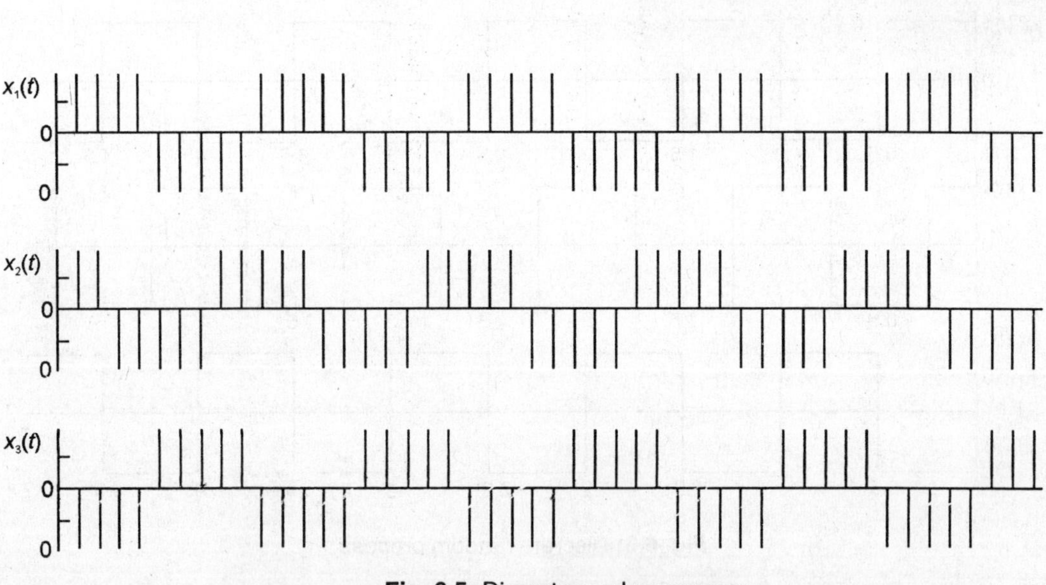

Fig. 6.5: Discrete random process

6.2 STATIONARITY AND INDEPENDENCE

6.2.1 Distribution Functions

Different orders of distributions are defined below:

i. The first order distribution function for random variable $X_1 = X(t_1)$ is

$$F_X(x_1; t_1) = P\{X(t_1) \leq x_1\} \quad ...(6.1)$$

ii. The second order distribution function for two random variables $X_1 = X(t_1)$ and $X_2 = X(t_2)$ is

$$F_X(x_1, x_2; t_1, t_2) = P\{X(t_1) \leq x_1, X(t_2) \leq x_2)\} \quad ...(6.2)$$

iii. The nth order distribution function is

$$F_X(x_1, x_2,...,x_n; t_1, t_2,...,t_n) = P\{X(t_1) \leq x_1, X(t_2) \leq x_2,..., X(t_n) \leq x_n\} \quad ...(6.3)$$

6.2.2 Density Functions

Different orders of density functions are defined below:

i. The first order probability density function for random variable $X_1 = X(t_1)$ is

$$f_X(x_1; t_1) = \frac{d[F_X(x_1; t_1)]}{dx_1} \quad ...(6.4)$$

ii. The second order distribution function for two random variables $X_1 = X(t_1)$ and $X_2 = X(t_2)$ is

$$f_X(x_1, x_2; t_1, t_2) = \frac{\partial^2 [F_X(x_1, x_2; t_1, t_2)]}{\partial x_1 \partial x_2} \quad ...(6.5)$$

iii. The nth order distribution function is

$$f_X(x_1, x_2,...,x_n; t_1, t_2,...,t_n) = \frac{\partial^n [F_X(x_1, x_2,...,x_n; t_1, t_2,...,t_n)]}{\partial x_1 \partial x_2 ... \partial x_n} \quad ...(6.6)$$

6.2.3 Statistical Independence

Two processes $X(t)$ and $Y(t)$ are statistically independent if the random variable group $X(t_1), X(t_2),..., X(t_n)$ is independent of $Y(t_1'), Y(t_2'),..., Y(t_m')$ for any choice of times $t_1, t_2,..., t_n, t_1', t_2',..., t_m'$. Independence requires that the joint density be factorable by groups:

$$\begin{aligned} & f_{XY}(x_1, x_2,..., x_n, y_1, y_2,..., y_m; t_1, t_2,..., t_n, t_1', t_2',..., t_m') \\ & = f_X(x_1, x_2,..., x_n; t_1, t_2,..., t_n) f_Y(y_1, y_2,..., y_m; t_1', t_2',..., t_m'). \end{aligned} \quad ...(6.7)$$

6.2.4 Stationary Random Process

The concept of stationarity plays an important role in solving practical problems involving random processes. Just like time invariance is an important characteristic of many deterministic systems, stationarity describes certain time invariant property of a random process. It also leads to frequency domain description of a random process. It refers to time invariance of some, or all, the statistics of a random process, e.g. mean, autocorrelation, nth order distribution, etc. We define two types of stationarity, viz. strict sense stationarity (SSS) and wide sense stationarity (WSS).

Depending on the density functions of the random variables in the process, there are several levels of stationarity.

First Order Stationary Processes

A random process is called stationary to order one if its first order density functions does not change with a shift in time origin. In other words,

$$f_X(x; t_1) = f_X(x; t_1 + c) \qquad ...(6.8)$$

must be true for any t and any real number c.

Theorem: Show that the mean of the first order stationary random process is constant.

Proof: Let us consider a random process $X(t)$.

Assume the mean of the random process $X(t)$ at t_1 is $E[X(t_1)] = \bar{X}$.

The mean of the random process at $t = t_2$ is

$$E[X(t_2)] = \int_{-\infty}^{\infty} xf(x, t_2)dx \qquad ...(6.9)$$

By letting $t_2 = t_1 + c$ in Eq. (6.9),

$$E[X(t_2)] = \int_{-\infty}^{\infty} xf(x, t_1 + c)dx \qquad ...(6.10)$$

Since it is a first order stationary random process

$$\therefore \quad f_X(x; t_1) = f_X(x; t_1 + c) \qquad ...(6.11)$$

From Eqs. (6.10) and (6.11),

$$E[X(t_2)] = \int_{-\infty}^{\infty} xf(x, t_1 + c)dx = \int_{-\infty}^{\infty} xf(x, t_1)dx = E[X(t_1)] = \bar{X}$$

From the above, the mean of the random process at t_1 is equal to the mean at t_2, i.e. mean is independent of time.

Hence, for first order stationary random process the mean is constant.

Second Order and Wide Sense Stationary Processes

A random process is called stationary to order two if its second order density functions does not change with a shift in time origin. In other words,

$$f_X(x_1, x_2; t_1, t_2) = f_X(x_1, x_2; t_1 + c, t_2 + c) \qquad ...(6.12)$$

for all t_1, t_2 and c. A second order stationary process is also first order stationary because the second order density function determines the lower, first order density.

Now the correlation $E[X_1X_2] = E[X(t_1)X(t_2)]$ of a random process will, in general, be a function of t_1 and t_2. Let us denote this function by $R_{XX}(t_1, t_2)$ and call it the autocorrelation function of the random process $X(t)$.

$$R_{XX}(t_1, t_2) = E[X(t_1)X(t_2)]. \qquad ...(6.13)$$

Theorem: The autocorrelation function of a second order stationary random process is a function of time difference only and not of absolute time, i.e. if $\tau = t_2 - t_1$, then auto-correlation for second order stationary process is $R_{XX}(\tau)$.

Proof: Expectation of second order density function is called autocorrelation.

The autocorrelation random process at t_1 and t_2 is

$$R_{XX}(t_1, t_2) = R_{XX}(t_1, t_1 + \tau)$$

$$= R_{XX}(\tau) = E[X(t_1)X(t_2)] = \int_{-\infty}^{\infty}\int_{-\infty}^{-\infty} x_1 x_2 f(x_1, x_2; t_1, t_2)dx_1 dx_2 \quad ...(6.14)$$

Similarly, at t_3 and t_4

$$R_{XX}(t_3, t_4) = E[X(t_3)X(t_4)] = \int_{-\infty}^{\infty}\int_{-\infty}^{\infty} x_1 x_2 f(x_1, x_2; t_3, t_4)dx_1\, dx_2 \quad ...(6.15)$$

By letting $t_3 = t_1 + c$ and $t_4 = t_2 + c$, Eq. (6.15) can be written as

$$R_{XX}(t_3, t_4) = R_{XX}(t_1 + c, t_2 + c) = \int_{-\infty}^{-\infty}\int_{-\infty}^{\infty} x_1 x_2\, f(x_1, x_2; t_1 + c, t_2 + c)dx_1\, dx_2 \quad ...(6.16)$$

But it is a second order stationary random process

$$f_X(x_1, x_2; t_1, t_2) = f_X(x_1, x_2; t_1 + c, t_2 + c) \quad ...(6.17)$$

Substitute Eq. (6.16) in Eq. (6.15),

$$R_{XX}(t_3, t_4) = R_{XX}(t_1 + c, t_2 + c) = \int_{-\infty}^{\infty}\int_{-\infty}^{\infty} x_1 x_2\, f(x_1, x_2; t_1, t_2)dx_1\, dx_2 \quad ...(6.18)$$

From the above, $R_{XX}(t_3, t_4) = R_{XX}(t_1, t_2) = R_{XX}(\tau)$,

Hence for a second order stationary random process the autocorrelation depends on time shift τ only, which is a real number.

A random process $X(t)$ is said to be *wide sense stationary* random process, if it satisfies the following conditions:

i. $E[X(t)] = \bar{X}$ = a constant ...(6.19)

ii. $R_{XX}(\tau) = E[X(t)\, X(t_1 + \tau)]$ ...(6.20)

i.e. autocorrelation function depends only on τ but not on time origin.

n^{th} Order and Strict Sense Stationary (SSS) Processes

A random process is called stationary to order n if its nth order density functions does not change with a shift in time origin, i.e.

$$f_X(x_1, x_2, ..., x_n; t_1, t_2, ..., t_n) = f_X(x_1, x_2, ..., x_n; t_1 + c, t_2 + c, ..., t_n + c) \quad ...(6.21)$$

For any c, where the left side represents the joint density function of the random variables and the right side corresponds to the joint density function of the random variables, stationary of order n implies stationary to all orders $k \le N$.

A process $X(t)$ stationary to all orders $n = 1, 2, 3, ...$ is called strict sense stationary.

Some practical applications of wide sense stationary random process are the random binary transmission wave, random telegraphic signal and random sinusoidal signals. However, semirandom telegraphic wave is not wide sense stationary (it is second order stationary but it is not stationary in the mean, i.e. mean is a function of time) and the semirandom binary transmission process is also not wide sense stationary (it is stationary in the mean but it is not second order and higher order stationary).

Example 6.2.1: A random process $X(t)$ has periodic sample function as shown in Fig. 6.6 where B, T and $2t_0 \le T$ are constants but ϵ is a random variable uniformly distributed on the interval $(0, T)$.

(a) Classify the process.

(b) Find the first order distribution function of $X(t)$.

(c) Find the first order density function.
(d) Find $E[X(t)]$, $E[X^2(t)]$ and σ^2_{XX}.
(e) Determine whether $X(t)$ is wide sense stationary or not.

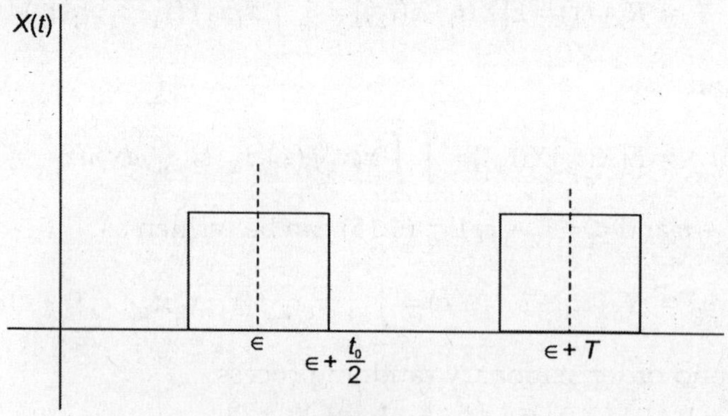

Fig. 6.6

Solution: (a) This is a discrete random process and it is deterministic since the future values of the sample function can be predicted from the past values.
(b) First order distribution function is nothing but the signal variation of sample function $X(t)$ in Fig. 6.6.

$$X(t) = F_X(x) = \begin{cases} A & \left(\in - \frac{t_0}{2}\right) < t < \left(\in + \frac{t_0}{2}\right) \\ 0 & \left(\in + \frac{t_0}{2}\right) < t < \left(\in + \frac{3t_0}{2}\right) \end{cases}$$

$$\text{Period} = \left(\in + \frac{3t_0}{2}\right) - \left(\in - \frac{t_0}{2}\right) = 2t_0 = T$$

(c) $f_X(x) = \dfrac{dF_X(x)}{dt}$,
$$f_X(x) = \begin{cases} A\delta\left(t - \left(\in - \frac{t_0}{2}\right)\right) & \left(\in - \frac{t_0}{2}\right) < t < \left(\in + \frac{t_0}{2}\right) \\ 0 & \left(\in + \frac{t_0}{2}\right) < t < \left(\in + \frac{3t_0}{2}\right) \end{cases}$$

(d) $\in$ is a random variable uniformly distributed over $(0, T)$ where T is one cycle period.

$$\text{PDF} = f_X(\in) = \frac{1}{T}; \qquad 0 < t < T$$

$$E[X(t)] = \int_{-\infty}^{\infty} x f_X(\in) d\in = \int_0^T x(t) f_X(\in) d\in$$

$$= \int_{-T/4}^{T/4} A \frac{1}{T} d\in = \frac{A}{T} \int_{-T/4}^{T/4} d\in$$

$$= \frac{A}{T} \in \Big|_{-T/4}^{T/4} = \frac{A}{T}\left(\frac{T}{4} + \frac{T}{4}\right)$$

$$= \frac{A}{2}$$

$$E[X^2(t)] = \int_{-\infty}^{\infty} x^2 f_X(\in) d\in = \int_0^T x^2(t) f_X(\in) d\in$$

$$= \int_{-T/4}^{T/4} A^2 \frac{1}{T} d\in = \frac{A^2}{T} \int_{-T/4}^{T/4} d\in$$

$$= \frac{A^2}{T} \in\Big|_{-T/4}^{T/4} = \frac{A^2}{T}\left(\frac{T}{4} + \frac{T}{4}\right)$$

$$= \frac{A^2}{2}$$

$$\sigma_X^2 = E[X^2(t)] - (E[X(t)])^2$$

$$\sigma_X^2 = \frac{A^2}{2} - \left(\frac{A}{2}\right)^2$$

$$= \frac{A^2}{4}$$

$$R_{XX}(\tau) = E[X(t)\, X(t+\tau)]$$

$$= \int_0^T x(t) x(t+\tau) f_X(\in) d\in$$

$$= \int_{-T/4}^{T/4} A^2 \frac{1}{T} d\in = \frac{A^2}{T} \int_{-T/4}^{T/4} d\in$$

$$= \frac{A^2}{T} \in\Big|_{-T/4}^{T/4} = \frac{A^2}{T}\left(\frac{T}{4} + \frac{T}{4}\right)$$

$$= \frac{A^2}{2}$$

Autocorrelation and mean values are independent of time. Hence, it is wide sense stationary.

6.3 CORRELATION AND COVARIANCE FUNCTION

When we take the expected value, or average, of a random process, we measure several important characteristics about how the process behaves in general. This proves to be a very important observation. However, suppose we have several random processes measuring different aspects of a system. The relationship between these different processes will also be an important observation. Covariance and correlation are the two important tools in finding these relationships. Now we will discuss as to what these words mean and how these tools are helpful. Note that much of the following discussion refers to just random variables, but keep in mind that these variables can represent random signals or random processes. Correlation and covariance are techniques for measuring the similarity of one signal to another.

6.3.1 Correlation

For anyone who has any kind of statistical background, you should be able to see that the idea of dependence/independence among variables and signals plays an important role when dealing with random processes. Because of this, the correlation of two variables provides us with a measure of how the two variables affect one another.

Definition: Correlation is a measure of how much one random variable depends upon the other. This measure of association between the variables will provide us with a clue as to how well the value of one variable can be predicted from the value of the other.

Autocorrelation and its Properties

The expected value of the product of a random variable or signal realization with a time-shifted version of itself is called autocorrelation function.

Mathematically, it is defined as

$$R_{XX}(t_1, t_2) = E[X(t_1)X(t_2)] = \int_{-\infty}^{\infty}\int_{-\infty}^{\infty} x_1 x_2 f(x_1, x_2; t_1, t_2) dx_1 dx_2 \quad ...(6.22)$$

The above equation is valid for stationary and nonstationary random processes. For stationary processes, we can generalize this expression a little further. Given a wide sense stationary processes, it can be proven that the expected values from our random process will be independent of the origin of our time function. Therefore, we can say that our autocorrelation function will depend on the time difference and not on absolute time. For this discussion, we will let $\tau = t_2 - t_1$, and thus we generalize our autocorrelation expression as

$$R_{XX}(t_1, t_2) = R_{XX}(t_1, t_1 + \tau) = R_{XX}(\tau)$$

$$R_{XX}(\tau) = E[X(t_1)X(t_1 + \tau)] \quad ...(6.23)$$

for the continuous time case. In most DSP course we will be more interested in dealing with real signal sequences, and thus we want to look at the discrete time case of the autocorrelation function. The formula below will prove to be more common and useful than Eq. (6.22):

$$R_{XX}(n, n + m) = \sum_{n=-\infty}^{\infty} (X[n]X[n+m]) \quad ...(6.24)$$

And again we can generalize the notation for our autocorrelation function as

$$R_{XX}(n, n + m) = R_{XX}(m) = E[X[n]X[n+m]] \quad ...(6.25)$$

Properties of Autocorrelation

We will now look at several properties of the autocorrelation function that hold for stationary random processes.

i. Autocorrelation is an even function for τ

$$R_{XX}(\tau) = R_{XX}(-\tau)$$

Proof: $\quad R_{XX}(\tau) = E[X(t_1)X(t_1 + \tau)]$

Let $\quad \tau = -\tau$

$$R_{XX}(-\tau) = E[X(t_1)X(t_1-\tau)]$$

Assume $\quad t-\tau = k$

$$t = k+\tau$$

$$R_{XX}(-\tau) = E[X(k+\tau)X(k)]$$

$$= E[X(k)X(k+\tau)] = R_{XX}(\tau)$$

$$\therefore \quad R_{XX}(\tau) = R_{XX}(-\tau) \qquad ...(6.26)$$

ii. The mean square value can be found by evaluating the autocorrelation where $\tau = 0$, which gives us

$$R_{XX}(0) = \overline{X^2}$$

Proof: $\quad R_{XX}(\tau) = E[X(t_1)X(t_1+\tau)]$

Substitute $\tau = 0$

$$R_{XX}(0) = E[X(t_1)X(t_1)]$$

$$= \overline{X^2}$$

$$\therefore \quad R_{XX}(0) = \overline{X^2} \qquad ...(6.27)$$

iii. $\quad R_{XX}(0) \geq |R_{XX}(\tau)|$

Proof: $\quad [X(t_1)+X(t_1+\tau)]^2 \geq 0$

$$E\left[(X(t_1)+X(t_1+\tau))^2\right] \geq 0$$

$$E\left[X^2(t_1)+X^2(t_1+\tau)+2X(t_1)X(t_1+\tau)\right] \geq 0$$

$$E\left[X^2(t_1)\right]+E\left[X^2(t_1+\tau)\right]+2E\left[X(t_1)X(t_1+\tau)\right] \geq 0$$

$$R_{XX}(0) + R_{XX}(0) + 2R_{XX}(\tau) \geq 0$$

$$R_{XX}(0) \geq -R_{XX}(\tau)$$

$$\therefore \quad R_{XX}(0) \geq |R_{XX}(\tau)| \qquad ...(6.28)$$

iv. If $X(t)$ has a periodic component, then $R_{XX}(\tau)$ will also have periodic component with the same frequency.

Proof: Let $X(t)$ has a periodic component T, i.e.

$$X(t+T) = X(t)$$

$$R_{XX}(\tau) = E[X(t)X(t+\tau)]$$

Let $\tau = T+\tau$

$$R_{XX}(T+\tau) = E[X(t)X(t+T+\tau)] = E[X(t)X(t+T+\tau)] \qquad \text{but } X(t) = X(t+T)$$

$$= E[X(t+T)X(t+T+\tau)] = R_{XX}(\tau)$$

$$\therefore \quad R_{XX}(T+\tau) = R_{XX}(\tau) \qquad ...(6.29)$$

v. If $E[X(t)] = \overline{X} \neq 0$ and $X(t)$ is ergodic with no periodic components then

$$\lim_{|\tau|\to\infty} R_{XX}(\tau) = \overline{X}^2$$

Proof: $R_{XX}(\tau) = E[X(t)X(t+\tau)]$

as $|\tau| \to \infty$, $X(t)$ and $X(t+\tau)$ are independent

$$\lim_{|\tau|\to\infty} R_{XX}(\tau) = E[X(t)]E[X(t+\tau)]$$

$$= \overline{X}\overline{X} = \overline{X}^2$$

$$\therefore \quad \lim_{|\tau|\to\infty} R_{XX}(\tau) = \overline{X}^2 \qquad ...(6.30)$$

vi. If $X(t)$ is ergodic, zero mean has no periodic components, then

$$\lim_{|\tau|\to\infty} R_{XX}(\tau) = 0$$

Proof: $R_{XX}(\tau) = E[X(t)X(t+\tau)]$

as $|\tau| \to \infty$, $X(t)$ and $X(t+\tau)$ are independent

$$\lim_{|\tau|\to\infty} R_{XX}(\tau) = E[X(t)]E[X(t+\tau)]$$

$$= 0 \cdot 0 = 0$$

$$\therefore \quad \lim_{|\tau|\to\infty} R_{XX}(\tau) = 0 \qquad ...(6.31)$$

vii. $R_{XX}(\tau)$ cannot have an arbitrary shape.

6.3.2 Cross-correlation and Its Properties

The cross-correlation function describes the general dependency of $X(t)$ with another random process $Y(t+\tau)$, delayed by a time delay τ.

The expected value of the product of a random variable or signal realization with a time-shifted version of other is called *cross-correlation function*.

Mathematically, it is defined as

$$R_{XY}(t_1, t_2) = E[X(t_1)Y(t_2)] = \int_{-\infty}^{\infty}\int_{-\infty}^{\infty} x_1 y_2 f(x_1, y_2; t_1, t_2)\,dx_1 dx_2 \qquad ...(6.32)$$

Let $\tau = t_2 - t_1$, and thus we generalize our cross-correlation expression as

$$R_{XY}(t_1, t_2) = R_{XY}(t_1, t_1 + \tau) = R_{XY}(\tau)$$

$$R_{XY}(\tau) = E[X(t_1)Y(t_1+\tau)] \qquad ...(6.33)$$

If

$$R_{XY}(t) = 0 \qquad ...(6.34)$$

then $X(t)$ and $Y(t)$ are called *orthogonal processes*. If the two random processes are statistically independent, the cross-correlation function becomes

$$R_{XY}(\tau) = E[X(t_1)]E[Y(t_1+\tau)] \qquad ...(6.35)$$

for the continuous time case. For the discrete time case of the cross-correlation function

$$R_{XY}(n, n+m) = \sum_{n=-\infty}^{\infty} (X[n]Y[n+m]) \qquad ...(6.36)$$

And again we can generalize the notation for our autocorrelation function as

$$R_{XY}(n, n+m) = R_{XY}(m) = E[X[n]Y[n+m]] \qquad ...(6.37)$$

Properties of Cross-correlation

1. Autocorrelation is an even function of τ.

$$R_{XY}(-\tau) = R_{YX}(\tau)$$

Proof: $R_{XY}(\tau) = E[X(t_1)Y(t_1+\tau)]$

Let $\tau = -\tau$

$$R_{XY}(-\tau) = E[X(t_1)Y(t_1-\tau)]$$

Assume $t-\tau = k$

$$t = k+\tau$$

$$R_{XY}(-\tau) = E[X(k+\tau)Y(k)]$$

$$= E[X(k)Y(k+\tau)] = R_{XY}(\tau)$$

$\therefore$ $R_{XY}(-\tau) = R_{YX}(\tau).$ ...(6.38)

2. $|R_{XY}(\tau)| \le \sqrt{R_{XX}(0)R_{YY}(0)}$

Proof: $[\alpha X(t_1) \pm Y(t_1+\tau)]^2 \ge 0$

$\alpha^2 X^2(t_1) + Y^2(t_1+\tau) \pm 2\alpha X(t_1) \pm Y(t_1+\tau) \ge 0$

It is a quadratic equation of order 2 and has maximum only one real root. Hence $b^2 - 4ac \le 0$.

$$E[\alpha^2 X^2(t_1)] + E[Y^2(t_1+\tau)] \pm E[2\alpha X(t_1) \pm Y(t_1+t)] \ge 0$$

$$(E[2X(t_1)Y(t_1+\tau)])^2 \mp 4E[X^2(t_1)]E[Y^2(t_1+\tau)] \le 0$$

$$4(R_{XY}(\tau))^2 \mp 4R_{XX}(0)R_{YY}(0) \le 0$$

$$4(R_{XY}(\tau))^2 \mp 4R_{XX}(0)R_{YY}(0) \le 0$$

$$(R_{XY}(\tau))^2 \le \pm R_{XX}(0)R_{YY}(0)$$

$$R_{XY}(\tau) \le \pm\sqrt{R_{XX}(0)R_{YY}(0)} \quad ...(6.39)$$

$\therefore$ $|R_{XY}(\tau)| \le \sqrt{R_{XX}(0)R_{YY}(0)}.$

3. $|R_{XY}(\tau)| \le \frac{1}{2}[R_{XX}(0) + R_{YY}(0)]$

Proof: $[X(t_1) + Y(t_1+\tau)]^2 \ge 0$

$$E[(X(t_1) + Y(t_1+\tau))^2] \ge 0$$

$$E[X^2(t_1) + Y^2(t_1+\tau) + 2X(t_1)Y(t_1+\tau)] \ge 0$$

$$E[X^2(t_1)] + E[Y^2(t_1+\tau)] + 2E[X(t_1)Y(t_1+\tau)] \ge 0$$

$$R_{YY}(0)+R_{XX}(0)+2R_{XY}(\tau)\geq 0$$

$$R_{XX}(0)+R_{YY}(0) \geq -2R_{XX}(\tau)$$

$\therefore$
$$|R_{XY}(\tau)| \leq \frac{1}{2}[R_{XX}(0)+R_{YY}(0)] \quad ...(6.40)$$

4. If two random processes $X(t)$ and $Y(t_1)$ are independent random processes, then

$$R_{XY}(\tau) = \bar{X}\bar{Y}.$$

Proof: $R_{XY}(\tau) = E[X(t)Y(t+\tau)]$

$X(t)$ and $Y(t_1)$ are independent random processes

$$E[X(t)Y(t+\tau)] = E[X(t)]E[Y(t+\tau)]$$

$$= \bar{X}\bar{Y}$$

$\therefore$
$$R_{XY}(\tau) = \bar{X}\bar{Y} \quad ...(6.41)$$

5. If $X(t)$ and $Y(t_1)$ are two random processes with zero means, then

$$\lim_{|\tau|\to\infty} R_{XY}(\tau) = 0.$$

Proof: $R_{XY}(\tau) = E[X(t)Y(t+\tau)]$

as $|\tau| \to \infty$, $X(t)$ and $Y(t+\tau)$ are independent

$$\lim_{|\tau|\to\infty} R_{XY}(\tau) = E[X(t)]E[Y(t+\tau)]$$

$$= 0 \cdot 0 = 0$$

$\therefore$
$$\lim_{|\tau|\to\infty} R_{XY}(\tau) = 0 \quad ...(6.42)$$

6. If a random process $Z(t) = X(t) + Y(t)$ where $X(t)$ and $Y(t_1)$ are two random processes, then

$$R_{ZZ}(\tau) = R_{XX}(\tau) + R_{XY}(\tau) + R_{YX}(\tau) + R_{YY}(\tau)$$

Proof: $R_{ZZ}(\tau) = E[Z(t)Z(t+\tau)]$

But $Z(t) = X(t) + Y(t)$

$$R_{ZZ}(\tau) = E[\{X(t)+Y(t)\}\{X(t+\tau)+Y(t+\tau)\}]$$

$$= E[X(t)X(t+\tau)+Y(t)X(t+\tau)+Y(t+\tau)X(t)+Y(t+\tau)Y(t)]$$

$$= E[X(t)X(t+\tau)]+E[Y(t)X(t+\tau)]+E[Y(t+\tau)X(t)]+E[Y(t+\tau)Y(t)]$$

$$= R_{XX}(\tau)+R_{XY}(\tau)+R_{YX}(\tau)+R_{YY}(\tau)$$

$\therefore$
$$R_{ZZ}(\tau) = R_{XX}(\tau) + R_{XY}(\tau) + R_{YX}(\tau) + R_{YY}(\tau) \quad ...(6.43)$$

7. If a random process $Z(t) = X(t) + K$, where mean of $X(t)$ is zero and K is constant, then

$$R_{ZZ}(\tau) = R_{XX}(\tau) + K^2$$

Proof: $R_{ZZ}(\tau) = E[Z(t)Z(t+\tau)]$

But $Z(t) = X(t) + K$

$$R_{ZZ}(\tau) = E\left[\{X(t)+K\}\{X(t+\tau)+K\}\right]$$

$$= E\left[X(t)X(t+\tau)+KX(t+\tau)+KX(t)+K^2\right]$$

$$= E[X(t)X(t+\tau)]+E[KX(t+\tau)]+E[KX(t)]+E[K^2]$$

$$= E[X(t)X(t+\tau)]+KE[X(t+\tau)]+KE[X(t)]+K^2$$

$$= E[X(t)X(t+\tau)]+K\cdot 0+K\cdot 0+K^2 \qquad (\because \text{ mean of } X(t) \text{ is zero})$$

$$\therefore \quad R_{ZZ}(\tau) = R_{XX}(\tau) + K^2 \qquad ...(6.44)$$

6.3.3 Covariance Functions

Autocovariance function: $C_{XX}(t, t_1)$ is defined as the covariance between the two time samples $X(t)$ and $X(t_1)$. Autocovariance of a random process is defined as

$$C_{XX}(t, t_1) = \mathrm{E}\left[\{X(t)-E[X(t)]\}\{X(t_1)-E[X(t_1)]\}\right] \qquad ...(6.45)$$

Substitute $t_1 = t + \tau$

$$C_{XX}(t, t+\tau) = E\left[\{X(t)-E[X(t)]\}\{X(t+\tau)-E[X(t+\tau)]\}\right]$$

$$C_{XX}(t, t+\tau) = R_{XX}(t,t+\tau)-E\left[\{X(t)X(t+\tau)\}\right] \qquad ...(6.46)$$

For the process that is at least wide sense stationary, Eq. (6.43) will be

$$C_{XX}(\tau) = R_{XX}(\tau)-\overline{X}^2 \qquad ...(6.47)$$

6.3.4 Cross-covariance Function

The variance of a random process is given by Eq. (6.43) with $\tau = 0$,

$$\sigma_{XX}^2 = R_{XX}(0)-\overline{X}^2 \qquad ...(6.48)$$

$C_{XY}(t, t_1)$ is defined as the covariance between the two time samples $X(t)$ and $Y(t_2)$. Cross-covariance of a random process is defined as

$$C_{XY}(t, t_1) = E\left[\{X(t)-E[X(t)]\}\{Y(t_1)-E[Y(t_1)]\}\right] \qquad ...(6.49)$$

Substitute $t_1 = t + \tau$

$$C_{XY}(t, t+\tau) = E\left[\{X(t)-E[X(t)]\}\{Y(t+\tau)-E[Y(t+\tau)]\}\right]$$

$$C_{XY}(t, t+\tau) = R_{XY}(t,t+\tau)-E\left[(X(t)Y(t+\tau)\right] \qquad ...(6.50)$$

For the processes that are at least wide sense stationary, Eq. (6.48) will be

$$C_{XY}(\tau) = R_{XY}(\tau)-\overline{X}^2 \qquad ...(6.51)$$

For two random processes, if

$$C_{XY}(\tau) = 0 \qquad ...(6.52)$$

they are uncorrelated. From Eq. (6.48) it means that

$$R_{XY}(t, t+\tau) = E[X(t)Y(t+\tau)] \qquad ...(6.53)$$

Since this result is same as Eq. (6.35), which applies to independent processes, we conclude that independent processes are uncorrelated.

6.3.5 Discrete Time Processes and Sequences

The discrete time processes and sequences are defined only at the "sample" times nT_s. Therefore, time offset, denoted above generally by τ, can have only discrete values, say kT_s.

For a DT process $X(nT_s)$, the mean, autocorrelation and autocovariance functions are as follows:

$$\text{Mean} = E[X(nT_s)] \quad \text{...(6.54)}$$

$$\text{Autocorrelation} = R_{XX}(nT_s, nT_s + kT_s) = E[X(nT_s)X(nT_s + kT_s)] \quad \text{...(6.55)}$$

$$\text{Autocovariance} = C_{XX}(nT_s, nT_s + kT_s)$$

$$= E[\{X(nT_s) - E[X(nT_s)]\}\{X(nT_s + kT_s) - E[X(nT_s + kT_s)]\}] \quad \text{...(6.56)}$$

$$= R_{XX}(nT_s, nT_s + kT_s) - E[X(nT_s)X(nT_s + kT_s)] \quad \text{...(6.57)}$$

Similarly, cross-correlation and cross-covariance functions are as follows:

Let $X(nT_s)$ and $Y(nT_s)$ are two discrete random processes

$$\text{Cross-correlation} = R_{XY}(nT_s, nT_s + kT_s) = E[X(nT_s)Y(nT_s + kT_s)] \quad \text{...(6.58)}$$

$$\text{Cross-covariance} = C_{XY}(nT_s, nT_s + kT_s)$$

$$= E[\{X(nT_s) - E[X(nT_s)]\}\{Y(nT_s + kT_s) - E[Y(nT_s + kT_s)]\}] \quad \text{...(6.59)}$$

$$= R_{XY}(nT_s, nT_s + kT_s) - E[X(nT_s)Y(nT_s + kT_s)] \quad \text{...(6.60)}$$

For processes that are at least jointly wide sense stationary, our expressions reduce to constants ($\overline{X}$ or $\overline{Y}$) and functions of only kT_s.

$$E[X(nT_s)] = \overline{X} \qquad E[Y(nT_s)] = \overline{Y} \quad \text{...(6.61)}$$

$$R_{XX}(nT_s, nT_s + kT_s) = R_{XX}(kT_s) \quad \text{...(6.62)}$$

$$R_{YY}(nT_s, nT_s + kT_s) = R_{YY}(kT_s) \quad \text{...(6.63)}$$

$$C_{XX}(nT_s, nT_s + kT_s) = R_{XX}(kT_s) - \overline{X}^2 \quad \text{...(6.64)}$$

$$C_{YY}(nT_s, nT_s + kT_s) = R_{YY}(kT_s) - \overline{Y}^2 \quad \text{...(6.65)}$$

$$R_{XY}(nT_s, nT_s + kT_s) = R_{XY}(kT_s) \quad \text{...(6.66)}$$

$$R_{YX}(nT_s, nT_s + kT_s) = R_{YX}(kT_s) \quad \text{...(6.67)}$$

$$C_{XY}(nT_s, nT_s + kT_s) = R_{XY}(kT_s) - \overline{X}\,\overline{Y} \quad \text{...(6.68)}$$

$$C_{YX}(nT_s, nT_s + kT_s) = R_{YX}(kT_s) - \overline{Y}\,\overline{X} \quad \text{...(6.69)}$$

Finally, we note that for DT sequences, where the dependence on T_s is very small so that it is dropped, the above results apply if T_s is everywhere omitted and the functional notation is everywhere replaced by brackets. For example, $R_{XY}(nT_s, nT_s + kT_s)$ becomes $R_{XY}[n, n + k]$ while $E[X(nT_s)] = E\{X[n]\}$.

Example 6.3.1: Suppose $X(t) = A\sin(\omega_c t + \theta)$, where A and ω_c are constants and θ is a random variable distributed uniformly on the interval $[-\pi, \pi]$. Define a new random process $Y(t) = X^2(t)$.

(a) Find the autocorrelation function of $Y(t)$.

(b) Find the cross-correlation function of $X(t)$ and $Y(t)$.

(c) Are $X(t)$ and $Y(t)$ jointly wide sense stationary?

Solution: Autocorrelation function of $Y(t)$ is

$$R_{YY}(t_1, t_2) = R_{YY}(\tau) = E[Y(t_1)Y(t_2)]$$

We know that $\quad E[X(t)] = \int_{-\infty}^{\infty} xf(\theta)d\theta$

Here θ is uniformly distributed on the interval $[-\pi, \pi]$, the density function is

$$f(\theta) = \frac{1}{\pi-(-\pi)} = \frac{1}{2\pi}$$

$$E[X(t)] = \int_{-\infty}^{\infty} xf(\theta)d\theta = \int_{-\pi}^{\pi} A\sin(\omega_c t+\theta)\frac{1}{2\pi}d\theta$$

$$= \frac{-A}{2\pi}\cos(\omega_c t+\theta)\Big|_{-\pi}^{\pi}$$

$$= \frac{-A}{2\pi}[\cos(\omega_c t+\pi)-\cos(\omega_c t-\pi)]$$

$$= \frac{-A}{2\pi}(0) = 0$$

$$E[Y(t)] = \int_{-\infty}^{\infty} X^2 f(\theta)d\theta$$

$$= \int_{-\pi}^{\pi} A^2\sin^2(\omega_c t+\theta)\frac{1}{2\pi}d\theta$$

$$= \int_{-\pi}^{\pi} A^2\left(\frac{1-\cos(2\omega_c t+2\theta)}{2}\right)\frac{1}{2\pi}d\theta$$

$$= \frac{A^2}{2\pi}\left(\int_{-\pi}^{\pi}\frac{1}{2}d\theta \int_{-\pi}^{\pi}\frac{1}{2}\cos(2\omega_c t+2\theta)d\theta\right)$$

$$= \frac{A^2}{4\pi}\theta\Big|_{-\pi}^{\pi} + \frac{A^2}{4\pi}\frac{\cos(2\omega_c t+2\theta)}{2}\Big|_{-\pi}^{\pi}$$

$$= \frac{A^2}{4\pi}2\pi + \frac{A^2}{4\pi}\left(\frac{\cos(2\omega_c t+2\pi)}{2} - \frac{\cos(2\omega_c t-2\pi)}{2}\right)$$

$$= \frac{A^2}{2} + 0 = \frac{A^2}{2}$$

$$R_{YY}(\tau) = E[Y(t)Y(t+\tau)]$$

$$R_{YY}(\tau) = E[Y(t)Y(t+\tau)] = \int_{-\infty}^{\infty} Y(t)Y(t+\tau)f(\theta)d\theta$$

$$= \int_{-\infty}^{\infty} [A^2\sin^2(\omega_c t+\theta)][A^2\sin^2(\omega_c t+\omega_c\tau+\theta)]f(\theta)d\theta$$

$$= \int_{-\pi}^{\pi} [A^4\sin^2(\omega_c t+\omega_c\tau+\theta)\sin^2(\omega_c t+\theta)]f(\theta)d\theta$$

$$= \frac{A^4}{8\pi}\int_{-\pi}^{\pi} [1-\cos(2\omega_c t+2\theta)][(1-\cos(2\omega_c t+\omega_c\tau+2\theta)]d\theta$$

$$= \frac{A^4}{8\pi}\int_{-\pi}^{\pi} [1-\cos(2\omega_c t+2\omega_c\tau+2\theta)-\cos(2\omega_c t+2\theta)+\cos(2\omega_c t+2\theta)\cos(2\omega_c t+2\omega_c\tau+2\theta)]d\theta$$

$$= \frac{A^4}{8\pi}\left[\int_{-\pi}^{\pi} 1d\theta - \int_{-\pi}^{\pi}\cos(2\omega_c t+2\omega_c\tau+2\theta)d\theta - \int_{-\pi}^{\pi}\cos(2\omega_c t+2\theta)d\theta + \int_{-\pi}^{\pi}\cos(2\omega_c t+2\theta)\cos(2\omega_c t+2\omega_c\tau+2\theta)d\theta\right]$$

$$= \frac{A^4}{8\pi}\left[\theta\Big|_{-\pi}^{\pi} - \frac{\sin(2\omega_c t+2\omega_c\tau+2\theta)}{2}\Big|_{-\pi}^{\pi} - \frac{\sin(2\omega_c t+2\theta)}{2}\Big|_{-\pi}^{\pi} + \frac{1}{2}\int_{-\pi}^{\pi}\cos(4\omega_c t+2\omega_c\tau+4\theta)d\theta + \frac{1}{2}\int_{-\pi}^{\pi}\cos(2\omega_c\tau)d\theta\right]$$

$$= \frac{A^4}{8\pi}\left[2\pi - \frac{\sin(2\omega_c t+2\omega_c\tau+2\theta)}{2}\Big|_{-\pi}^{\pi} - \frac{\sin(2\omega_c t+2\theta)}{2}\Big|_{-\pi}^{\pi} + \frac{1}{2}\frac{\sin(4\omega_c t+2\omega_c\tau+4\theta)}{4}\Big|_{-\pi}^{\pi} + \frac{1}{2}\cos(2\omega_c\tau)\theta\Big|_{-\pi}^{\pi}\right]$$

$$= \frac{A^4}{8\pi}\left[2\pi-0+0-0+\frac{1}{2}\cos(2\omega_c\tau)2\pi\right]$$

$$= \frac{A^4}{8}[2+\cos(2\omega_c\tau)].$$

Cross-correlation function of $Y(t)$ and $X(t)$ is

$$R_{XY}(\tau) = E[X(t)Y(t+\tau)]$$

$$R_{XY}(\tau) = E[X(t)Y(t+\tau)] = \int_{-\infty}^{\infty} X(t)Y(t+\tau)f(\theta)d\theta$$

$$= \int_{-\infty}^{\infty} [A\sin(\omega_c t+\theta)][A^2\sin^2(\omega_c t+\omega_c\tau+\theta)]f(\theta)d\theta$$

$$= \int_{-\pi}^{\pi} \left[A^3 \sin^2(\omega_c t + \omega_c \tau + \theta)\sin(\omega_c t + \theta) \right] f(\theta) d\theta$$

$$= \frac{A^3}{4\pi} \int_{-\pi}^{\pi} \sin(\omega_c t + \theta)\left[1 - \cos(2\omega_c t + \omega_c \tau + 2\theta)\right] d\theta$$

$$= \frac{A^3}{4\pi} \int_{-\pi}^{\pi} \left[\sin(\omega_c t + \theta) - \sin(\omega_c t + \theta)\cos(2\omega_c t + 2\omega_c \tau + 2\theta)\right] d\theta$$

$$= \frac{A^3}{4\pi} \left[\int_{-\pi}^{\pi} \sin(\omega_c t + \theta) d\theta - \int_{-\pi}^{\pi} \sin(3\omega_c t + 2\omega_c \tau + 3\theta) d\theta - \int_{-\pi}^{\pi} \sin(-\omega_c t - 2\omega_c \tau - \theta) d\theta \right]$$

$$= \frac{A^3}{4\pi} \left[-\frac{\cos(\omega_c t + \theta)}{1}\Big|_{-\pi}^{\pi} - \frac{\cos(3\omega_c t + 2\omega_c \tau + 3\theta)}{6}\Big|_{-\pi}^{\pi} + \frac{1}{2}\cos(\omega_c t + 2\omega_c \tau + \theta)\Big|_{-\pi}^{\pi} \right]$$

$$= \frac{A^3}{4\pi} \left[-\{\cos(\omega_c t + \pi) - \cos(\omega_c t - \pi)\} - \left(\frac{\{\cos(3\omega_c t + 2\omega_c \tau + 3\pi) - \cos(3\omega_c t + 2\omega_c \tau - 3\pi)\}}{6} \right) \right.$$

$$\left. + \frac{1}{2}\{\cos(\omega_c t + 2\omega_c \tau + \pi) - \cos(\omega_c t + 2\omega_c \tau - \pi)\} \right]$$

$$= \frac{A^4}{8\pi}(0) = 0$$

Here mean value of $X(t)$ and $Y(t)$ is constant and also the cross-correlation $X(t)$ and $Y(t)$ is independent of t. Hence, it is jointly wide sense stationary.

6.4 TIME AVERAGES AND ERGODICITY

6.4.1 Sample Function

At particular time interval of a function is called sample function. A collection of sample functions is called *ensemble function*. Ensemble average is called mean value.

6.4.2 Time Averages

The time average of a function $g(x(t))$ of a continuous random process $X(t)$ is defined by

$$<g(X(t))>_T = \frac{1}{2T}\int_{-T}^{T} g(X(t))dt \qquad ...(6.70)$$

Similarly, the time average $g(X[n])$ of a discrete random process $X[n]$ is

$$<g(X[n])>_T = \frac{1}{2N+1}\sum_{i=-N}^{N} x_i \qquad ...(6.71)$$

The above definitions are in contrast to the corresponding ensemble average defined by

$$E[g(X(t))] = \int_{-\infty}^{\infty} g(X) f_X(x;t) dx \text{ for continuous}$$

$$= \sum_{i \in R_{X(t)}} g(x_i) p(x_i;t) \text{ for discrete}$$

The time average of a sample function is defined as

$$A[\,] = \frac{1}{2T}\int_{-T}^{T} [\,]dt \qquad \text{...(6.72)}$$

where A is used to denote time average in a manner analogous to expectation for the statistical average. Time average is taken over all time because, as applied to random process, sample function, samples functions of processes are presumed to exist for all time.

Specific averages of interest are the mean value $\overline{x} = A[x(t)]$ of a sample function (a lower case letter is used to imply a sample function), and the time autocorrelation function denoted by

$$\mathfrak{R}_{XX}(\tau) = A[x(t)x(t+\tau)]$$

$$\overline{x} = A[x(t)] = \langle \mu_X \rangle_T = \frac{1}{2T}\int_{-T}^{T} x(t)\,dt \quad \text{(for continuous case)} \qquad \text{...(6.73)}$$

$$\overline{x} = A[x(n)] = \langle \mu_X \rangle_N = \frac{1}{2N+1}\sum_{i=-N}^{N} x_i \quad \text{(for discrete case)} \qquad \text{...(6.74)}$$

$$\mathfrak{R}_{XX}(\tau) = \langle R_{XX}(\tau) \rangle_T = \frac{1}{2T}\int_{-T}^{T} x(t)x(t+\tau)\,dt \; \text{(for continuous case)} \qquad \text{...(6.75)}$$

$$\mathfrak{R}_{XX}(m) = \langle R_{XX}(m) \rangle = \frac{1}{2N+1}\sum_{i=-N}^{N} x_i\, x_{i+m} \quad \text{(for discrete case)} \qquad \text{...(6.76)}$$

The above mean, autocorrelations of time averages are defined only for a single sample function of the process $X(t)$. However, when all sampled functions are considered, $\overline{x}$ and $\mathfrak{R}_{XX}(\tau)$ are actual random variables. By taking the expected value on both sides of equation from Eq. (6.73) to Eq. (6.76), we will get the mean and autocorrelation of a random process $X(t)$.

6.4.3 Mean and Variance of the Time Averages Random Process

Let us consider the simplest case of the time averaged mean of a random process of a discrete time WSS random process $\{X_n\}$ given by

$$\langle \mu_X \rangle_N = \frac{1}{2N+1}\sum_{i=-N}^{N} x_i$$

The mean of $\langle \mu_X \rangle_N$ is

$$E\left[\langle \mu_X \rangle_N\right] = E\left[\frac{1}{2N+1}\sum_{i=-N}^{N} x_i\right] = \frac{1}{2N+1}E\left[\sum_{i=-N}^{N} x_i\right] = \overline{X} \qquad \text{...(6.77)}$$

and the variance is

$$E\left[\left(\langle\mu_X\rangle_N - \overline{X}\right)^2\right] = E\left[\left(\frac{1}{2N+1}\sum_{i=-N}^{N} x_i - \overline{X}\right)^2\right] = \left(\frac{1}{2N+1}\right)^2 E\left[\left(\sum_{i=-N}^{N} x_i - \overline{X}\right)^2\right]$$

$$= \left(\frac{1}{2N+1}\right)^2 \sum_{i=-N}^{N} E\left(x_i - \overline{X}\right)^2$$

$$= \left(\frac{1}{2N+1}\right)^2 \sum_{i=-N}^{N} \sigma_X^2 = \frac{\sigma_X^2}{2N+1}$$

From the above, we observe that as $N \to \infty$, variance is zero.

$$\lim_{N\to\infty} E\left[\left(\langle\mu_X\rangle_N - \overline{X}\right)^2\right] = 0$$

Let us consider the time averaged mean for the continuous case, we have

$$\langle\mu_X\rangle_T = \frac{1}{2T}\int_{-T}^{T} x(t)\,dt$$

$$E\left[\langle\mu_X\rangle_T\right] = E\left[\frac{1}{2T}\int_{-T}^{T} x(t)\,dt\right] = \frac{1}{2T}\int_{-T}^{T} E[x(t)]\,dt = \overline{X} \qquad \text{...(6.78)}$$

The variance is

$$E\left[\left(\langle\mu_X\rangle_T - \overline{X}\right)^2\right] = E\left[\left(\frac{1}{2T}\int_{-T}^{T} x(t)\,dt - \overline{X}\right)^2\right] = \left(\frac{1}{2T}\right)^2 E\left[\left(\int_{-T}^{T} x(t)\,dt - \overline{X}\right)^2\right]$$

$$= \left(\frac{1}{2T}\right)^2 \int_{-T}^{T} E\left[\left(x(t) - \overline{X}\right)^2\right] dt$$

$$= \left(\frac{1}{2T}\right)^2 \int_{-T}^{T}\int_{-T}^{T} E\left[\left(x(t_1) - \overline{X}\right)\left(x(t_2) - \overline{X}\right)\right] dt_1\,dt_2$$

$$= \left(\frac{1}{2T}\right)^2 \int_{-T}^{T}\int_{-T}^{T} C_{XX}(t_1, t_2)\,dt_1\,dt_2$$

The above double integral is evaluated on the square area bounded by $t_1 = \pm T$. We divide the square region into sum of trapezoidal stripes parallel to $t_1 - t_2 = 0$. Putting $t_1 - t_2 = \tau$ and noting that the differential area between $t_1 - t_2 = \tau$ and $t_1 - t_2 = \tau + d\tau$ is $(2T - |\tau|)d\tau$, the above integral is converted to a single integral as follows:

$$E\left[\left(\langle\mu_X\rangle_T - \overline{X}\right)^2\right] = \frac{1}{4T^2}\int_{-T}^{T}\int_{-T}^{T} C_{XX}(t_1 - t_2)\,dt_1\,dt_2 \qquad \text{..(6.79)}$$

where $\quad t_1 - t_2 = \tau$

From Eq. (6.79), the integrand depends only on the difference $t_1 - t_2$ (not absolute t_1 and/or t_2). Since the integrand C_{XX} is a function of one quantity, i.e. $t_1 - t_2$, one should suspect that the double integral could be reduced to a single integral over this one variable. These suspicions are correct. Equation (6.79) can be simplified by the transformation

$$\tau = t_1 - t_2$$
$$\tau' = t_1 + t_2$$

with inverse

$$t_1 = \frac{1}{2}(\tau + \tau') \text{ and } t_2 = \frac{1}{2}(\tau - \tau')$$

As shown in Fig. 6.7, $t_1 - t_2$ plane points P_1, P_2, P_3 and P_4 map to $\tau - \tau'$ plane points P_1', P_2', P_3' and P_4' respectively.

The jacobian is

$$\frac{\partial(t_1, t_2)}{\partial(\tau, \tau')} = \det\begin{bmatrix} \dfrac{\partial(t_1)}{\partial(\tau)} & \dfrac{\partial(t_1)}{\partial(\tau')} \\ \dfrac{\partial(t_2)}{\partial(\tau)} & \dfrac{\partial(t_2)}{\partial(\tau')} \end{bmatrix} = \det\begin{bmatrix} \dfrac{1}{2} & \dfrac{1}{2} \\ -\dfrac{1}{2} & \dfrac{1}{2} \end{bmatrix} = \frac{1}{2}$$

In (τ, τ') plane, as τ goes from $-2T$ to 0, τ' traverses from $-2T - \tau$ to $2T + \tau$, as can be seen from the dotted vertical line in Fig. 6.7. In a similar manner, as τ goes from 0 to $2T$, τ' traverses $-2T + \tau$ to $2T - \tau$. Hence, Eq. (6.79) can be expressed as

$$\frac{1}{4T^2}\int_{-T}^{T}\int_{-T}^{T} C_{XX}(t_1 - t_2)dt_1 dt_2 = \frac{1}{4T^2}\iint_{\Re_2} C_{XX}(\tau)\left|\frac{\partial(t_1, t_2)}{\partial(\tau, \tau')}\right| d\tau\, d\tau'$$

where $\Re_2$ is the "rotated square" region in the (τ, τ') plane shown on the right hand side of Fig. 6.7.

$$= \frac{1}{4T^2}\int_{-2T}^{0}\int_{-2T-\tau}^{2T+\tau} \frac{1}{2} C_{XX}(\tau)d\tau\, d\tau' + \frac{1}{4T^2}\int_{0}^{2T}\int_{-2T+\tau}^{2T-\tau} \frac{1}{2} C_{XX}(\tau)d\tau\, d\tau'$$

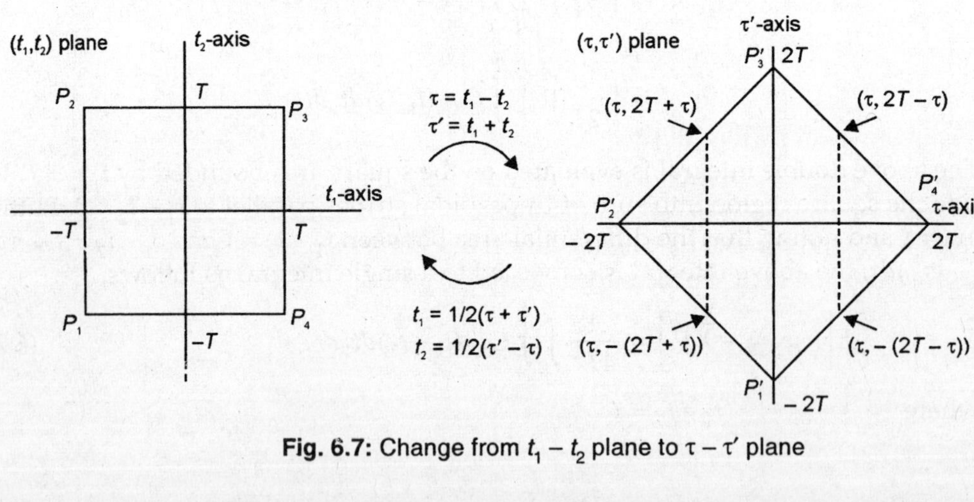

Fig. 6.7: Change from $t_1 - t_2$ plane to $\tau - \tau'$ plane

$$= \frac{1}{4T^2}\int_{-2T}^{0} C_{XX}(\tau)(2T+\tau)d\tau + \frac{1}{4T^2}\int_{0}^{2T} C_{XX}(\tau)(2T-\tau)d\tau$$

$$\therefore \quad \frac{1}{4T^2}\int_{-T}^{T}\int_{-T}^{T} C_{XX}(t_1-t_2)dt_1dt_2 = \frac{1}{2T}\int_{-2T}^{2T}\left(1-\frac{|\tau|}{2T}\right)C_{XX}(\tau)d\tau \qquad ...(6.80)$$

Second Method

The above equation gives a trapezoidal shown in Fig. 6.8.

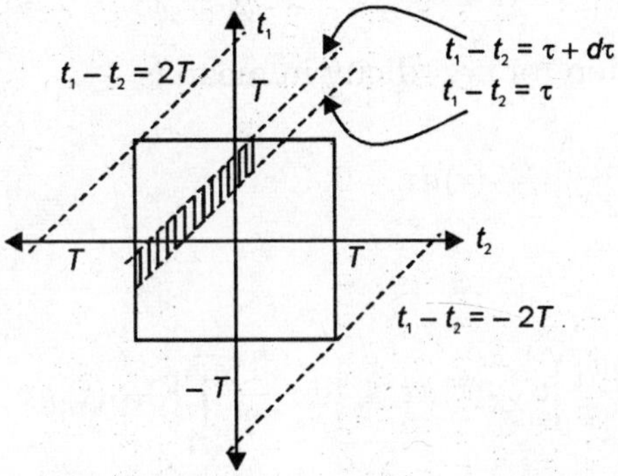

Fig. 6.8: Closed region of integration applicable to Eq. (6.79)

The area is calculated as explained below.

t_1, t_2 varies from $-T$ to T. Moreover $R_{XX}(\tau)$ is constant over the shaded region in Fig. 6.8, whose area is given by

$$\int_{2T-\tau-d\tau}^{2T-\tau} \tau d\tau = \frac{\tau^2}{2}\Bigg|_{2T-\tau-d\tau}^{2T-\tau} = \frac{1}{2}(2T-\tau)^2 - \frac{1}{2}(2T-\tau-d\tau)^2$$

$$= (2T-\tau)\,d\tau \qquad (\because \tau \text{ is small so that } d\tau \text{ is very small and } (d\tau)^2 \text{ is neglected})$$

Hence the above integral reduces to

$$E\left[\left(\langle\mu_X\rangle_T - \overline{X}\right)^2\right] = \frac{1}{4T^2}\int_{-2T}^{2T}(2T-|\tau|)C_{XX}(\tau)d\tau$$

$$E\left[\left(\langle\mu_X\rangle_T - \overline{X}\right)^2\right] = \frac{1}{2T}\int_{-2T}^{2T}\left(1-\frac{|\tau|}{2T}\right)C_{XX}(\tau)d\tau. \qquad ...(6.81)$$

6.5 ERGODICITY PRINCIPLE

Random processes are said to be ergodic processes when their time averages and ensemble averages are equal or interchangeable. We can define ergodicity with respect to the various statistical parameters like mean, variance and correlation, etc. and hence we have mean ergodic process, autocorrelation ergodic, etc.

6.5.1 Mean Ergodic Process

A WSS process $X(t)$ is said to be ergodic in mean, if $\langle\mu_X\rangle_T \xrightarrow{\text{m.s}} \mu_X$ as $T \to \infty$. Thus, for a mean ergodic process $X(t)$,

$$\lim_{T\to\infty} E\langle\mu_X\rangle_T = \mu_X$$

and
$$\lim_{T\to\infty} \text{Var}\langle\mu_X\rangle_T = 0$$

We have earlier shown that

$$E\langle\mu_X\rangle_T = \mu_X$$

and
$$\text{Var}\langle\mu_X\rangle_T = \frac{1}{2T}\int_{-2T}^{2T}\left(1-\frac{|\tau|}{2T}\right)C_{XX}(\tau)d\tau$$

Therefore, the condition for ergodicity in mean is

$$\lim_{T\to\infty}\frac{1}{2T}\int_{-2T}^{2T}\left(1-\frac{|\tau|}{2T}\right)C_{XX}(\tau)d\tau = 0 \quad \text{...(6.82)}$$

Further,

$$\lim_{T\to\infty}\frac{1}{2T}\int_{-2T}^{2T}\left(1-\frac{|\tau|}{2T}\right)C_{XX}(\tau)d\tau \le \lim_{T\to\infty}\frac{1}{2T}\int_{-2T}^{2T}|C_{XX}(\tau)|d\tau \quad \text{...(6.83)}$$

Therefore, a sufficient condition for mean ergodicity is

$$\int_{-2T}^{2T}|C_{XX}(\tau)|d\tau < \infty.$$

Example 6.5.1: Consider the random binary waveform $X(t)$. The process has the auto-covariance function given by

$$C_{XX}(\tau) = \begin{cases} 1-\dfrac{|\tau|}{T_P} & |\tau|\le T_P \\ 0 & \text{otherwise} \end{cases}$$

Find if it is mean ergodic or not.

Solution: Here $\displaystyle\int_{-2T}^{2T}|C_{XX}(\tau)|d\tau = 2\int_{0}^{2T}|C_{XX}(\tau)|d\tau$

But our function exists only on the interval $(-T_P, T_P)$

$$= 2\int_0^{T_P}\left(1-\frac{\tau}{T_P}\right)d\tau$$

$$= 2\tau\Big|_0^{T_P} - \frac{2}{T_P}\frac{\tau^2}{2}\Big|_0^{T_P} = 2T_P - \frac{2}{T_P}\frac{T_P^2}{2} = T_P$$

$$\therefore \quad \int_{-2T}^{2T}|C_{XX}(\tau)|d\tau < \infty$$

Hence $X(t)$ is mean ergodic.

6.5.2 Autocorrelation Ergodicity

Analogous to a mean ergodic process, a stationary continuous process $X(t)$ with autocorrelation function $R_{XX}(\tau)$ is called autocorrelation ergodic or ergodic in the autocorrelation, if and only if for all τ

$$\langle R_{XX}(\tau)\rangle_T = \frac{1}{2T}\int_{-T}^{T} x(t)x(t+\tau)\,dt$$

We consider $Z(t) = X(t)X(t + \lambda)$ so that $\mu_Z = R_{XX}(\lambda)$, where λ is a time offset. Then $X(t)$ will be autocorrelation ergodic if $Z(t)$ is mean ergodic.

$$E[Z(t)] = E[X(t)X(t+\lambda)] = R_{XX}(\lambda) \quad \text{...(6.84)}$$

$$\begin{aligned} R_{ZZ}(\lambda) &= E[W(t)W(t+\tau)] \\ &= E[X(t)X(t+\lambda)X(t+\tau)X(t+\tau+\lambda)] \end{aligned}$$

$$C_{ZZ}(t) = R_{ZZ}(\tau) - \{E[Z(t)]\}^2$$

$$C_{ZZ}(t) = R_{ZZ}(\tau) - (R_{XX}(\lambda))^2 \quad \text{...(6.85)}$$

Thus, $X(t)$ will be autocorrelation ergodic if

$$\lim_{T\to\infty} \frac{1}{2T}\int_{-2T}^{2T}\left(1-\frac{|\tau_1|}{2T}\right)C_{ZZ}(\tau_1)\,d\tau_1 = 0 \quad \text{...(6.86)}$$

$C_{ZZ}(\tau_1)$ involves fourth order moment.

A simpler condition for autocorrelation ergodicity of a jointly Gaussian process can be found.

Example 6.5.2: Consider a random-phased sinusoid given by $X(t) = A\cos(\omega_c t + \theta)$ where A and ω_c are constants and θ is random variable uniformly distributed on the interval $[0, 2\pi]$. Determine whether the random process is mean and autocorrelation ergodic or not.

Solution: For any particular realization $X(t) = A\cos(\omega_c t + \theta)$

$$\langle \mu_X\rangle_T = \frac{1}{2T}\int_{-T}^{T} x(t)\,dt$$

$$\langle \mu_X\rangle_T = \frac{1}{2T}\int_{-T}^{T} A\cos(\omega_c t+\theta)\,dt$$

$$\begin{aligned} \langle \mu_X\rangle_T &= \frac{A}{2T}\frac{\sin(\omega_c t+\theta)}{\omega_c}\Bigg|_{-T}^{T} \\ &= \frac{A}{2T}\frac{\sin(\omega_c T+\theta)}{\omega_c} - \frac{A}{2T}\frac{\sin(-\omega_c T+\theta)}{\omega_c} \\ &= \frac{A}{T\omega_c}\sin(\omega_c T+\theta) \end{aligned}$$

and $$\langle R_{XX}(\tau)\rangle_T = \frac{1}{2T}\int_{-T}^{T} x(t)x(t+\tau)\,dt$$

$$\langle R_{XX}(\tau)\rangle_T = \frac{1}{2T}\int_{-T}^{T} A\cos(\omega_c t+\theta)A\cos(\omega_c t+\omega_c\tau+\theta)dt$$

$$= \frac{A^2}{2T}\int_{-T}^{T} \cos(\omega_c t+\theta)\cos(\omega_c t+\omega_c\tau+\theta)dt$$

$$= \frac{A^2}{4T}\int_{-T}^{T} 2\cos(\omega_c t+\theta)\cos(\omega_c t+\omega_c\tau+\theta)dt$$

$$= \frac{A^2}{4T}\int_{-T}^{T} [\cos(2\omega_c t+\omega_c\tau+2\theta)+\cos(\omega_c\tau)]dt$$

$$= \frac{A^2}{4T}\int_{-T}^{T} \cos(2\omega_c t+\omega_c\tau+2\theta)dt+\frac{A^2}{4T}\int_{-T}^{T} \cos(\omega_c\tau)dt$$

$$= \frac{A^2}{4T}\frac{\sin(2\omega_c t+\omega_c\tau+2\theta)}{2\omega_c}\bigg|_{-T}^{T}+\frac{A^2}{4T}\cos(\omega_c\tau)t\big|_{-T}^{T}$$

$$= \frac{A^2}{4T}\left(\frac{\sin(2\omega_c T+\omega_c\tau+2\theta)}{2\omega_c}-\frac{\sin(2\omega_c-T+\omega_c\tau+2\theta)}{2\omega_c}\right)+\frac{A^2}{4T}\cos(\omega_c\tau)(T+T)$$

$$= \frac{A^2}{4T}\frac{\sin(2\omega_c T+\omega_c\tau+2\theta)}{\omega_c}+\frac{A^2}{2}\cos(\omega_c\tau)$$

We see that as $T\to\infty$, $\langle\mu_X\rangle_T\to 0$ and $\langle R_{XX}(\tau)\rangle_T\to\frac{A^2}{2}\cos(\omega_c\tau)$

We know that mean

$$E[X(t)] = \int_{-\infty}^{\infty} X(t)f(\theta)d\theta$$

Here θ is uniformly distributed on the interval $[0, 2\pi]$, the density function is

$$f(\theta) = \frac{1}{2\pi-0}=\frac{1}{2\pi}$$

$$\bar{X} = E[X(t)]=\int_0^{2\pi} A\cos(\omega_c t+\theta)\frac{1}{2\pi}d\theta$$

$$= \frac{A}{2\pi}\int_0^{2\pi}\cos(\omega_c t+\theta)d\theta$$

$$= \frac{A}{2\pi}\sin(\omega_c t+\theta)\big|_0^{2\pi} = 0$$

$$\therefore \quad E[X(t)] = 0$$

$$R_{XX}(\tau) = E[X(t)X(t+\tau)]$$

$$= E[\{A\cos(\omega_c t+\theta)\}\{A\cos(\omega_c t+\omega_c\tau+\theta)\}]$$

$$= E[A^2\cos(\omega_c t+\theta)\cos(\omega_c t+\omega_c\tau+\theta)]$$

$$R_{XX}(\tau) = E[X(t)X(t+\tau)] = \int_{-\infty}^{\infty} X(t)X(t+\tau)f(\theta)d\theta$$

$$= E[A^2\cos(\omega_c t+\theta)\cos(\omega_c t+\omega_c\tau+\theta)]$$

$$R_{XX}(\tau) = A^2E[\cos(\omega_c t+\theta)\cos(\omega_c t+\omega_c\tau+\theta)]$$

$$= A^2E[\cos(\omega_c t+\theta)\cos(\omega_c t+\omega_c\tau+\theta)]$$

$$R_{XX}(\tau) = A^2\int_{-\infty}^{\infty} \cos(\omega_c t+\theta)\cos(\omega_c t+\omega_c\tau+\theta)\, f(\theta)d\theta$$

$$R_{XX}(\tau) = A^2\int_{0}^{2\pi} \cos(\omega_c t+\theta)\cos(\omega_c t+\omega_c\tau+\theta)\frac{1}{2\pi}d\theta$$

$$= \frac{A^2}{4\pi}\int_{0}^{2\pi} 2\cos(\omega_c t+\theta)\cos(\omega_c t+\omega_c\tau+\theta)d\theta$$

$$= \frac{A^2}{4\pi}\int_{0}^{2\pi} \cos(\omega_c\tau)+\cos(2\omega_c t+\omega_c\tau+2\theta)d\theta$$

$$= \frac{A^2}{4\pi}\cos(\omega_c\tau)\,\theta\Big|_0^{2\pi} + \frac{A^2}{4\pi}\frac{\cos(2\omega_c t+\omega_c\tau+2\theta)}{2}\Bigg|_0^{2\pi}$$

$$= \frac{A^2}{4\pi}\cos(\omega_c\tau)\,2\pi + \frac{A^2}{4\pi}\left[\frac{\cos(2\omega_c t+\omega_c\tau)}{2} - \frac{\cos(2\omega_c t+\omega_c\tau)}{2}\right]$$

$$= \frac{A^2}{2}\cos(\omega_c\tau)$$

$$\therefore \quad R_{XX}(\tau) = \frac{A^2\cos(\omega_c\tau)}{2}$$

From the above, we conclude that

$$\lim_{T\to\infty}\langle\mu_X\rangle_T = E[X(t)]$$

and $\quad \lim_{T\to\infty}\langle R_{XX}(\tau)\rangle_T = R_{XX}(\tau).$

For each realization, both the time averaged mean and the time averaged autocorrelation function converge to the corresponding ensemble averages. Thus, the random phased sinusoid is ergodic in both mean and autocorrelation.

Remark

A random process $X(t)$ is ergodic if its ensemble averages converge in the m.s. sense to the corresponding time averages. This is a stronger requirement than stationarity—the ensemble averages of all orders of such a process are independent of time. This implies that *an ergodic process is necessarily stationary in the strict sense.* The converse is not true—there are stationary random processes which are not ergodic.

Figure 6.9 shows a hierarchical classification of random processes.

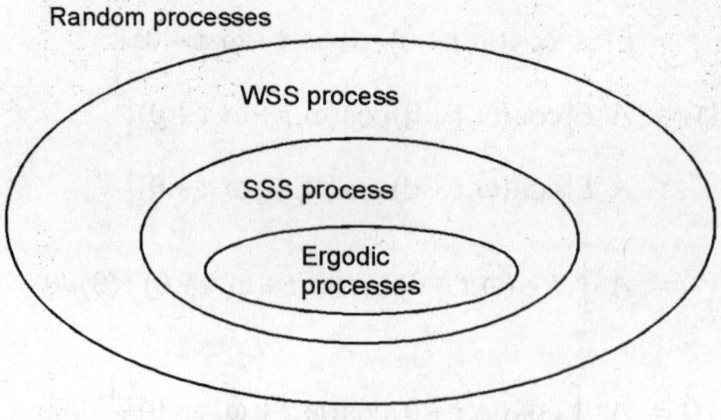

Fig. 6.9: Hierarchical classification of random processes

6.6 GAUSSIAN AND POISSON RANDOM PROCESSES

6.6.1 Gaussian Random Process

Consider a continuous random process $X(t)$ and define N random variables $X_1 = X(t_1)$, $X_2 = X(t_2),\ldots, X_i = X(t_i),\ldots,X_N = X(t_N)$ corresponding to N time instants $t_1, t_2,\ldots, t_i,\ldots,t_N$. If for any $N = 1, 2, 3\ldots$ and any times $t_1, t_2,\ldots,t_i,\ldots,t_N$, these random variables are jointly Gaussian with the joint density function given by

$$f_{X(t_1),\,X(t_2),\ldots,X(t_n)}(x_1,x_2,\ldots,x_n) = \frac{\exp\left\{\frac{-1}{2}[X-\bar{X}]^t[C_x]^{-1}[X-\bar{X}]\right\}}{\sqrt{(2\pi)^N \, |C_x|}} \quad \ldots(6.87)$$

where

$$[X-\bar{X}] = \begin{bmatrix} X(t_1)-\bar{X} \\ X(t_2)-\bar{X} \\ \vdots \\ X(t_n)-\bar{X} \end{bmatrix} \quad \ldots(6.88)$$

The mean values are

$$\bar{X}_i = E(X) = [E(X_1)E(X_2)\ldots E(X_N)] = E(X_i) = E[X(t_i)] \quad \ldots(6.89)$$

The elements of covariance matrix $[C_x]$ are

$$C_{ik} = C_{X_iX_k} = E\left[(X_i-\bar{X}_i)(X_k-\bar{X}_k)\right] = E\left[\{X(t_i)-E[X(t_i)]\}\{X(t_k)-E[X(t_k)]\}\right]$$

$$= C_{XX}(t_i, t_k)$$

$$C_{XX}(t_i, t_k) = R_{XX}(t_i, t_k) - E[X(t_i)]E[X(t_k)] \quad \ldots(6.90)$$

If the Gaussian process is not stationary, the mean and autocovariance functions will depend on absolute time.

However, if the process is WSS, the mean will be constant

$$\bar{X}_i = E[X(t_i)] = \bar{X} \quad \text{(constant)}$$

while the autocovariance and autocorrelation functions will depend only on time differences and absolute time.

$$C_{XX}(t_i, t_k) = C_{XX}(t_k - t_i)$$
$$R_{XX}(t_i, t_k) = R_{XX}(t_k - t_i)$$

$$C_{XX} = \begin{bmatrix} C_X(0)C_X(t_1 - t_1)...C_X(t_1 - t_n) \\ C_X(t_2 - t_1)C_X(0)...C_X(t_2 - t_n) \\ \vdots \\ C_X(t_n - t_1)C_X(t_n - t_2)...C_X(0) \end{bmatrix} \quad ...(6.91)$$

We see that $f_{X(t_1), X(t_2),...,X(t_n)}(x_1, x_2,...,x_n)$ depends only on the time lags. Thus, for a Gaussian random process, WSS implies strict sense stationarity, because this process is completely described by the mean and the autocorrelation functions.

Two random processes $X(t)$ and $Y(t)$ are said to be jointly Gaussian if the random variables $X_1 = X(t_1), X_2 = X(t_2),...,X_N = X(t_N), Y_1 = Y(t_1'), Y_2 = Y(t_2'),...,Y_M = Y(t_M')$ defined at times $t_1, t_2,...,t_i,...,t_N$ for $X(t)$ and $t_1', t_2',...,t_M'$ for $Y(t)$, are jointly Gaussian for any N, $t_1, t_2,...,t_i,...,t_N$, any M, $t_1', t_2',...,t_M'$.

6.6.2 Poisson Random Process

Poisson random process is an example of discrete random process. It describes the number of times that some event has occurred as a function of time, where the events occur at random times. The event might be the arrival of a customer at a bank or supermarket check-out register, the occurrence of a lightning strike within some prescribed area, the failure of some component in a system, or the emission of an electron from the surface of a light sensitive material. In each of these examples, a single event occurs at a random time and the process amounts to counting the number of such occurrences with time. For this reason, the process is also known as Poisson counting process.

To visualize the Poisson process, let $X(t)$ represent the number of event occurrences with time; then $X(t)$ has integer valued, nondecreasing sample functions, as illustrated in Fig. 6.10a for the random occurrence times of Fig. 6.10b. For convenience, we take $X(t) = 0$ at $t = 0$; for $t > 0$, $X(t)$ is the number of occurrences in the interval $(t, 0)$; for $t < 0$, $X(t)$ is the negative of the number of occurrences in the interval $(t, 0)$.

To define the Poisson process we shall require two conditions. First, we require that only one event can occur at a time. This condition does not prevent the times of occurrence of events from being very close together, only that do not coincide. Second, we require that occurrence times be statistically independent so that the number that occurs in any given time interval is independent of the number in any other non-overlapping time interval.

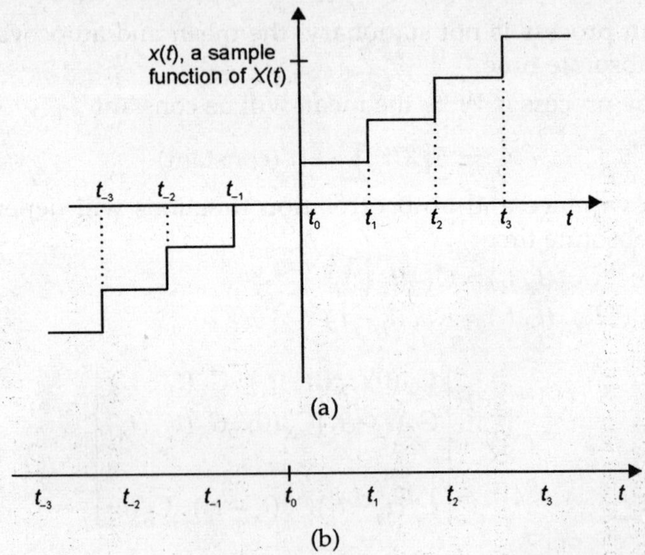

Fig. 6.10: (a) A sample function of random process (b) The random times of occurrence of events being counted to form the process

Probability Density Function

The probability of exactly k occurrences over a time interval $(0, t)$ with $b = \lambda t$ is

$$P[X(t) = k] = \frac{(\lambda t)^k e^{-\lambda t}}{k!} \quad ...(6.92)$$

and the probability density of the number of occurrence is

$$f_X(x) = \sum_{k=0}^{\infty} \frac{(\lambda t)^k e^{-\lambda t}}{k!} \delta(x-k) \quad ...(6.93)$$

Mean and Variance of the Poisson Random Process

The mean of the Poisson RP $X(t)$ is given by

$$E[X(t)] = \int_{-\infty}^{\infty} x f_X(x) dx$$

$$= \int_{-\infty}^{\infty} x \sum_{k=0}^{\infty} \frac{(\lambda t)^k e^{-\lambda t}}{k!} \delta(x-k) dx$$

$$E[X(t) = k] = \int_{-\infty}^{\infty} k \sum_{k=0}^{\infty} \frac{(\lambda t)^k e^{-\lambda t}}{k!} \delta(k-k) dx$$

$$= \sum_{k=0}^{\infty} k \frac{(\lambda t)^k e^{-\lambda t}}{k!} \qquad \left(\because \int_{-\infty}^{\infty} \delta(x) = 1\right)$$

$$= 0 + \lambda t \sum_{k=1}^{\infty} \frac{(\lambda t)^{(k-1)} e^{-\lambda t}}{(k-1)!}$$

$$= \lambda t e^{-\lambda t} \sum_{k=1}^{\infty} \frac{(\lambda t)^{(k-1)}}{(k-1)!}$$

$$= \lambda t e^{-\lambda t}\left(1 + \frac{\lambda t}{1!} + \frac{(\lambda t)^2}{2!} + \dots\right)$$

$$= \lambda t e^{-\lambda t} e^{\lambda t}$$

$$= \lambda t$$

Second moment is

$$E[X^2(t)] = \int_{-\infty}^{\infty} x^2 f_X(x)\,dx$$

$$= \int_{-\infty}^{\infty} x^2 \sum_{k=0}^{\infty} \frac{(\lambda t)^k e^{-\lambda t}}{k!} \delta(x-k)\,dx$$

$$E[X^2(t) = k^2] = \int_{-\infty}^{\infty} k^2 \sum_{k=0}^{\infty} \frac{(\lambda t)^k e^{-\lambda t}}{k!} \delta(k-k)\,dx$$

$$= \sum_{k=0}^{\infty} k^2 \frac{(\lambda t)^k e^{-\lambda t}}{k!} \qquad \left(\because \int_{-\infty}^{\infty} \delta(x) = 1\right)$$

$$= e^{-\lambda t} \sum_{k=0}^{\infty} k \frac{(\lambda t)^k}{(k-1)!}$$

$$= e^{-\lambda t} \sum_{k=0}^{\infty} (k-1+1) \frac{(\lambda t)^k}{(k-1)!}$$

$$= e^{-\lambda t} \sum_{k=1}^{\infty} (k-1) \frac{(\lambda t)^k}{(k-1)!} + e^{-\lambda t} \sum_{k=1}^{\infty} \frac{(\lambda t)^k}{(k-1)!}$$

$$= e^{-\lambda t} \sum_{k=2}^{\infty} \frac{(\lambda t)^k}{(k-2)!} + e^{-\lambda t} \sum_{k=1}^{\infty} \frac{(\lambda t)^k}{(k-1)!}$$

$$= e^{-\lambda t} (\lambda t)^2 \sum_{k=1}^{\infty} \frac{(\lambda t)^{k-2}}{(k-2)!} + e^{-\lambda t} \lambda t \sum_{k=1}^{\infty} \frac{(\lambda t)^{k-1}}{(k-1)!}$$

$$= e^{-\lambda t} (\lambda t)^2 e^{\lambda t} + e^{-\lambda t} \lambda t e^{\lambda t}$$

$$= (\lambda t)^2 + \lambda t = \lambda t\,(1 + \lambda t)$$

Variance is $\sigma_X^2 = E[X^2(t)] - E[X(t)]^2$

$\therefore \quad \sigma_X^2 = \lambda t(1 + \lambda t) - (\lambda t)^2 = \lambda t$

Joint Probability Density

To determine the joint probability density function for the Poisson process at times $0 < t_1 < t_2$, first observe that the probability of k_1 event occurrences over $(0, t_1)$ is

$$P[X(t_1) = k_1] = \frac{(\lambda t_1)^{k_1} e^{-\lambda t}}{k_1!} \qquad k_1 = 0, 1, 2, \ldots \qquad \ldots(6.94)$$

The conditional probability of k_2 occurrences over, $(0, t_2)$, given that k_1 events occurred over $(0, t_1)$, is just the probability that $(k_2 - k_1)$ events occurred over (t_1, t_2), which is

$$P\left[X(t_2) = k_2 / X(t_1) = k_1\right] = \frac{[\lambda(t_2 - t_1)]^{k_2 - k_1} e^{-\lambda(t_2 - t_1)}}{(k_2 - k_1)!} \quad \text{for } k_2 \geq k_1 \qquad \ldots(6.95)$$

The joint probability of k_2 occurrences at time t_2 and k_1 occurrences at time t_1 is

$$P(k_1, k_2) = P\left[X(t_2) = k_2 / X(t_1) = k_1\right] P[X(t_1) = k_1]$$

$$P(k_1, k_2) = \frac{(\lambda t_1)^{k_1} [\lambda(t_2 - t_1)]^{k_2 - k_1} e^{-\lambda(t_2)}}{k_1!(k_2 - k_1)!} \quad \text{for } k_2 \geq k_1 \qquad \ldots(6.96)$$

The joint density function is

$$f(x_1, x_2) = \sum_{k_1=0}^{\infty} \sum_{k_2=0}^{\infty} P(k_1, k_2)\delta(x_1 - k_1)\delta(x_2 - k_2) \qquad \ldots(6.97)$$

Example 6.6.1: A WSS Gaussian random process with mean $\overline{X} = 4$ and autocorrelation function $R_{XX}(\tau) = 25e^{-3|\tau|}$. Find the covariance matrix for three random variables $X(t_i)$, $i = 1, 2, 3$, defined at times $t_i = t_0 + [(i - 1)/2]$ with t_0 a constant.

Solution: Here $t_k - t_i = [(k - i)/2]$, i and $k = 1, 2, 3, \ldots$

$$R_{XX}(t_k - t_i) = 25e^{-3|k-i|/2}$$

$$C_{XX}(t_k - t_i) = 25e^{-3|k-i|/2} - 16$$

Thus

$$C_{XX} = \begin{bmatrix} (25-16) & (25e^{-3/2} - 16) & (25e^{-6/2} - 16) \\ (25e^{-3/2} - 16) & (25-16) & (25e^{-3/2} - 16) \\ (25e^{-6/2} - 16) & (25e^{-3/2} - 16) & (25-16) \end{bmatrix}$$

Example 6.6.2: A petrol pump serves on an average 30 cars per hour. Find the probability that during a period of 5 minutes: (i) no car comes to the station, (ii) exactly 3 cars come to the station and (iii) more than 3 cars come to the station.

Solution: Average arrival = 30 cars/hr

$$= \frac{1}{2} \text{ car/min}$$

Probability of no car in 5 minutes

$$P[X(t_1) = k_1] = \frac{(\lambda t_1)^{k_1} e^{-\lambda t}}{k_1!}$$

i. $P[X(5)=0] = \dfrac{\left(\frac{1}{2}\right)^0 e^{-\frac{1}{2}\times 5}}{0!} = e^{-2.5} = 0.0821$

ii. $P[X(5)=3] = \dfrac{\left(\frac{1}{2}\right)^3 e^{-\frac{1}{2}\times 5}}{3!} = \dfrac{\left(\frac{1}{2}\right)^3 e^{-2.5}}{3!} = 0.2138$

iii. $P[X(5)>3] = 1 - P[X(5)\le 3]$

$= 1 - P[X(5)=0] - P[X(5)=1] - P[X(5)=2] - P[X(5)=3]$

$= 1 - 0.08 - 0.2052 - 0.2565 - 0.21 = 0.2424$

EXAMPLES

Example 6.1: Show that the random process $X(t) = A\cos(\omega_c t + \theta)$ is wide sense stationary where A and ω_c are constants and θ is a random variable distributed uniformly on the interval $[0, 2\pi]$.

Solution: Conditions for WSS are:

i. Mean of random process should be constant.

ii. Autocorrelation is only function of τ.

Mean is $E[X(t)] = \int_{-\infty}^{\infty} x f(x;t)dx$ where x is random variable

In this problem, random variable is θ and uniformly distributed on the interval $[0, 2\pi]$.

The density function is

$$f_X(\theta) = \frac{1}{2\pi - 0} = \frac{1}{2\pi}$$

For uniform random variable, PDF is $\dfrac{1}{b-a}$

where b is upper limit and a is lower limit.

$$E[X(t)] = \int_{-\infty}^{\infty} x f_X(\theta)d\theta$$

$$= \int_0^{2\pi} A\cos(\omega_c t+\theta)\frac{1}{2\pi}d\theta$$

$$= \frac{A}{2\pi}\int_0^{2\pi}\cos(\omega_c t+\theta)d\theta$$

$$= \frac{A}{2\pi}\sin(\omega_c t+\theta)\Big|_0^{2\pi}$$

$$= \frac{A}{2\pi}[\sin(\omega_c t+2\pi) - \sin(\omega_c t+0)]$$

$$= \frac{A}{2\pi}0 = 0$$

$\therefore\ E[X(t)] = 0 = \text{constant}$

$$R_{XX}(\tau) = E[X(t)X(t+\tau)]$$
$$= E\left[\{(A\cos(\omega_c t+\theta)\}\{A\cos(\omega_c t+\omega_c\tau+\theta)\}\right]$$
$$= E\left[A^2\cos(\omega_c t+\theta)\cos(\omega_c t+\omega_c\tau+\theta)\right]$$

$$R_{XX}(\tau) = E[X(t)X(t+\tau)] = \int_{-\infty}^{\infty} X(t)X(t+\tau)f(\theta)d\theta$$
$$= \int_{-\infty}^{\infty}\left[A^2\cos(\omega_c t+\theta)\cos(\omega_c t+\omega_c\tau+\theta)\right]f(\theta)d\theta$$
$$= \int_{0}^{2\pi}\left[A^2\cos(\omega_c t+\theta)\cos(\omega_c t+\omega_c\tau+\theta)\right]f(\theta)d\theta$$
$$= \frac{A^2}{2}\int_{0}^{2\pi}[\cos(2\omega_c t+2\theta+\omega_c\tau)+\cos(\omega_c\tau)]\frac{1}{2\pi}d\theta$$
$$= \frac{A^2}{2}\int_{0}^{2\pi}\cos(2\omega_c t+2\theta+\omega_c\tau)\frac{1}{2\pi}d\theta + \frac{A^2}{2}\int_{0}^{2\pi}\cos(\omega_c\tau)\frac{1}{2\pi}d\theta$$
$$= \frac{A^2}{4\pi}\int_{0}^{2\pi}\cos(2\omega_c t+2\theta+\omega_c\tau)d\theta + \frac{A^2}{4\pi}\cos(\omega_c\tau)\theta\Big|_0^{2\pi}$$
$$= \frac{A^2}{4\pi}\frac{\sin(2\omega_c t+2\theta+\omega_c\tau)}{2}\Bigg|_0^{2\pi} + \frac{A^2}{4\pi}\cos(\omega_c\tau)2\pi$$
$$= \frac{A^2}{4\pi}\left[\frac{\sin(2\omega_c t+\omega_c\tau)}{2} - \frac{\sin(2\omega_c t+\omega_c\tau)}{2}\right] + \frac{A^2}{2}\cos(\omega_c\tau)$$
$$= \frac{A^2}{4\pi}[0] + \frac{A^2}{2}\cos(\omega_c\tau).$$

$$\therefore \quad R_{XX}(\tau) = \frac{A^2}{2}\cos(\omega_c\tau).$$

Mean is constant, i.e. zero and autocorrelation is only function of τ. Hence, the random process is wide sense stationary.

Example 6.2: Consider a random process $X(t) = \cos(\omega t)$, where ω is a random variable distributed uniformly on the interval $[0, \omega_0]$. Determine whether $X(t)$ is stationary.

Solution: The density function is

$$f_X(\omega) = \frac{1}{\omega_0 - 0} = \frac{1}{\omega_0}$$

$$E[X(t)] = \int_{-\infty}^{\infty} x f_x(\omega)d\omega$$

$$= \int_0^{\omega_0} \cos \omega t \frac{1}{\omega_0} d\omega$$

$$= \frac{1}{\omega_0} \int_0^{\omega_0} \cos \omega t \, d\omega$$

$$= \frac{1}{\omega_0} \frac{\sin \omega t}{t} \Big|_0^{\omega_0}$$

$$= \frac{1}{\omega_0} \left[\frac{\sin(\omega_0 t) - \sin(0)}{t} \right]$$

$$= \frac{1}{\omega_0} \left[\frac{\sin(\omega_0 t)}{t} \right]$$

From the above, mean of the random process is not constant. Hence, the random process is not stationary.

Example 6.3: A random process $X(t)$ has periodic sample function as shown in Fig.6.11 where B, T and $4t_0 \le T$ are constants but $\in$ is a random variable uniformly distributed in the interval $(0, T)$.

(a) Classify the process.

(b) Find the first order distribution function of $X(t)$.

(c) Find the first order density function.

(d) Find $E[X(t)]$, $E[X^2(t)]$ and σ_{XX}^2.

(e) Determine whether $X(t)$ is first order stationary or not.

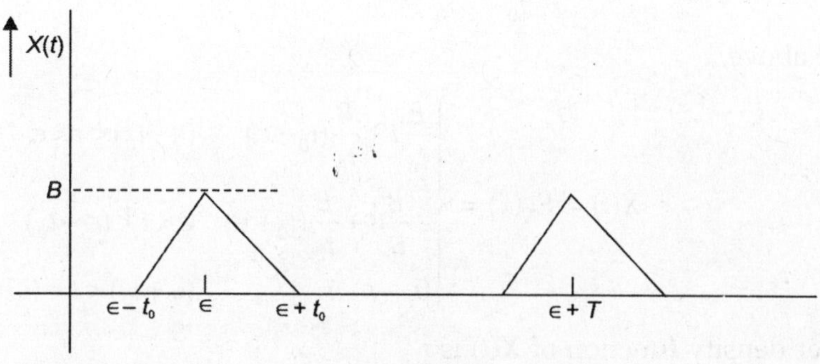

Fig. 6.11

Solution: (a) This is a continuous random process and it is deterministic since the future values of the sample function can be predicted from the past values.

(b) First order distribution function is nothing but the signal variation of sample function $X(t)$ in Fig. 6.11.

From Fig. 6.11, $X(t)$ has two different functions and is a periodic signal with period $(\in + 3t_0) - (\in - 3t_0) = 4t_0 = T$.

Here we are obtaining the function only for one interval because the same can be repeated after the time interval T.

First obtain the points.
These are $(\in - t_0, 0)$, $(\in, B)$ and $(\in + t_0, 0)$.
Time interval $\in - t_0 < t < \in$
We have two points $(\in - t_0, 0)$ and $(\in, B)$, on Y-axis $X(t)$ and on X-axis t.
Straight line equation using two points (x_1, y_1) and (x_2, y_2) is given by

$$y - y_1 = \frac{y_2 - y_1}{x_2 - x_1}(x - x_1)$$

Substitute the values in above equation,

$$X(t) - 0 = \frac{B-0}{\in - \in + t_0}[t - (\in - t_0)]$$

$$X(t) = \frac{B}{t_0}(t - \in + t_0)$$

$$= \frac{B}{t_0}t + \frac{B}{t_0}(t_0 - \in)$$

Time interval is $\in > t > \in + t_0$.
We have two points $(\in, B)$ and $(\in + t_0, 0)$

$$X(t) - B = \frac{0-B}{\in + t_0 - \in}(t - \in)$$

$$= B + \frac{-B}{t_0}(t - \in)$$

$$= -\frac{B}{t_0}t + \frac{B}{t_0}(t_0 + \in)$$

From the above,

$$X(t) = F_X(x) = \begin{cases} \frac{B}{t_0}t + \frac{B}{t_0}(t_0 - \in) & (\in - t_0) < t < \in \\ -\frac{B}{t_0}t + \frac{B}{t_0}(t_0 + \in) & \in < t < (\in + t_0) \\ 0 & (\in + t_0) < t < (\in + 3t_0) \end{cases}$$

(c) First order density function of $X(t)$ is

$$f_X(x) = \frac{dF_X(x)}{dt}$$

$$f_X(x) = \begin{cases} \frac{B}{t_0} & (\in - t_0) < t < \in \\ -\frac{B}{t_0}t & \in < t < (\in + t_0) \\ 0 & (\in + t_0) < t < (\in + 3t_0) \end{cases}$$

(d) $\in$ is a random variable uniformly distributed over $(0, T)$ where T is one cycle period.

$$\text{PDF} = f_X(\in) = \frac{1}{T}; \qquad 0 < t < T$$

$$E[X(t)] = \int_{-\infty}^{\infty} x f_x(\in) d\in = \int_0^T x(t)\, f_x(\in) d\in$$

$$= \frac{1}{T}\int_{-T/4}^{0}\left(\frac{B}{t_0}t + \frac{B}{t_0}(t_0 - \in)\right)d\in + \frac{1}{T}\int_0^{T/4}\left(-\frac{B}{t_0}t + \frac{B}{t_0}(t_0 + \in)\right)d\in$$

$$= \frac{1}{T}\int_{-T/4}^{0}\frac{B}{t_0}(t+t_0)\,d\in - \frac{B}{T}\int_{-T/4}^{0}\in d\in + \frac{1}{T}\int_0^{T/4}\frac{B}{t_0}(-t+t_0)\,d\in \frac{B}{T} + \frac{B}{T}\int_0^{T/4}\in d\in$$

$$= \frac{1}{T}\left[\left\{\frac{B(t+t_0)\in}{t_0} - \frac{B\in^2}{2t_0}\right\}_{-T/4}^{0} + \left\{\frac{B(-t+t_0)\in}{t_0} + \frac{B\in^2}{2t_0}\right\}_0^{T/4}\right]$$

$$= \frac{1}{T}\left[\left\{\frac{B(t+t_0)T}{4t_0} - \frac{B(0 - T^2/16)}{2t_0}\right\} + \left\{\frac{B(-t+t_0)T}{4t_0} + \frac{B(T^2/16) - 0)}{2t_0}\right\}\right]$$

$$= \frac{1}{T}\left[\frac{BtT}{4t_0} + \frac{BT}{4} + \frac{BT^2}{32t_0} - \frac{BtT}{4t_0} + \frac{BT}{4} + \frac{BT^2}{32t_0}\right]$$

$$= \frac{1}{T}\left[\frac{BT}{2} + \frac{BT^2}{16t_0}\right]$$

$$= \frac{B}{2}\left[1 + \frac{T}{8t_0}\right]$$

$$E[X(t)] = \frac{B}{2}\left[1 + \frac{T}{8t_0}\right]$$

$$E[X^2(t)] = \int_{-\infty}^{\infty} x^2 f_x(\in)\, d\in = \int_0^T x^2(t)\, f_x(\in) d\in$$

$$= \frac{B^2}{Tt_0^2}\int_{-T/4}^{0}(t + t_0 - \in)^2\, d\in + \frac{B^2}{Tt_0^{\,2}}\int_0^{T/4}(-t + t_0 - \in)^2\, d\in$$

$$= \frac{B^2}{Tt_0^{\,2}}\int_{-T/4}^{0}\left[(t+t_0)^2 + \in^2 - 2(t+t_0)\in\right]d\in$$

$$+ \frac{B^2}{Tt_0^{\,2}}\int_0^{T/4}\left[(-t+t_0)^2 + \in^2 + 2(-t+t_0)\in\right]d\in$$

$$= \frac{B^2}{Tt_0^{\,2}}\left[\left\{(t+t_0)^2 \in\right\}_{-T/4}^{0} + \left\{\frac{\in^3}{3}\right\}_{-T/4}^{0} - \left\{(t+t_0)\in^2\right\}_{-T/4}^{0}\right]$$

$$+ \frac{B^2}{Tt_0^2}\left[\left\{(-t+t_0)^2 \in\right\}_{0}^{T/4} + \left\{\frac{\in^3}{3}\right\}_{0}^{T/4} - \left\{(-t+t_0)\in^2\right\}_{0}^{T/4}\right]$$

$$= \frac{B^2}{Tt_0^{\,2}}\left[(t+t_0)^2\frac{T}{4} + \frac{T^3}{64\times 3} - (t+t_0)\frac{T^2}{16} + (-t+t_0)^2\frac{T}{4} + \frac{T^3}{64\times 3} + (-t+t_0)\frac{T^2}{16}\right]$$

$$= \frac{B^2}{Tt_0^{\,2}}\left[\frac{T^3}{32\times 3} - \frac{tT^2}{16} - \frac{t_0T^2}{16} - \frac{tT^2}{16} + \frac{t_0T^2}{16} + \frac{T}{4}\left\{(t+t_0)^2 + (-t+t_0)^2\right\}\right]$$

$$= \frac{B^2}{Tt_0^{\,2}}\left[\frac{T^3}{96} - \frac{tT^2}{8} + \frac{T}{2}\left\{t^2 + t_0^2\right\}\right]$$

$$= \frac{B^2}{t_0^2}\left[\frac{T^2}{96} - \frac{tT}{8} + \frac{t^2+t_0^2}{2}\right]$$

$$E[X^2(t)] = \frac{B^2}{2t_0^2}\left[\frac{T^2}{48} - \frac{tT}{4} + t^2 + t_0^2\right]$$

$$\sigma_X^2 = E[X^2(t)] - (E[X(t)])^2$$

$$\sigma_X^2 = \frac{B^2}{2t_0^2}\left[\frac{T^2}{48} - \frac{tT}{4} + t^2 + t_0^2\right] - \left(\frac{B}{2}\left[1 + \frac{T}{8t_0}\right]\right)^2$$

$$= \frac{B^2}{2t_0^2}\left[t^2 + t_0^2 + \frac{T^2}{48} - \frac{tT}{4} - \frac{t_0^2}{2} - \frac{T^2}{128} - \frac{t_0T}{8}\right]$$

$$\sigma_X^2 = \frac{B^2}{2t_0^2}\left[t^2 + \frac{5T^2}{384} - \frac{tT}{4} + \frac{t_0^2}{2} - \frac{t_0T}{8}\right]$$

(e) It is first order stationary since $E[X(t)]$ is constant.

Example 6.4: The autocorrelation function of a stationary RP $X(t)$ is given by $R_{XX}(\tau) = 25 + \frac{4}{1+6\tau^2}$. Find mean and variance.

Solution: According to autocorrelation property,

$$\lim_{\tau\to\infty} R_{XX}(\tau) = \bar{X}^2$$

$$\lim_{\tau\to\infty} R_{XX}(\tau) = \lim_{|\tau|\to\infty} 25 + \frac{4}{1+6\tau^2} = 25$$

We have also another autocorrelation property

$$R_{XX}(0) = E[X^2]$$

$$R_{XX}(0) = 25 + \frac{4}{1+0} = 29$$

$$\sigma_X^2 = E[X^2(t)] - (E[X(t)])^2$$

$$\sigma_X^2 = 29 - (5)^2 = 4.$$

Example 6.5: A random process is defined by $X(t) = A$, where A is a continuous random variable uniformly distributed in the interval (0, 1).
(a) Determine the form of sample function.
(b) Classify the process.
(c) Is it deterministic?

Solution: (a) Given random process $X(t) = A$, where A is a random variable uniformly distributed on the interval (0, 1).
Sample function is

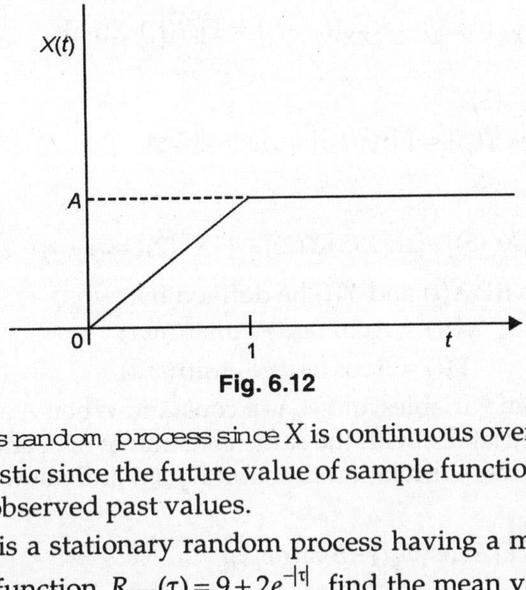

Fig. 6.12

(b) This is continuous random process since X is continuous over time t.
(c) It is nondeterministic since the future value of sample function cannot be predicted exactly from the observed past values.

Example 6.6: If $X(t)$ is a stationary random process having a mean value $E[X(t)] = 3$ and autocorrelation function $R_{XX}(\tau) = 9 + 2e^{-|\tau|}$, find the mean value and the variance of the random variable

$$Y = \int_0^2 X(t)\,dt$$

Solution: $E[X(t)] = 3$

$$R_{XX}(\tau) = 9 + 2e^{-|\tau|}$$

$$R_{XX}(0) = E[X^2(t)] = 9 + 2 = 11$$

$$\text{Mean} = E[Y(t)] = E\left[\int_0^2 X(t)\,dt\right] = \int_0^2 3\,dt = 3\times 2 = 6$$

$$\text{Variance } \sigma_Y^2 = E[Y^2(t)] - (E[Y(t)])^2$$

$$E[Y^2(t)] = E\left[\int_0^2 X(t)dt\int_0^2 X(t)dt\right] = \int_0^2\int_0^2 X^2(t)dt\,dt$$

$$= \int_0^2\int_0^2 11 dt\,dt = 44$$

$$\sigma_Y^2 = E[Y^2(t)] - (E[Y(t)])^2 = 44 - 36 = 8.$$

Example 6.7: If $X(t)$ is a WSS random process having autocorrelation function $R_{XX}(\tau) = Ae^{-\alpha|\tau|}$, find the second moment of the random variable $Y(t) = X(5) - X(2)$.

Solution: The second moment of $Y(t)$ is $E[Y^2(t)]$, i.e.

$$E[Y^2(t)] = E[(X(5) - X(2))^2]$$
$$= E[X^2(5) - 2X(5)\,X(2) + X^2(2)]$$
$$= E[X^2(5)] - 2E[X(5)\,X(2)] + E[X^2(2)]$$

We know that

$$R_{XX}(t_1, t_2) = R_{XX}(t_2, -t_1) = E[X(t_1)\,X(t_2)]$$

From the above

$$E[X(5)\,X(2)] = R_{XX}(3)$$
$$R_{XX}(0) = E[X^2(t_1)] = E[X^2(t_2)] = Ae^{-\alpha\times 0} = A$$
$$R_{XX}(3) = Ae^{-\alpha|3|}$$
$$E[Y^2(t)] = E[X^2(5)] - 2E[X(5)X(2)] + E[X^2(2)] = A + A - 2Ae^{-\alpha|3|} = 2\left(1 - e^{-\alpha|3|}\right)$$

Example 6.8: Let two RP $X(t)$ and $Y(t)$ be defined by

$$X(t) = A\cos(\omega_0 t) + B\sin(\omega_0 t)$$
$$Y(t) = B\cos(\omega_0 t) + A\sin(\omega_0 t)$$

where A, B are random variables and ω_0 is a constant. When A and B are uncorrelated zero mean random variables, with the same constraints of A and B, $Y(t)$ is wide sense stationary. Find the cross-correlation function $R_{XY}(\tau)$ and show that $X(t)$ and $Y(t)$ are jointly WSS.

Solution:

$$E[X(t)] = E[A\cos(\omega_0 t) + B\sin(\omega_0 t)]$$
$$= \cos(\omega_0 t)E[A] + \sin(\omega_0 t)[B]$$
$$= \cos(\omega_0 t)\cdot 0 + \sin(\omega_0 t)\cdot 0 = 0$$
$$E[Y(t)] = E[B\cos(\omega_0 t) - A\sin(\omega_0 t)]$$
$$= \cos(\omega_0 t)E[B] - \sin(\omega_0 t)E[A]$$
$$= \cos(\omega_0 t)\cdot 0 + \sin(\omega_0 t)\cdot 0 = 0$$
$$R_{XY}(\tau) = E[X(t)Y(t+\tau)]$$
$$= E[(A\cos(\omega_0 t) + B\sin(\omega_0 t))(B\cos(\omega_0 t + \omega_0\tau) - A\sin(\omega_0 t + \omega_0\tau))]$$
$$= E[AB\cos(\omega_0 t)\cos(\omega_0 t + \omega_0\tau) + B^2\sin(\omega_0 t)\cos(\omega_0 t + \omega_0\tau)$$
$$-A^2\cos(\omega_0 t)B\sin(\omega_0 t + \omega_0\tau) - AB\sin(\omega_0 t)\sin(\omega_0 t + \omega_0\tau)]$$

$$= E[AB\{\cos(\omega_0 t)\cos(\omega_0 t+\omega_0\tau)-\sin(\omega_0 t)\sin(\omega_0 t+\omega_0\tau)\}$$
$$+B^2\sin(\omega_0 t)\cos(\omega_0 t+\omega_0\tau)-A^2\cos(\omega_0 t)B\sin(\omega_0 t+\omega_0\tau)]$$
$$= E[AB\cos(\omega_0 t)\cos(2\omega_0 t+\omega_0\tau)$$
$$+B^2\sin(\omega_0 t)\cos(\omega_0 t+\omega_0\tau)-A^2\cos(\omega_0 t)B\sin(\omega_0 t+\omega_0\tau)]$$
$$= \cos(\omega_0 t)\cos(2\omega_0 t+\omega_0\tau)E[A]\,E[B]$$
$$+\sin(\omega_0 t)\cos(\omega_0 t+\omega_0\tau)E[B^2]-\cos(\omega_0 t)B\sin(\omega_0 t+\omega_0\tau)E[A^2]$$

A and B are uncorrelated and zero mean random variables.

$$E[AB] = E[A]\,E[B] = 0$$

$$R_{XY}(\tau) = 0+\sin(\omega_0 t)\cos(\omega_0 t+\omega_0\tau)E[B^2]-\cos(\omega_0 t)B\sin(\omega_0 t+\omega_0\tau)E[A^2]$$

We know that variance,

$$\sigma_X^2 = E[X^2]-(E[X])^2,$$
$$\sigma_Y^2 = E[Y^2]-(E[Y])^2$$

But $E[A] = 0$ and $E[B] = 0$

$$\sigma_X^2 = E[X^2],\ \sigma_Y^2 = E[Y^2]$$

Because all constraints are same, variance of $X(t)$ and $Y(t)$ is same.

Hence $\quad \sigma^2 = E[X^2] = E[Y^2]$

$$R_{XY}(\tau) = \sin(\omega_0 t)\cos(\omega_0 t+\omega_0\tau)\sigma^2-\cos(\omega_0 t)B\sin(\omega_0 t+\omega_0\tau)\sigma^2$$
$$R_{XY}(\tau) = \sigma^2[\sin(\omega_0 t)\cos(\omega_0 t+\omega_0\tau)-\cos(\omega_0 t)B\sin(\omega_0 t+\omega_0\tau)]$$
$$R_{XY}(\tau) = \sigma^2[\sin(-\omega_0\tau)]$$
$$R_{XY}(\tau) = -\sigma^2\sin(\omega_0\tau)$$

The mean and cross-correlation of $X(t)$ and $Y(t)$ are independent of time, so it is jointly wide sense stationary.

EXERCISE

1. $\bar{X}=6$ and $R_{XY}(t,t+\tau)=36+25\exp(-|\tau|)$ for a random process $X(t)$. Indicate which of the following statements are true based on what is known with certainty.
   (a) $X(t)$ is first order stationary
   (b) $X(t)$ has total average power of 61 W
   (c) $X(t)$ is ergodic
   (d) $X(t)$ is wide sense stationary
   (e) $X(t)$ has a periodic component
   (f) $X(t)$ has an AC power of 36 W.

   Hint: The DC power is given by $\overline{X(t)}^2$, AC power is given by $\sigma_X^2=\overline{X^2(t)}-\overline{X(t)}^2$ and total power is given by $\overline{X^2(t)}$.

2. A random process $Y(t) = X(t) - X(t + \tau)$ is defined in terms of a process $X(t)$ that is at least wide sense stationary.
   (a) Show that mean value of $Y(t)$ is 0 even if $X(t)$ has a nonzero mean value.
   (b) Show that $\sigma_Y^2 = 2[R_{XX}(0) + R_{XX}(\tau)]$.
   (c) If $Y(t) = X(t) + X(t + \tau)$, find $E[Y(t)]$ and σ_Y^2.
3. Find the time average mean and time autocorrelation function of the random process for the random process $X(t) = A\cos(\omega_0 + \theta)$, where A, ω_0 are constants and θ is a uniformly distributed random variable in the interval $(0, 2\pi)$.
4. An aircraft arrive at an airport according to a Poisson process at a rate of 12 per hour. All aircrafts are handled by one air traffic controller. If the controller takes a 2 minute coffee break, what is the probability that he will miss one or more arriving aircrafts?
5. Statistically independent zero mean random processes $X(t)$ and $Y(t)$ have auto-correlations functions $R_{XX}(\tau) = \exp(-|\tau|)$ and $R_{YY}(\tau) = \cos(2\pi\tau)$ respectively.
   (a) Find the autocorrelation function of the sum $W_1(t) = X(t) + Y(t)$.
   (b) Find the autocorrelation function of difference $W_2(t) = X(t) - Y(t)$.
   (c) Find the cross-correlation function of $W_1(t)$ and $W_2(t)$.
6. Let $X(t)$ be a stationary continuous random process that is differentiable. Denote its time derivative by $\overline{X}$.
   (a) Show that $E[\overline{X(t)}] = 0$.
   (b) Find $R_{\overline{X}\overline{X}}(\tau)$ in terms of $R_{XX}(\tau)$.
7. If the autocorrelation function of a stationary process is $R_{XX}(\tau) = 36 + \dfrac{4}{1+3\tau^2}$, find the mean and variance of the process.
8. A random process consists of three sample functions $x(t, s_1) = 2$, $x(t, s_2) = 2\cos t$ and $x(t, s_3) = 3\sin t$, each occurring with equal probability. Is the process stationary in any sense?
9. Let two random processes $X(t)$ and $Y(t)$ be defined by

$$X(t) = A\cos(\omega_0 t) + B\sin(\omega_0 t)$$
$$Y(t) = B\cos(\omega_0 t) - A\sin(\omega_0 t)$$

   where A and B are random variables and ω_0 is a constant. Assume A and B are uncorrelated, zero mean random variables with same variance. Find the cross-correlation function $R_{XY}(t, t + \tau)$ and show that $X(t)$ and $Y(t)$ are jointly wide sense stationary.

7 Random Processes: Spectral Characteristics

7.0 INTRODUCTION

All of the foregoing discussions concerning random processes have involved the time domain, that is, we have characterized processes by means of autocorrelation, cross-correlation and covariance functions without any consideration of spectral properties. It is well known, both time domain and frequency domain analysis methods exist for analyzing linear systems and deterministic waveforms. But what about random waveforms? In this chapter, we introduce the most important concepts that apply to characterizing random processes in the frequency domain.

The spectral description of a deterministic waveform is obtained by Fourier transforming the waveform, and the reader would be correct in concluding that Fourier transforms play an important role in the spectral characterization of random waveforms. However, the direct transformation approach is not attractive for random waveforms because the transform may not exist.

An appropriate spectrum to be associated with a random process is introduced below.

7.1 POWER DENSITY SPECTRUM AND ITS PROPERTIES

The spectral properties of a deterministic signal $x(t)$ are contained in its *Fourier transform* $X(\omega)$ given by

$$X(\omega) = F[x(t)] = \int_{-\infty}^{\infty} x(t)e^{-j\omega t}dt \qquad ...(7.1)$$

where $e^{-j\omega t} = \cos \omega t + j \sin \omega t$ is the complex exponential.

The Fourier transform $X(\omega)$ exists if $x(t)$ satisfies the following conditions:

1. $x(t)$ is absolutely integrable, that is,

$$\int_{-\infty}^{\infty} |x(t)|dt < \infty$$

2. $x(t)$ has only a finite number of discontinuties in any finite interval.
3. $x(t)$ has only finite number of maxima and minima within any finite interval.

The signal $x(t)$ can be obtained from $X(\omega)$ by the inverse Fourier transform (IFT) as follows:

$$x(t) = F^{-1}[X(\omega)] = \frac{1}{2\pi}\int_{-\infty}^{\infty} X(\omega)e^{j\omega t}d\omega \quad ...(7.2)$$

The existence of the inverse Fourier transform implies that we can represent a function $x(t)$ as a superposition of continuum of complex sinusoids. The Fourier transform $X(\omega)$ is the strength of the sinusoid of frequency ω present in the signal. If $x(t)$ is a voltage signal measured in volt, $X(\omega)$ has the unit of volt/radian. The function $X(\omega)$ is also called the spectrum of $x(t)$.

7.1.1 The Power Density Spectrum

For a random process $X(t)$, let $x_T(t)$ be defined as that portion of a sample function $x(t)$ that exists between $-T$ and T, i.e.

$$X_T(t) = \begin{cases} x(t) & -T < t < T \\ 0 & \text{otherwise} \end{cases} \quad ...(7.3)$$

The energy contained $x(t)$ in the interval is $(-T, T)$ is

$$E(T) = \int_{-T}^{T} x_T^2(t)dt = \int_{-T}^{T} x^2(t)dt \quad ...(7.4)$$

As long as T is finite, we presume that $x_T(t)$ has bounded variation. Hence $x_T(t)$ is Fourier transformable, its energy must also be related to $X_T(\omega)$ by Parseval's theorem. Thus,

$$E(T) = \int_{-T}^{T} x^2(t)dt = \frac{1}{2\pi}\int_{-\infty}^{\infty} |X_T(\omega)|^2 d\omega \quad ...(7.5)$$

By dividing the above expression by $2T$, we obtain the average power in $x(t)$ over the interval $(-T, T)$

$$P(T) = \frac{E(T)}{2T} = \frac{1}{2T}\int_{-T}^{T} x^2(t)dt = \frac{1}{2\pi}\int_{-\infty}^{\infty} \frac{|X_T(\omega)|^2}{2T}d\omega \quad ...(7.6)$$

Here $P(T)$ is actually a random variable with respect to random process. By taking the limit as $T \to \infty$ and expected value of Eq. (7.6) to obtain a suitable power density spectrum for the random process.

$$P_{XX} = \lim_{T\to\infty}\frac{1}{2T}\int_{-T}^{T} E[x^2(t)]dt = \lim_{T\to\infty}\frac{1}{2\pi}\int_{-\infty}^{\infty} \frac{E\left[|X_T(\omega)|^2\right]}{2T}d\omega$$

$$P_{XX} = \lim_{T\to\infty}\frac{1}{2T}\int_{-T}^{T} E[x^2(t)]dt = \frac{1}{2\pi}\int_{-\infty}^{\infty} \lim_{T\to\infty}\frac{E\left[|X_T(\omega)|^2\right]}{2T}d\omega \quad ...(7.7)$$

From the above Eq. (7.7), average power P_{XX} in random process $X(t)$ is given by the time average of its second moment

$$P_{XX} = \lim_{T\to\infty}\frac{1}{2T}\int_{-T}^{T} E[x^2(t)]dt = A\{E[X^2(t)]\} \quad ...(7.8)$$

From the above Eq. (7.7), average power P_{XX} in random process $X(t)$ can be obtained by frequency domain integration. If we define the *power density spectrum* for the random process by

$$\mathcal{S}_{XX}(\omega) = \lim_{T\to\infty} \frac{E\left[\left|X_T(\omega)\right|^2\right]}{2T} \qquad \text{...(7.9)}$$

then from Eqs. (7.8) and (7.9),

$$\text{Power } P_{XX} = \lim_{T\to\infty} \frac{1}{2T} \int_{-T}^{T} E[x^2(t)]dt = \frac{1}{2\pi} \int_{-\infty}^{\infty} \mathcal{S}_{XX}(\omega) d\omega \qquad \text{...(7.10)}$$

Properties of the Power Density Spectrum

i. $\mathcal{S}_{XX}(\omega) \geq 0$

Proof: Power density spectrum is defined as

$$\mathcal{S}_{XX}(\omega) = \lim_{T\to\infty} \frac{E\left[\left|X_T(\omega)\right|^2\right]}{2T} \qquad \text{...(7.11a)}$$

We know that expectation of a positive function is positive. Thus, right hand side of Eq. (7.11a) is positive or equal to the zero. Hence, power density spectrum is always greater or equal to zero.

ii. $\frac{1}{2\pi} \int_{-\infty}^{\infty} \mathcal{S}_{XX}(\omega) d\omega = A\{E[X^2(t)]\}$

Proof: From Eq. (7.10),

$$P_{XX} = \lim_{T\to\infty} \frac{1}{2T} \int_{-T}^{T} E[x^2(t)]dt = \frac{1}{2\pi} \int_{-\infty}^{\infty} \mathcal{S}_{XX}(\omega) d\omega \qquad \text{...(7.11b)}$$

But we have the time averaged autocorrelation function as

$$\lim_{T\to\infty} \frac{1}{2T} \int_{-T}^{T} E[x^2(t)]dt = A\{E[X^2(t)]\} \qquad \text{...(7.11c)}$$

From Eqs. (7.11b) and (7.11c),

$$\frac{1}{2\pi} \int_{-\infty}^{\infty} \mathcal{S}_{XX}(\omega) d\omega = A\{E[X^2(t)]\}$$

iii. $\mathcal{S}_{XX}(\omega)$ is real

Proof: We know that

$$\mathcal{S}_{XX}(\omega) = \lim_{T\to\infty} \frac{E\left[\left|X_T(\omega)\right|^2\right]}{2T}$$

where $|X_T(\omega)|^2 = X_T(\omega)X_T^*(\omega)$, is always real because the product of any function and its conjugate is always real and expectation of a real function is real only.

Hence $\mathcal{S}_{XX}(\omega)$ is real.

iv. $A[R_{XX}(t,t+\tau)] = \frac{1}{2\pi}\int_{-\infty}^{\infty} \mathcal{S}_{XX}(\omega)e^{j\omega\tau}d\omega$

$$\mathcal{S}_{XX}(\omega) = \int_{-\infty}^{\infty} A[R_{XX}(t,t+\tau)]e^{-j\omega\tau}d\tau$$

Proof: The above relation will be proved in the next section, which is called Wiener–Khintchine relation.

If $X(t)$ is at least wide sense stationary, i.e. $A[R_{XX}(t,t+\tau)] = R_{XX}(\tau)$

$$\mathcal{S}_{XX}(\omega) = \int_{-\infty}^{\infty} R_{XX}(\tau)e^{-j\omega\tau}d\tau \quad \text{...(7.11d)}$$

$$R_{XX}(\tau) = \frac{1}{2\pi}\int_{-\infty}^{\infty} \mathcal{S}_{XX}(\omega)e^{j\omega\tau}d\omega \quad \text{...(7.11e)}$$

v. The zero frequency value of the power density spectrum of a WSS random process is equal to the total area under the graph of autocorrelation function, i.e.

$$\mathcal{S}_{XX}(0) = \int_{-\infty}^{\infty} R_{XX}(\tau)d\tau$$

Proof: If the random process is WSS, then we have

$$\mathcal{S}_{XX}(\omega) = \int_{-\infty}^{\infty} R_{XX}(\tau)e^{-j\omega\tau}d\tau$$

Substitute $\omega = 0$.

Then
$$\mathcal{S}_{XX}(0) = \int_{-\infty}^{\infty} R_{XX}(\tau)e^{-j0\cdot\tau}d\tau$$

$\therefore$
$$\mathcal{S}_{XX}(0) = \int_{-\infty}^{\infty} R_{XX}(\tau)d\tau$$

vi. $\mathcal{S}_{XX}(\omega) = \mathcal{S}_{XX}(-\omega)$; $X(t)$ is real.

Proof:

First method: From Eq. (7.11c), we have

$$\mathcal{S}_{XX}(\omega) = \int_{-\infty}^{\infty} R_{XX}(\tau)e^{-j\omega\tau}d\tau$$

By substituting $\omega = -\omega$,

$$\mathcal{S}_{XX}(-\omega) = \int_{-\infty}^{\infty} R_{XX}(\tau)e^{j\omega\tau}d\tau$$

But we have $R_{XX}(\tau) = R_{XX}(-\tau)$, i.e. autocorrelation is a function of $-\tau$.
The above equation can be written as

$$\mathscr{S}_{XX}(-\omega) = \int_{-\infty}^{\infty} R_{XX}(-\tau)e^{-j\omega(-\tau)}d\tau = \int_{-\infty}^{\infty} R_{XX}(\tau)e^{-j\omega(\tau)}d\tau$$

$$\therefore \quad \mathscr{S}_{XX}(\omega) = \mathscr{S}_{XX}(-\omega).$$

Second method: From the definition of PSD,

$$\mathscr{S}_{XX}(\omega) = \lim_{T\to\infty}\frac{E\left[|X_T(\omega)|^2\right]}{2T}$$

where $X_T(\omega)$ is Fourier transform of $x_T(t)$, and is a real function.

$$X_T(\omega) = \int_{-\infty}^{\infty} x_T(t)e^{-j\omega t}dt$$

$$\mathscr{S}_{XX}(\omega) = \lim_{T\to\infty}\frac{E\left[|X_T(\omega)|^2\right]}{2T} = \lim_{T\to\infty}\frac{E\left[X_T^*(\omega)X_T(\omega)\right]}{2T}$$

Substitute $\omega = -\omega$,

$$X_T(-\omega) = \int_{-\infty}^{\infty} x_T(t)e^{j\omega t}dt = X_T^*(\omega)$$

$$|X_T(\omega)|^2 = X_T^*(\omega)X_T(\omega) = X_T(\omega)X_T^*(\omega)$$

$$\mathscr{S}_{XX}(-\omega) = \lim_{T\to\infty}\frac{E\left[|X_T(-\omega)|^2\right]}{2T} = \lim_{T\to\infty}\frac{E\left[X_T(\omega)\,X_T^*(\omega)\right]}{2T} = \mathscr{S}_{XX}(\omega)$$

$$\therefore \quad \mathscr{S}_{XX}(\omega) = \mathscr{S}_{XX}(-\omega).$$

vii. $\mathscr{S}_{\overline{X}\overline{X}}(\omega) = \omega^2\mathscr{S}_{XX}(\omega)$, $\overline{X}$ denotes differentiation of X.

Proof:
First method: From Eq. (7.11c), we have

$$\mathscr{S}_{XX}(\omega) = \int_{-\infty}^{\infty} R_{XX}(\tau)e^{-j\omega\tau}d\tau$$

$$\mathscr{S}_{\overline{X}\overline{X}}(\omega) = \int_{-\infty}^{\infty} \frac{d^2R_{XX}(\tau)}{d\tau^2}e^{-j\omega\tau}d\tau$$

And we have the property,

$$F\left(\frac{d^n[x(t)]}{dt^n}\right) = (-j\omega)^n X(\omega)$$

$$\mathcal{S}_{\overline{XX}}(\omega) = \int_{-\infty}^{\infty} (-j\omega)(j\omega) R_{XX}(\tau) e^{-j\omega\tau} d\tau \text{ (because one is complex conjugate of other)}$$

$$\mathcal{S}_{\overline{XX}}(\omega) = \int_{-\infty}^{\infty} (\omega)^2 R_{XX}(\tau) e^{-j\omega\tau} d\tau = \omega^2 \int_{-\infty}^{\infty} R_{XX}(\tau) e^{-j\omega\tau} d\tau$$

$$\therefore \quad \mathcal{S}_{\overline{XX}}(\omega) = \omega^2 \mathcal{S}_{XX}(\omega).$$

Second method: From $\mathcal{S}_{XX}(\omega) = \lim_{T\to\infty} \frac{E\left[|X_T(\omega)|^2\right]}{2T}$

where $X_T(\omega)$ is Fourier transform of $X_T(t)$.
And we have the property,

$$F\left(\frac{d^n[x(t)]}{dt^n}\right) = (-j\omega)^n X(\omega)$$

From that

$$F\left(\frac{d[x_T(t)]}{dt}\right) = (-j\omega) X_T(\omega)$$

$$F\left[\frac{d}{dt}\left(\frac{d[x_T(t)]}{dt}\right)\right] = (-j\omega)(j\omega) X_T(\omega)$$

But $|X_T(\omega)|^2 = X_T^*(\omega) X_T(\omega) = X_T(\omega) X_T^*(\omega)$

$$\mathcal{S}_{\overline{XX}}(\omega) = \lim_{T\to\infty} \frac{E[X_T^*(\omega) X_T(\omega)]}{2T} = \lim_{T\to\infty} \frac{E[(-j\omega)(j\omega)|X_T(\omega)|^2]}{2T}$$

$$= \lim_{T\to\infty} \frac{\omega^2 E\left[|X_T(\omega)|^2\right]}{2T} = \omega^2 \mathcal{S}_{XX}(\omega)$$

$$\therefore \quad \mathcal{S}_{\overline{XX}}(\omega) = \omega^2 \mathcal{S}_{XX}(\omega).$$

7.2 RELATION BETWEEN THE AUTOCORRELATION FUNCTION AND PSD: WIENER-KHINTCHINE RELATION

Statement: The inverse Fourier transform of the power density spectrum is the time average of the autocorrelation function, i.e.

$$\frac{1}{2\pi} \int_{-\infty}^{\infty} \mathcal{S}_{XX}(\omega) e^{j\omega\tau} d\omega = A[R_{XX}(t, t+\tau)] \qquad \text{...(7.12)}$$

The definition of power spectrum density is

$$\mathcal{S}_{XX}(\omega) = \lim_{T\to\infty} \frac{E\left[|X_T(\omega)|^2\right]}{2T}$$

where $X_T(\omega)$ is Fourier transform of $x_T(t)$.
$X_T(\omega)$ is defined as

$$X_T(\omega) = \int_{-T}^{T} x_T(t)e^{-j\omega t}dt$$

$$\mathscr{S}_{XX}(\omega) = \lim_{T\to\infty}\frac{E\left[\left|X_T(\omega)\right|^2\right]}{2T} = \lim_{T\to\infty}\frac{1}{2T}E\left[\int_{-T}^{T} x_T(t_1)e^{j\omega t_1}dt_1\int_{-T}^{T} x_T(t_2)e^{-j\omega t_2}dt_2\right]$$

$$= \lim_{T\to\infty}\frac{1}{2T}E\left[\int_{-T}^{T}\int_{-T}^{T} x_T(t_1)x_T(t_2)e^{j\omega t_1}e^{-j\omega t_2}dt_2dt_1\right]$$

$$= \lim_{T\to\infty}\frac{1}{2T}\int_{-T}^{T}\int_{-T}^{T} E[x_T(t_1)x_T(t_2)]e^{-j\omega(t_2-t_1)}dt_2dt_1 \qquad ...(7.13)$$

$$E[x_T(t_1)x_T(t_2)] = R_{XX}(t_1, t_2) \qquad -T < (t_1 \text{ and } t_2) < T \qquad ...(7.14)$$

Thus, Eq. (7.13) becomes

$$\mathscr{S}_{XX}(\omega) = \lim_{T\to\infty}\frac{1}{2T}\int_{-T}^{T}\int_{-T}^{T} R_{XX}(t_1,t_2)e^{-j\omega(t_2-t_1)}dt_2\,dt_1 \qquad ...(7.15)$$

Substitute Eq. (7.15) on left side of Eq. (7.12).
∴ Equation (7.12) becomes

$$\frac{1}{2\pi}\int_{-\infty}^{\infty}\mathscr{S}_{XX}(\omega)e^{j\omega\tau}d\omega = \frac{1}{2\pi}\int_{-\infty}^{\infty}\lim_{T\to\infty}\frac{1}{2T}\int_{-T}^{T}\int_{-T}^{T} R_{XX}(t_1,t_2)e^{-j\omega(t_2-t_1)}e^{j\omega\tau}dt_2\,dt_1\,d\omega$$

$$= \frac{1}{2\pi}\int_{-\infty}^{\infty}\lim_{T\to\infty}\frac{1}{2T}\int_{-T}^{T}\int_{-T}^{T} R_{XX}(t_1,t_2)e^{-j\omega(t_2-t_1-\tau)}d\omega\,dt_2\,dt_1$$

Now exchange the integrations,

$$\frac{1}{2\pi}\int_{-\infty}^{\infty}\mathscr{S}_{XX}(\omega)e^{j\omega\tau}d\omega = \lim_{T\to\infty}\frac{1}{2T}\int_{-T}^{T}\int_{-T}^{T} R_{XX}(t_1,t_2)\frac{1}{2\pi}\left(\int_{-\infty}^{\infty} e^{j\omega\,(\tau+t_1-t_2)}d\omega\right)dt_2\,dt_1 \qquad ...(7.16)$$

But we have $\frac{1}{2\pi}\int_{-\infty}^{\infty} e^{j\omega x}d\omega = \delta(x)$ and $\frac{1}{2\pi}\int_{-\infty}^{\infty} e^{-j\omega x}dx = \delta(\omega)$

Substitute the above relation in Eq. (7.16)

$$\frac{1}{2\pi}\int_{-\infty}^{\infty}\mathscr{S}_{XX}(\omega)e^{j\omega\tau}d\omega = \lim_{T\to\infty}\frac{1}{2T}\int_{-T}^{T}\int_{-T}^{T} R_{XX}(t_1,t_2)\delta(\tau+t_1-t_2)dt_2\,dt_1 \qquad ...(7.17)$$

It can be written as $\delta(\tau + t_1 - t_2) = \delta(t_2 - \tau - t_1)$

$$\frac{1}{2\pi}\int_{-\infty}^{\infty} \mathcal{S}_{XX}(\omega)e^{j\omega\tau}d\omega = \lim_{T\to\infty}\frac{1}{2T}\int_{-T}^{T}\int_{-T}^{T} R_{XX}(t_1,t_2)\delta(t_2-\tau-t_1)dt_2\,dt_1 \quad ...(7.18)$$

The impulse function property

$$\int_{-\infty}^{\infty} \phi(x)\delta(x-x_0)dx = \phi(x_0) \quad ...(7.19)$$

Compare Eqs. (7.18) and (7.19),

$$\phi(x) = R_{XX}(t_1,t_2),\ \delta(x-x_0)= \delta(t_2-\tau-t_1),\ x = t_2 \text{ and } x_0 = \tau + t_1$$

Equation (7.18) becomes

$$\frac{1}{2\pi}\int_{-\infty}^{\infty} \mathcal{S}_{XX}(\omega)e^{j\omega\tau}d\omega = \lim_{T\to\infty}\frac{1}{2T}\int_{-T}^{T} R_{XX}(t_1,t_1+\tau)dt_1 \quad ...(7.20)$$

Substitute $t_1 = t$, Eq. (7.20) becomes

$$\frac{1}{2\pi}\int_{-\infty}^{\infty} \mathcal{S}_{XX}(\omega)e^{j\omega\tau}d\omega = \lim_{T\to\infty}\frac{1}{2T}\int_{-T}^{T} R_{XX}(t,t+\tau)dt \quad ...(7.21)$$

But time averaged autocorrelation is

$$A[R_{XX}(t,t+\tau)] = \lim_{T\to\infty}\frac{1}{2T}\int_{-T}^{T} R_{XX}(t,t+\tau)dt \quad ...(7.22)$$

From Eqs. (7.21) and (7.22),

$$\therefore \quad \frac{1}{2\pi}\int_{-\infty}^{\infty} \mathcal{S}_{XX}(\omega)e^{j\omega\tau}d\omega = A[R_{XX}(t,t+\tau)]$$

If the random process $X(t)$ is WSS, i.e. $A[R_{XX}(t, t + \tau)] = R_{XX}(\tau)$, then

$$\mathcal{S}_{XX}(\omega) = \int_{-\infty}^{\infty} R_{XX}(\tau)e^{-j\omega\tau}d\tau \quad ...(7.23)$$

and

$$R_{XX}(\tau) = \frac{1}{2\pi}\int_{-\infty}^{\infty} \mathcal{S}_{XX}(\omega)e^{j\omega\tau}d\omega \quad ...(7.24)$$

Example 7.2.1: The autocorrelation function of a WSS process $X(t)$ is given by

$$R_{XX}(\tau) = a^2e^{-b|\tau|} \qquad b > 0$$

Find the power spectral density of the process.

Solution: $\mathcal{S}_{XX}(\omega) = \int_{-\infty}^{\infty} R_{XX}(\tau)e^{-j\omega\tau}d\tau$

$$= \int_{-\infty}^{\infty} a^2e^{-b|\tau|}e^{-j\omega\tau}d\tau$$

$$= \int_{-\infty}^{0} a^2 e^{b\tau} e^{-j\omega\tau} d\tau + \int_{0}^{\infty} a^2 e^{-b\tau} e^{-j\omega\tau} d\tau$$

$$= \int_{-\infty}^{0} a^2 e^{-(j\omega - b)\tau} d\tau + \int_{0}^{\infty} a^2 e^{-(j\omega + b)\tau} d\tau$$

$$= a^2 \left. \frac{e^{-(j\omega - b)\tau}}{-(j\omega - b)} \right|_{-\infty}^{0} + a^2 \left. \frac{e^{-(j\omega + b)\tau}}{-(j\omega + b)} \right|_{0}^{\infty}$$

$$= \frac{a^2}{b - j\omega} + \frac{a^2}{b + j\omega}$$

$$= \frac{2a^2 b}{b^2 + \omega^2}.$$

The autocorrelation function and the PSD are shown in Fig. 7.1.

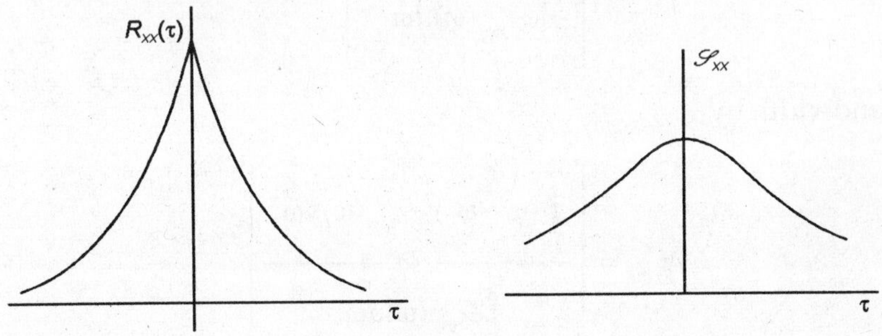

Fig. 7.1

7.3 BANDWIDTH OF POWER DENSITY SPECTRUM

Except for the fact that area of $\mathscr{S}_{XX}(\omega)$ is not necessarily unity, $\mathscr{S}_{XX}(\omega)$ has characteristics similar to probability density function $f_X(\omega)$. By dividing $\mathscr{S}_{XX}(\omega)$ by its area, we get a new function similar to probability density function (PDF), i.e.

$$f_X(\omega) = \frac{\mathscr{S}_{XX}(\omega)}{\int_{-\infty}^{\infty} \mathscr{S}_{XX}(\omega)\, d\omega} \quad \text{...(7.25)}$$

$f_X(\omega) \geq 0$ for all ω and the total area under the function $f_X(\omega)$ is unity. Hence, normalized form of power spectral density behaves similar to a PDF. The normalized power spectrum is a measure of its "spread" and this spread is called "rms bandwidth" (root mean squared bandwidth). It is denoted by W_{rms} and its unit is rad/s.

The rms bandwidth for a WSS random process $X(t)$ is defined as

$$W_{rms} = \left[\int_{-\infty}^{\infty} \omega^2 f_X(\omega)\, d\omega \right]^{1/2}$$

$$W_{rms} = \left[\frac{\int_{-\infty}^{\infty} \omega^2 \mathcal{S}_{XX}(\omega)\, d\omega}{\int_{-\infty}^{\infty} \mathcal{S}_{XX}(\omega)\, d\omega}\right]^{1/2} \qquad \text{...(7.26)}$$

This is valid only for low pass random process $X(t)$, i.e. its spectral components are clustered near $\omega = 0$. Such processes are also called *low pass processes*. The above concept readily extend to a process that has a bandpass form of power spectrum, i.e. its significant spectral components cluster near some frequencies $\bar{\omega}_0$ and $-\bar{\omega}_0$. If we assume that the process $X(t)$ is real, $\mathcal{S}_{XX}(\omega)$ will be real and have even symmetry about $\omega = 0$. With this assumption, we define a *mean frequency* $\bar{\omega}_0$ by

$$\bar{\omega}_0 = \left[\frac{\int_{-\infty}^{\infty} \omega \mathcal{S}_{XX}(\omega)\, d\omega}{\int_{0}^{\infty} \mathcal{S}_{XX}(\omega)\, d\omega}\right]^{1/2} \qquad \text{...(7.27)}$$

and rms bandwidth by

$$W_{rms} = \left[\frac{\int_{-\infty}^{\infty} (\omega - \omega_0)^2 \mathcal{S}_{XX}(\omega)\, d\omega}{\int_{0}^{\infty} \mathcal{S}_{XX}(\omega)\, d\omega}\right]^{1/2} \qquad \text{...(7.28)}$$

7.4 CROSS POWER SPECTRUM DENSITY AND ITS PROPERTIES

Consider a random process $Z(t)$ which is sum of two real jointly WSS random processes $X(t)$ and $Y(t)$, i.e. $Z(t) = X(t) + Y(t)$ ...(7.29)

$$R_{ZZ}(\tau) = R_{XX}(\tau) + R_{XY}(\tau) + R_{YX}(\tau) + R_{YY}(\tau) \qquad \text{...(7.30)}$$

If we take the Fourier transform of both sides,

$$F[R_{ZZ}(\tau)] = F[R_{XX}(\tau)] + F[R_{XY}(\tau)] + F[R_{YX}(\tau)] + F[R_{YY}(\tau)]$$

$$\mathcal{S}_{ZZ}(\omega) = \mathcal{S}_{XX}(\omega) + \mathcal{S}_{XY}(\omega) + \mathcal{S}_{YX}(\omega) + \mathcal{S}_{YY}(\omega) \qquad \text{...(7.31)}$$

where $F(\cdot)$ stands for the Fourier transform.

Thus, we see that $\mathcal{S}_{ZZ}(\omega)$ includes contribution from Fourier transform of the cross-correlation functions $R_{XY}(\tau)$ and $R_{YX}(\tau)$. These Fourier transforms represent *cross power spectral densities*.

7.4.1 Cross Power Spectrum Density

For two random processes $X(t)$ and $Y(t)$, let $x_T(t)$ and $y_T(t)$ be defined as that portion of sample functions $x(t)$ and $y(t)$ that exists between $-T$ and T, i.e.

$$x_T(t) = \begin{cases} x(t) & -T < t < T \\ 0 & \text{otherwise} \end{cases} \qquad ...(7.32)$$

and
$$y_T(t) = \begin{cases} y(t) & -T < t < T \\ 0 & \text{otherwise} \end{cases} \qquad ...(7.33)$$

As long as T is finite, we presume that $x_T(t)$ and $y_T(t)$ has bounded variation. Hence, $x_T(t)$ and $y_T(t)$ are Fourier transformable, its power must also be related to $X_T(\omega)$ and $Y_T(\omega)$ by Parseval's theorem. Thus,

$$P(T) = \frac{1}{2T}\int_{-T}^{T} x_T(t)y_T(t)dt = \frac{1}{2T}\int_{-T}^{T} x(t)y(t)dt \qquad ...(7.34)$$

Here $P(T)$ is actually a random variable with respect to random process. By taking the limit as $T \to \infty$ and expected value of Eq. (7.34) to obtain a suitable power density spectrum for the random process, we get

$$P_{XY} = \lim_{T\to\infty}\frac{1}{2T}\int_{-T}^{T} E[x(t)y(t)]dt = \lim_{T\to\infty}\frac{1}{2\pi}\int_{-\infty}^{\infty}\frac{E\left[X_T^*(\omega)Y_T(\omega)\right]}{2T}d\omega$$

$$P_{XY} = \lim_{T\to\infty}\frac{1}{2T}\int_{-T}^{T} E[x(t)y(t)]dt = \frac{1}{2\pi}\int_{-\infty}^{\infty}\lim_{T\to\infty}\frac{E\left[X_T^*(\omega)Y_T(\omega)\right]}{2T}d\omega \qquad ...(7.35)$$

From the above Eq. (7.35), average power P_{XX} in random process $X(t)$ is given by the time average of its second moment:

$$P_{XY} = \lim_{T\to\infty}\frac{1}{2T}\int_{-T}^{T} E[x(t)y(t)]dt = A\{E[x(t)y(t)]\} = \lim_{T\to\infty}\frac{1}{2T}\int_{-T}^{T} R_{XY}(t,t)dt \qquad ...(7.36)$$

From the above Eq. (7.36), average power P_{XX} in random process $X(t)$ can be obtained by frequency domain integration. If we define the *power density spectrum* for the random process by

$$\mathcal{S}_{XY}(\omega) = \lim_{T\to\infty}\frac{E\left[X_T^*(\omega)Y_T(\omega)\right]}{2T} \qquad ...(7.37)$$

then from Eqs. (7.36) and (7.37),

$$\text{Power } P_{XY} = \lim_{T\to\infty}\frac{1}{2T}\int_{-T}^{T} E[x(t)y(t)]dt = \frac{1}{2\pi}\int_{-\infty}^{\infty}\mathcal{S}_{XY}(\omega)d\omega. \qquad ...(7.38)$$

7.4.2 Properties of Cross Power Density Spectrum

The CPSD is a complex function of the frequency ω. Some properties of the CPSD of two jointly WSS processes $X(t)$ and $Y(t)$ are listed below:

i. $\mathcal{S}_{XY}(\omega) = \mathcal{S}_{YX}(-\omega) = \mathcal{S}_{YX}^*(\omega)$

Proof: $\mathscr{S}_{XY}(\omega) = \int_{-\infty}^{\infty} R_{XY}(\tau)e^{-j\omega\tau}d\tau$

But we have $R_{XY}(\tau) = R_{XY}(-\tau)$, i.e. autocorrelation is a function of $-\tau$.
The above equation can be written as

$$\mathscr{S}_{XY}(\omega) = \int_{-\infty}^{\infty} R_{YX}(-\tau)e^{j\omega(-\tau)}d\tau = \int_{-\infty}^{\infty} R_{YX}(\tau)e^{j\omega(\tau)}d\tau = \mathscr{S}^*_{YX}(\omega) \quad \text{...(7.39)}$$

$$\mathscr{S}_{XY}(\omega) = \int_{-\infty}^{\infty} R_{YX}(\tau)e^{-j\omega\tau}d\tau$$

But we have $R_{YX}(\tau) = R_{YX}(-\tau)$, i.e. autocorrelation is a function of $-\tau$.
The above equation can be written as

$$\mathscr{S}_{YX}(\omega) = \int_{-\infty}^{\infty} R_{YX}(-\tau)e^{j\omega(-\tau)}d\tau = \int_{-\infty}^{\infty} R_{XY}(\tau)e^{j\omega(\tau)}d\tau$$

By substituting $\omega = -\omega$,

$$\mathscr{S}_{YX}(-\omega) = \int_{-\infty}^{\infty} R_{XY}(\tau)e^{-j\omega\tau}d\tau = \mathscr{S}_{XY}(\omega) \quad \text{...(7.40)}$$

ii. $\text{Re}(\mathscr{S}_{XY}(\omega))$ is an even function of ω and $\text{Img}(\mathscr{S}_{XY}(\omega))$ is an odd function of ω.

We have $\quad \mathscr{S}_{XY}(\omega) = \int_{-\infty}^{\infty} R_{XY}(\tau)e^{-j\omega\tau}d\tau$

$$\mathscr{S}_{XY}(\omega) = \int_{-\infty}^{\infty} R_{XY}(\tau)(\cos\omega\tau + j\sin\omega\tau)d\tau$$

$$\mathscr{S}_{XY}(\omega) = \int_{-\infty}^{\infty} R_{XY}(\tau)\cos\omega\tau\, d\tau - \int_{-\infty}^{\infty} R_{XY}(\tau)\, j\sin\omega\tau\, d\tau$$

$$\mathscr{S}_{XY}(\omega) = \text{Re}(\mathscr{S}_{XY}(\omega)) - \text{Img}(\mathscr{S}_{XY}(\omega))$$

where $\text{Re}(\mathscr{S}_{XY}(\omega))$ is $\int_{-\infty}^{\infty} R_{XY}(\tau)\cos\omega\tau\, d\tau$.

By substituting $\omega = -\omega$,

$$\mathscr{S}_{XY}(-\omega) = \int_{-\infty}^{\infty} R_{XY}(\tau)\cos(-\omega)\tau\, d\tau = \int_{-\infty}^{\infty} R_{XY}(\tau)\cos\omega\tau\, d\tau = \text{Re}(\mathscr{S}_{XY}(\omega))$$

$\therefore$ $\text{Re}(\mathscr{S}_{XY}(\omega))$ is an even function of ω.

$\text{Img}(\mathscr{S}_{XY}(\omega))$ is $-\int_{-\infty}^{\infty} R_{XY}(\tau)\sin\omega\tau\, d\tau$

By substituting $\omega = -\omega$,

$$\mathcal{S}_{XY}(\omega) = -\int_{-\infty}^{\infty} R_{XY}(\tau)\sin(-\omega)\tau\, d\tau = \int_{-\infty}^{\infty} R_{XY}(\tau)\sin\omega\tau\, d\tau = -\text{Img}(\mathcal{S}_{XY}(\omega))$$

$\therefore$ $\text{Img}(\mathcal{S}_{XY}(\omega))$ is an odd function of ω.

iii. If $X(t)$ and $Y(t)$ are uncorrelated and have constant means $\overline{X}$ and $\overline{Y}$, then

$\mathcal{S}_{XY}(\omega) = \mathcal{S}_{YX}(\omega) = \overline{X}\,\overline{Y}\,\delta(\omega)$, where $\delta(\omega)$ is the dirac delta function.

Proof: Observe that

$$\begin{aligned} R_{XY}(\tau) &= E[x(t)y(t+\tau)] \\ &= E[x(t)]E[y(t+\tau)] \\ &= \overline{X}\,\overline{Y} \\ &= R_{YX}(\tau) \end{aligned}$$

$$\begin{aligned} \mathcal{S}_{XY}(\omega) &= \int_{-\infty}^{\infty} R_{XY}(\tau)e^{-j\omega\tau}d\tau = \int_{-\infty}^{\infty} \overline{X}\,\overline{Y}e^{-j\omega\tau}d\tau \\ &= \overline{X}\,\overline{Y}\delta(\omega) \qquad \because F(1) = \delta(\omega) \end{aligned}$$

Similarly, $$\mathcal{S}_{YX}(\omega) = \overline{X}\,\overline{Y}\,\delta(\omega)$$

$\therefore$ $$\mathcal{S}_{XY}(\omega) = \mathcal{S}_{YX}(\omega) = \overline{X}\,\overline{Y}\,\delta(\omega).$$

iv. If $X(t)$ and $Y(t)$ are orthogonal, then $\mathcal{S}_{XY}(\omega) = \mathcal{S}_{YX}(\omega) = 0$.

Proof: If $X(t)$ and $Y(t)$ are orthogonal, we have

$$R_{XY}(\tau) = E[x(t)y(t+\tau)] = 0$$

Similarly, $$R_{YX}(\tau) = E[x(t+\tau)\,y(t)] = 0$$

$\therefore$ $$\mathcal{S}_{XY}(\omega) = \mathcal{S}_{YX}(\omega) = 0.$$

v. $$A[R_{XY}(t,t+\tau)] = \frac{1}{2\pi}\int_{-\infty}^{\infty} \mathcal{S}_{XY}(\omega)e^{j\omega\tau}d\omega$$

$$\mathcal{S}_{XY}(\omega) = \int_{-\infty}^{\infty} A[R_{XY}(t,t+\tau)]e^{-j\omega\tau}d\tau$$

Proof: The above relation will be proved in the next section, which is called Wiener–Khintchine relation.

If $X(t)$ is at least wide sense stationary, i.e.

$$A[R_{XX}(t,t+\tau)] = R_{XX}(\tau)$$

then $$\mathcal{S}_{XY}(\omega) = \int_{-\infty}^{\infty} R_{XY}(\tau)e^{-j\omega\tau}d\tau \qquad ...(7.41)$$

$$R_{XY}(\tau) = \frac{1}{2\pi}\int_{-\infty}^{\infty} \mathcal{S}_{XY}(\omega)e^{j\omega\tau}d\omega \qquad ...(7.42)$$

7.5 RELATION BETWEEN THE CROSS-CORRELATION AND CROSS POWER SPECTRUM DENSITY

Statement: The inverse Fourier transform of the cross power density spectrum is the time average of the cross-correlation function, i.e.

$$\frac{1}{2\pi}\int_{-\infty}^{\infty} \mathcal{S}_{XY}(\omega)e^{j\omega\tau}d\omega = A[R_{XY}(t,t+\tau)] \qquad \text{...(7.43)}$$

The definition of cross power spectrum density is

$$\mathcal{S}_{XY}(\omega) = \lim_{T\to\infty}\frac{E\left[X_T^*(\omega)Y_T(\omega)\right]}{2T}$$

where $X_T(\omega)$ and $Y_T(\omega)$ are Fourier transform of $x_T(t)$ and $y_T(t)$.

$X_T(\omega)$ is defined as $X_T(\omega) = \int_{-T}^{T} x_T(t)e^{j\omega t}dt$

$Y_T(\omega)$ is defined as $Y_T(\omega) = \int_{-T}^{T} y_T(t)e^{-j\omega t}dt$

$$\mathcal{S}_{XY}(\omega) = \lim_{T\to\infty}\frac{E\left[X_T^*(\omega)Y_T(\omega)\right]}{2T} = \lim_{T\to\infty}\frac{1}{2T}E\left[\int_{-T}^{T} x_T(t_1)e^{j\omega t_1}dt_1\int_{-T}^{T} y_T(t_2)e^{-j\omega t_2}dt_2\right]$$

$$= \lim_{T\to\infty}\frac{1}{2T}E\left[\int_{-T}^{T}\int_{-T}^{T} x_T(t_1)y_T(t_2)e^{j\omega t_1}e^{-j\omega t_2}\,dt_2 dt_1\right]$$

$$= \lim_{T\to\infty}\frac{1}{2T}\int_{-T}^{T}\int_{-T}^{T} E[x_T(t_1)y_T(t_2)]e^{-j\omega(t_2-t_1)}dt_2\,dt_1 \qquad \text{...(7.44)}$$

$$E[x_T(t_1)\,y_T(t_2)] = R_{XY}(t_1, t_2) \qquad -T < (t_1 \text{ and } t_2) < T \qquad \text{...(7.45)}$$

Thus, Eq. (7.44) becomes

$$\mathcal{S}_{XY}(\omega) = \lim_{T\to\infty}\frac{1}{2T}\int_{-T}^{T}\int_{-T}^{T} R_{XY}(t_1,t_2)e^{-j\omega(t_2-t_1)}dt_2\,dt_1 \qquad \text{...(7.46)}$$

Substitute Eq. (7.46) on the left side of Eq. (7.43), then Eq. (7.43) becomes

$$\frac{1}{2\pi}\int_{-\infty}^{\infty}\mathcal{S}_{XY}(\omega)e^{j\omega\tau}d\omega = \frac{1}{2\pi}\int_{-\infty}^{\infty}\lim_{T\to\infty}\frac{1}{2T}\int_{-T}^{T}\int_{-T}^{T} R_{XY}(t_1,t_2)e^{-j\omega(t_2-t_1)}e^{j\omega\tau}dt_2\,dt_1\,d\omega$$

$$= \frac{1}{2\pi}\int_{-\infty}^{\infty}\lim_{T\to\infty}\frac{1}{2T}\int_{-T}^{T}\int_{-T}^{T} R_{XY}(t_1,t_2)e^{-j\omega(t_2-t_1-\tau)}d\omega\,dt_2\,dt_1$$

Now exchange the integrations,

$$\frac{1}{2\pi}\int_{-\infty}^{\infty}\mathcal{S}_{XY}(\omega)e^{j\omega\tau}d\omega = \lim_{T\to\infty}\frac{1}{2T}\int_{-T}^{T}\int_{-T}^{T}R_{XY}(t_1,t_2)\frac{1}{2\pi}\left(\int_{-\infty}^{\infty}e^{j\omega(\tau+t_1-t_2)}d\omega\right)dt_2\,dt_1 \quad ...(7.47)$$

But we have $\frac{1}{2\pi}\int_{-\infty}^{\infty}e^{j\omega x}d\omega = \delta(x)$ and $\frac{1}{2\pi}\int_{-\infty}^{\infty}e^{-j\omega x}dx = d(\omega)$

Substitute the above relation in Eq. (7.47),

$$\frac{1}{2\pi}\int_{-\infty}^{\infty}\mathcal{S}_{XY}(\omega)e^{j\omega\tau}d\omega = \lim_{T\to\infty}\frac{1}{2T}\int_{-T}^{T}\int_{-T}^{T}R_{XY}(t_1,\ t_2)\delta(\tau+t_1-t_2)dt_2\,dt_1 \quad ...(7.48)$$

It can be written as $\delta(\tau + t_1 - t_2) = d(t_2 - \tau - t_1)$

$$\frac{1}{2\pi}\int_{-\infty}^{\infty}\mathcal{S}_{XY}(\omega)e^{j\omega\tau}d\omega = \lim_{T\to\infty}\frac{1}{2T}\int_{-T}^{T}\int_{-T}^{T}R_{XY}(t_1,t_2)\delta(t_2-\tau-t_1)dt_2\,dt_1 \quad ...(7.49)$$

The impulse function property is

$$\int_{-\infty}^{\infty}\phi(x)\delta(x-x_0)dx = \phi(x_0) \quad ...(7.50)$$

Compare Eqs. (7.49) and (7.50),

$$\phi(x) = R_{XY}(t_1,t_2), \delta(x-x_0)= \delta(t_2-\tau-t_1),\ x = t_2 \text{ and } x_0 = \tau + t_1$$

Equation (7.49) becomes

$$\frac{1}{2\pi}\int_{-\infty}^{\infty}\mathcal{S}_{XY}(\omega)e^{j\omega\tau}d\omega = \lim_{T\to\infty}\frac{1}{2T}\int_{-T}^{T}R_{XY}(t_1,t_1+\tau)dt_1 \quad ...(7.51)$$

Substitute $t_1 = t$ in Eq. (7.51), it becomes

$$\frac{1}{2\pi}\int_{-\infty}^{\infty}\mathcal{S}_{XY}(\omega)e^{j\omega\tau}d\omega = \lim_{T\to\infty}\frac{1}{2T}\int_{-T}^{T}R_{XY}(t,t+\tau)dt \quad ...(7.52)$$

But time averaged cross-correlation is

$$A[R_{XY}(t,t+\tau)] = \lim_{T\to\infty}\frac{1}{2T}\int_{-T}^{T}R_{XY}(t,t+\tau)dt \quad ...(7.53)$$

From Eqs. (7.52) and (7.53), we have

$$\frac{1}{2\pi}\int_{-\infty}^{\infty}\mathcal{S}_{XY}(\omega)e^{j\omega\tau}d\omega = A[R_{XY}(t,t+\tau)]$$

If random processes are WSS, i.e. $A[R_{XY}(t, t + \tau)] = R_{XY}(\tau)$, then

$$\mathcal{S}_{XY}(\omega) = \int_{-\infty}^{\infty}R_{XY}(\tau)e^{-j\omega\tau}d\tau \quad ...(7.54)$$

and

$$R_{XY}(\tau) = \frac{1}{2\pi}\int_{-\infty}^{\infty}\mathcal{S}_{XY}(\omega)e^{j\omega\tau}d\omega \quad ...(7.55)$$

Example 7.5.1: Consider the random process given by $Z(t) = X(t) + Y(t)$. Here $Z(t)$ is the sum of two jointly WSS orthogonal random processes $X(t)$ and $Y(t)$.

We have, $R_{ZZ}(\tau) = R_{XX}(\tau) + R_{XY}(\tau) + R_{YX}(\tau) + R_{YY}(\tau)$

Taking Fourier transform of both sides,

$$\mathcal{S}_{ZZ}(\omega) = \mathcal{S}_{XX}(\omega)+\mathcal{S}_{XY}(\omega)+\mathcal{S}_{YX}(\omega)+\mathcal{S}_{YY}(\omega) \quad ...(7.56)$$

But average power is

$$P_{XX} = \frac{1}{2\pi}\int_{-\infty}^{\infty} \mathcal{S}_{XX}(\omega)d\omega$$

By integrating both sides of Eq. (7.56),

$$\int_{-\infty}^{\infty} \mathcal{S}_{ZZ}(\omega)d\omega = \int_{-\infty}^{\infty} \mathcal{S}_{XX}(\omega)d\omega + \int_{-\infty}^{\infty} \mathcal{S}_{XY}(\omega)d\omega + \int_{-\infty}^{\infty} \mathcal{S}_{YX}(\omega)d\omega + \int_{-\infty}^{\infty} \mathcal{S}_{YY}(\omega)d\omega$$

Therefore, $P_{ZZ} = P_{XX} + P_{XY} + P_{YX} + P_{YY}$. ...(7.57)

7.6 POWER SPECTRUM FOR DISCRETE TIME PROCESSES AND SEQUENCES

7.6.1 Discrete Time Processes

Let $X(t)$ be a band limited random process for which samples are taken at times nT_s, $n = 0, \pm 1, \pm 2,...$ to form a discrete time process $X(nT_s)$. When sampling is the result of multiplying $X(t)$ by a periodic sampling pulse train consisting of rectangular

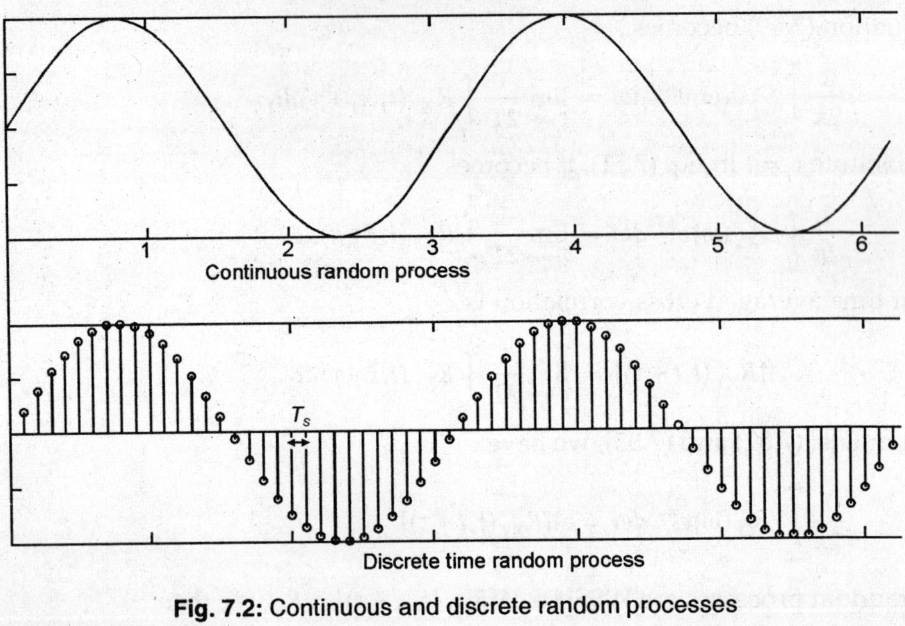

Fig. 7.2: Continuous and discrete random processes

pulses of short duration T_p and amplitude $1/T_p$ occurring each T_s seconds. The process autocorrelation function $R_{XX}(\tau)$ has a sampled representation denoted by $R_{X_sX_s}(\tau)$, that is given by

$$R_{X_sX_s}(\tau) = R_{XX}(\tau)\frac{1}{T_p}\sum_{n=-\infty}^{\infty} \text{rect}\left(\frac{\tau - nT_s}{T_p}\right)$$

$$\approx R_{XX}(\tau)\sum_{n=-\infty}^{\infty}\delta(\tau-nT_s)$$

$$= \sum_{n=-\infty}^{\infty} R_{XX}(nT_s)\delta(\tau-nT_s) \qquad ...(7.58)$$

The second sum is an approximation to the first sum that results from use of narrow sample pulses (T_p small). The third sum derives from the second by use of the equation given below.

For a function $g(x)$,

$$g(x)\,\delta(x-x_0) = g(x_0)\,\delta(x) \qquad ...(7.59)$$

$$\mathcal{S}_{X_sX_s}(\omega) = \sum_{n=-\infty}^{\infty} R_{X_sX_s}(nT_s)e^{-jn\omega T_s} \qquad ...(7.60)$$

It is a periodic in the variable ω with period $\omega_s = \dfrac{2\pi}{T_s}$.

$$\mathcal{S}_{X_sX_s}(\omega) = \sum_{n=-\infty}^{\infty} R_{XX}(nT_s)\delta(\tau-nT_s)e^{-jn\omega T_s} \qquad ...(7.61)$$

But we have

$$\sum_{n=-\infty}^{\infty}\delta(\tau-nT_s) = \frac{1}{T_s}\sum_{n=-\infty}^{\infty} e^{jn\omega_s\tau} \qquad ...(7.62)$$

From Eqs. (7.61) and (7.62),

$$\mathcal{S}_{X_sX_s}(\omega) = \frac{1}{T_s}\sum_{n=-\infty}^{\infty} R_{XX}(nT_s)e^{jn\omega_s\tau}e^{-jn\omega T_s}$$

$$= \frac{1}{T_s}\sum_{n=-\infty}^{\infty} R_{XX}(nT_s)e^{-jn(\omega T_s-\omega_s\tau)}$$

$$\tau = nT_s$$

$$= \frac{1}{T_s}\sum_{n=-\infty}^{\infty} R_{XX}(nT_s)e^{-jn(\omega-n\omega_s)T_s}$$

$$\mathcal{S}_{X_sX_s}(\omega) = \frac{1}{T_s}\sum_{n=-\infty}^{\infty} \mathcal{S}_{XX}(\omega-n\omega_s) \qquad ...(7.63)$$

If we write a rectangular function $\text{rect}(\omega/\omega_s)$ on both sides to select out $\mathcal{S}_{XX}(\omega)$, we have

$$\mathcal{S}_{XX}(\omega) = T_s\,\text{rect}(\omega/\omega_s)\mathcal{S}_{X_sX_s}(\omega) \qquad ...(7.64)$$

7.6.2 Discrete Time Sequence

Since computers treat samples as simply as a sequence of numbers, they are not about the time separation (T_s) used between samples. In recognition of this fact, the explicit dependence on T_s and $X(nT_s)$ is written as $X[n]$, a function of index n denoted by brackets rather than parenthesis. Two other changes must also be recognized.

First, a new variable, Ω is called the *discrete frequency*, and defined as $\Omega = \omega T_s$.

Second, we can define power spectrum density by using the Z-transform of the discrete time process, whose autocorrelation is $R_{XX}[n]$, by the relation

$$\mathcal{S}_{XX}(z) = \sum_{n=-\infty}^{\infty} R_{XX}[n]z^{-n}$$

where z is a complex variable.

$\mathcal{S}_{XX}(\omega)$ is related to $\mathcal{S}_{XX}(z)$ by

$$\mathcal{S}_{XX}(\omega) = \mathcal{S}_{XX}(z)\Big|_{z=e^{j\Omega}} \quad \text{...(7.65)}$$

Example 7.6.1: Suppose $R_{XX}[n] = 2^{-|n|}$ where $n = 0, \pm 1, \pm 2, \pm 3...$ Then find $\mathcal{S}_{XX}(\omega)$.

Solution: $\mathcal{S}_{XX}(\omega) = \sum_{n=-\infty}^{\infty} R_{XX}[n]e^{-jn\Omega}$

$$= \sum_{n=-\infty}^{-1} 2^n e^{-jn\Omega} + \sum_{n=0}^{\infty} 2^{-n} e^{-jn\Omega} \qquad \because \sum_{n=N_1}^{N_2} W^n = \frac{W^{N_1} + W^{N_2+1}}{1-W}$$

$$= \sum_{n=-\infty}^{-1} \left(2e^{-j\Omega}\right)^n + \sum_{n=0}^{\infty} \left(2e^{j\Omega}\right)^{-n}$$

$$= \frac{1}{1-2e^{-j\Omega}} + \frac{1}{1-2^{-1}e^{-j\Omega}} = \frac{3}{5-4\cos\omega}$$

The plot of the autocorrelation sequence and the power spectral density is shown in Fig. 7.3.

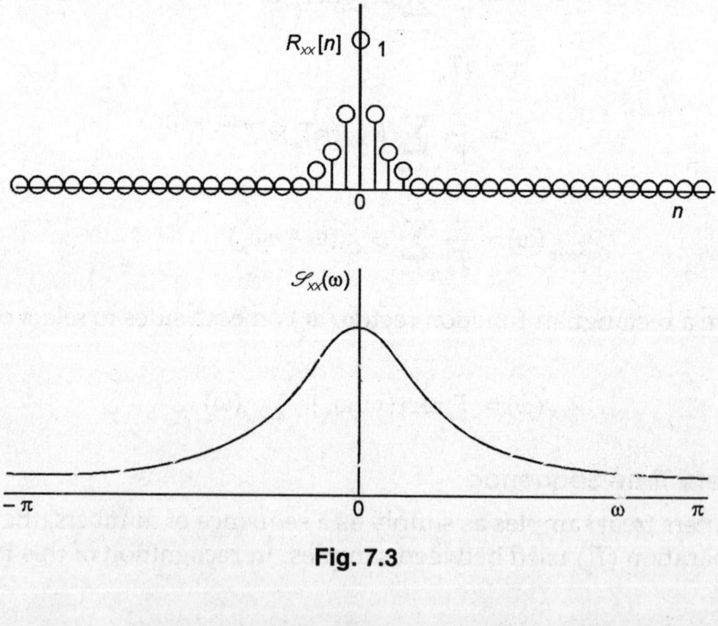

Fig. 7.3

EXAMPLES

Example 7.1: Determine which of the following functions can and cannot be valid power density spectrum. For those which are not, explain why.

i. $\mathcal{S}_{XX}(\omega)=\dfrac{1}{\sqrt{1-3\omega^2}}$ ii. $\mathcal{S}_{XX}(\omega)=\dfrac{\omega^2}{\sqrt{\omega^6+3\omega^2+3}}$

Solution: Power spectrum density is valid if it satisfies three properties. These are:

(a) $\mathcal{S}_{XX}(\omega)\geq 0$ (b) $\mathcal{S}_{XX}(\omega)=\mathcal{S}_{XX}(-\omega)$ (c) $\mathcal{S}_{XX}(\omega)$ is real

i. $\mathcal{S}_{XX}(\omega)=\dfrac{1}{\sqrt{1-3\omega^2}}$

For all values of ω, $\mathcal{S}_{XX}(\omega)>0$ but imaginary, i.e. the third property is not satisfied. Hence, this function cannot be a valid power spectrum.

ii. $\mathcal{S}_{XX}(\omega)=\dfrac{\omega^2}{\sqrt{\omega^6+3\omega^2+3}}$

This function is obeying all the three properties. Hence, it can be a valid power spectrum.

Example 7.2: A stationary RP with sample function $X(t)$ has a spectral density of $\mathcal{S}_{XX}(\omega)=\dfrac{16}{\omega^2+16}$ and an independent stationary random process with sample function $Y(t)$ has a spectral density $\mathcal{S}_{YY}(\omega)=\dfrac{\omega^2}{\omega^2+16}$. A new random variable is formed from $U(t) = X(t) + Y(t)$. Find:

i. The spectral of $U(t)$.

ii. The cross spectrum density of $\mathcal{S}_{XY}(\omega)$.

iii. The cross spectrum density of $\mathcal{S}_{XU}(\omega)$.

Solution: i. We have $\mathcal{S}_{XX}(\omega)=\dfrac{16}{\omega^2+16}$ and $\mathcal{S}_{YY}(\omega)=\dfrac{\omega^2}{\omega^2+16}$.

A new random process is defined as $U(t) = X(t) + Y(t)$.

Power spectrum density of $U(t)$ is $\mathcal{S}_{UU}(\omega)=\mathcal{S}_{XX}(\omega)+\mathcal{S}_{XY}(\omega)+\mathcal{S}_{YX}(\omega)+\mathcal{S}_{YY}(\omega)$

If $X(t)$ and $Y(t)$ are independent random processes, then $R_{XY}(\tau)=R_{YX}(\tau)=0$. Hence,

$$\mathcal{S}_{XY}(\omega) = \mathcal{S}_{YX}(\omega)=0$$

$$\mathcal{S}_{UU}(\omega) = \mathcal{S}_{XX}(\omega)+\mathcal{S}_{YY}(\omega)$$

$$= \frac{16}{\omega^2+16}+\frac{\omega^2}{\omega^2+16}=1$$

Power spectrum density of $U(t)$ is $\mathcal{S}_{UU}(\omega)=1$

ii. $\mathscr{S}_{XU}(\omega) = F[R_{XU}(\tau)]$

$$R_{XU}(\tau) = E[X(t)U(t+\tau)]$$
$$= E[X(t)X(t+\tau) + X(t)Y(t+\tau)]$$
$$= E[X(t)X(t+\tau)] + E[X(t)Y(t+\tau)]$$
$$= R_{XX}(\tau) + R_{XY}(\tau)$$
$$= R_{XX}(\tau)$$
$$\mathscr{S}_{XU}(\omega) = F[R_{XX}(\tau)]$$
$$= \mathscr{S}_{XX}(\omega)$$
$$\mathscr{S}_{XU}(\omega) = \frac{16}{\omega^2+16}.$$

Example 7.3: Suppose $X(t) = A + B\sin(\omega_c t + \theta)$ where A is a constant bias and θ is a random variable distributed uniformly on the interval $[0, 2\pi]$. Find $R_{XX}(\tau)$ and $\mathscr{S}_{XX}(\omega)$.

Solution: $R_{XX}(\tau) = E[X(t)X(t+\tau)]$

$$= E\left[\{A + B\sin(\omega_c t + \theta)\}\{A + B\sin(\omega_c t + \omega_c\tau + \theta)\}\right]$$

$$= E\left[A^2 + AB\sin(\omega_c t + \omega_c\tau + \theta) + AB\,\sin(\omega_c t + \theta) + AB\sin(\omega_c t + \omega_c\tau + \theta)\sin(\omega_c t + \theta)\right]$$

We know that

$$E[X(t)] = \int_{-\infty}^{\infty} x f(\theta)\,d\theta$$

Here θ is uniformly distributed on the interval $[0, 2\pi]$, the density function is

$$f(\theta) = \frac{1}{2\pi - 0} = \frac{1}{2\pi}$$

$$R_{XX}(t) = E[X(t)X(t+\tau)] = \int_{-\infty}^{\infty} X(t)X(t+\tau)f(\theta)\,d\theta$$

$$= \int_{-\infty}^{\infty} [A + B\sin(\omega_c t + \theta)][A + B\sin(\omega_c t + \omega_c\tau + \theta)]f(\theta)\,d\theta$$

$$= \int_{-\infty}^{\infty} \left[A^2 + AB\sin(\omega_c t + \omega_c\tau + \theta) + AB\,\sin(\omega_c t + \theta) + AB\,\sin(\omega_c t + \omega_c\tau + \theta)\sin(\omega_c t + \theta)\right]f(\theta)\,d\theta$$

$$= \int_0^{2\pi} A^2 \frac{1}{2\pi}d\theta + \int_0^{2\pi} AB\sin(\omega_c t + \omega_c\tau + \theta)\frac{1}{2\pi}d\theta + \int_0^{2\pi} AB\,\sin(\omega_c t + \theta)\frac{1}{2\pi}d\theta$$

$$+ \int_0^{2\pi} B^2\sin(\omega_c t + \omega_c\tau + \theta)\sin(\omega_c t + \theta)\frac{1}{2\pi}d\theta$$

$$= \frac{A^2}{2\pi}\int_0^{2\pi} d\theta + \frac{AB}{2\pi}\int_0^{2\pi} \sin(\omega_c t + \omega_c \tau + \theta)d\theta + \frac{AB}{2\pi}\int_0^{2\pi} \sin(\omega_c t + \theta)d\theta$$

$$+\frac{B^2}{2\pi}\int_0^{2\pi} \sin(\omega_c t + \omega_c \tau + \theta)\sin(\omega_c t + \theta)d\theta$$

$$= \frac{A^2}{2\pi}\cdot 2\pi + \frac{AB}{2\pi}\cos(\omega_c t + \omega_c \tau + \theta)\Big|_0^{2\pi} + \frac{AB}{2\pi}\cos(\omega_c t + \theta)\Big|_0^{2\pi}$$

$$+\frac{B^2}{4\pi}\int_0^{2\pi} 2\sin(\omega_c t + \omega_c \tau + \theta)\sin(\omega_c t + \theta)d\theta$$

$$= A^2 + \frac{AB}{2\pi}[\cos(\omega_c t + \omega_c \tau) - \cos(\omega_c t + \omega_c \tau)] + \frac{AB}{2\pi}[\cos\omega_c t - \cos\omega_c t]$$

$$+\frac{B^2}{4\pi}\int_0^{2\pi} [\cos(\omega_c \tau) - \cos(2\omega_c t + \omega_c \tau + 2\theta)]d\theta$$

$$= A^2 + \frac{AB}{2\pi}\cdot 0 + \frac{AB}{2\pi}\cdot 0 + \frac{AB}{4\pi}\cos(\omega_c \tau)\cdot 2\pi - \frac{AB}{4\pi}\frac{\cos(2\omega_c t + \omega_c \tau + 2\theta)}{2}\Bigg|_0^{2\pi}$$

$$= A^2 + \frac{B^2}{4\pi}\cos(\omega_c \tau)\cdot 2\pi - \frac{B^2}{4\pi}\left[\frac{\sin(2\omega_c t + \omega_c \tau)}{2} - \frac{\sin(2\omega_c t + \omega_c \tau)}{2}\right]$$

$$= A^2 + \frac{B^2}{2}\cos(\omega_c \tau)$$

We know that

$$\mathcal{S}_{XX}(\omega) = \int_{-\infty}^{\infty} A[R_{XX}(\tau)]e^{-j\omega\tau}d\tau$$

$$A[R_{XX}(t, t+\tau)] = \lim_{T\to\infty}\frac{1}{2T}\int_{-T}^{T} R_{XX}(t, t+\tau)dt$$

$$= \lim_{T\to\infty}\frac{1}{2T}\int_{-T}^{T}\left(A^2 + \frac{B^2}{2}\cos(\omega_c \tau)\right)dt$$

$$= \lim_{T\to\infty}\frac{1}{2T}A^2\, t\Big|_{-T}^{T} + \lim_{T\to\infty}\frac{1}{2T}\frac{B^2}{2}\cos(\omega_c \tau)t\Big|_{-T}^{T}$$

$$= A^2 + \frac{B^2}{2}\cos(\omega_c \tau)$$

$$\therefore \qquad A[R_{XX}(\tau)] = R_{XX}(\tau)$$

$$\mathcal{S}_{XX}(\omega) = \int_{-\infty}^{\infty}\left[A^2 + \frac{B^2}{2}\cos(\omega_c \tau)\right]e^{-j\omega\tau}d\tau$$

$$= \int_{-\infty}^{\infty} A^2 e^{-j\omega\tau} d\tau + \int_{-\infty}^{\infty} \frac{B^2}{2} \cos(\omega_c \tau) e^{-j\omega\tau} d\tau$$

$$= A^2 \int_{-\infty}^{\infty} e^{-j\omega\tau} d\tau + \frac{B^2}{2} \int_{-\infty}^{\infty} \frac{(e^{j\omega_c\tau} + e^{-j\omega_c\tau})}{2} e^{-j\omega\tau} d\tau$$

$$= A^2 2\pi\delta(\omega) + \frac{B^2}{2} \int_{-\infty}^{\infty} \frac{e^{-j(\omega-\omega_c)\tau}}{2} d\tau + \frac{B^2}{2} \int_{-\infty}^{\infty} \frac{e^{-j(\omega-\omega_c)\tau}}{2} d\tau \quad \left[\because \int_{-\infty}^{\infty} e^{-j\omega\tau} d\tau = 2\pi\delta(\omega)\right]$$

$$= A^2 2\pi\delta(\omega) + \frac{B^2}{2} \pi\delta(\omega - \omega_c) + \frac{B^2}{2} \pi\delta(\omega + \omega_c)$$

$$\mathscr{S}_{XX}(\omega) = A^2 2\pi\delta(\omega) + \frac{B^2}{2} \pi[\delta(\omega - \omega_c) + \delta(\omega + \omega_c)]$$

The power spectrum of random process $X(t)$ is shown in Fig. 7.4.

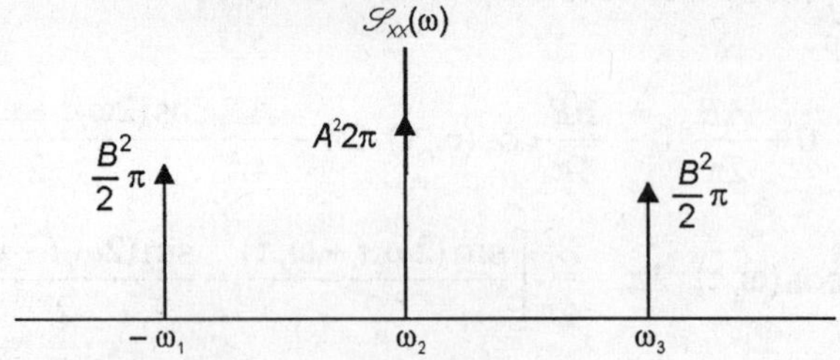

Fig. 7.4

Example 7.4: Find the PSD of the amplitude modulated random phase sinusoid, $X(t) = M(t)\cos(\omega_c t + \theta)$, where θ is random variable uniformly distributed on the interval $[0, 2\pi]$, and $M(t)$ is a WSS process independent of θ.

Solution: $R_{XX}(\tau) = E[X(t)\, X(t + \tau)]$

$$= E\left[\{M(t)\cos(\omega_c t + \theta)\}\{M(t+\tau)\cos(\omega_c t + \omega_c \tau + \theta)\}\right]$$

$$= E\left[M(t)M(t+\tau)\cos(\omega_c t + \theta)\cos(\omega_c t + \omega_c \tau + \theta)\right]$$

We know that $E[X(t)] = \int_{-\infty}^{\infty} x f(\theta) d\theta$.

Here θ is uniformly distributed on the interval $[0, 2\pi]$, the density function is

$$f(\theta) = \frac{1}{2\pi - 0} = \frac{1}{2\pi}$$

$$R_{XX}(\tau) = E[X(t)X(t+\tau)] = \int_{-\infty}^{\infty} X(t)X(t+\tau) f(\theta) d\theta$$

$$= E\left[M(t)M(t+\tau)\cos(\omega_c t + \theta)\cos(\omega_c t + \omega_c \tau + \theta)\right]$$

$M(t)$ is independent of θ.

$$R_{XX}(\tau) = E\left[M(t)M(t+\tau)\right]E\left[\cos(\omega_c t + \theta)\cos(\omega_c t + \omega_c \tau + \theta)\right]$$

$$= R_{MM}(\tau) E\left[\cos(\omega_c t + \theta)\cos(\omega_c t + \omega_c \tau + \theta)\right]$$

$$R_{XX}(\tau) = R_{MM}(\tau)\int_{-\infty}^{\infty} \cos(\omega_c t+\theta)\cos(\omega_c t+\omega_c\tau+\theta)\, f(\theta)\,d\theta$$

$$R_{XX}(\tau) = R_{MM}(\tau)\int_{0}^{2\pi} \cos(\omega_c t+\theta)\cos(\omega_c t+\omega_c\tau+\theta)\frac{1}{2\pi}\,d\theta$$

$$= R_{MM}(\tau)\frac{1}{4\pi}\int_{0}^{2\pi} 2\cos(\omega_c t+\theta)\cos(\omega_c t+\omega_c\tau+\theta)\,d\theta$$

$$= R_{MM}(\tau)\frac{1}{4\pi}\int_{0}^{2\pi} \cos(\omega_c\tau)+\cos(2\omega_c t+\omega_c\tau+2\theta)\,d\theta$$

$$= R_{MM}(\tau)\frac{1}{4\pi}\cos(\omega_c\tau)\,\theta\Big|_0^{2\pi} + R_{MM}(\tau)\frac{1}{4\pi}\frac{\cos(2\omega_c t+\omega_c\tau+2\theta)}{2}\Bigg|_0^{2\pi}$$

$$= R_{MM}(\tau)\frac{1}{4\pi}\cos(\omega_c\tau)2\pi + R_{MM}(\tau)\frac{1}{4\pi}\left[\frac{\cos(2\omega_c t+\omega_c\tau)}{2} - \frac{\cos(2\omega_c t+\omega_c\tau)}{2}\right]$$

$$= R_{MM}(\tau)\frac{1}{2}\cos(\omega_c\tau)$$

$$= \frac{R_{MM}(\tau)\cos(\omega_c\tau)}{2}$$

We know that $\mathscr{S}_{XX}(\omega) = \displaystyle\int_{-\infty}^{\infty} R_{XX}(\tau)e^{-j\omega\tau}d\tau$

$$= \int_{-\infty}^{\infty} \frac{R_{MM}(\tau)\cos(\omega_c\tau)}{2}e^{-j\omega\tau}d\tau$$

$$= \int_{-\infty}^{\infty} \frac{R_{MM}(\tau)}{2}\frac{(e^{j\omega_c\tau}+e^{-j\omega_c\tau})}{2}e^{-j\omega\tau}d\tau$$

$$= \int_{-\infty}^{\infty} \frac{R_{MM}(\tau)}{2}\frac{e^{-j(\omega-\omega_c)\tau}}{2}d\tau + \int_{-\infty}^{\infty} \frac{R_{MM}(\tau)}{2}\frac{e^{-j(\omega+\omega_c)\tau}}{2}d\tau$$

$$\therefore \quad \mathscr{S}_{XX}(\omega) = \frac{1}{4}\left[\mathscr{S}_{MM}(\omega-\omega_c)+\mathscr{S}_{MM}(\omega+\omega_c)\right]$$

where $\mathscr{S}_{MM}(\omega)$ is the PSD of $M(t)$.

Example 7.5: The PSD of a noise process is given by

$$\mathscr{S}_{NN}(\omega) = \begin{cases} \dfrac{N_0}{2} & |\omega \pm \omega_c| \le \dfrac{W}{2} \\ 0 & \text{otherwise} \end{cases}$$

Find the autocorrelation of the process.

Solution: $R_{XX}(\tau) = \dfrac{1}{2\pi}\displaystyle\int_{-\infty}^{\infty} \mathscr{S}_{XX}(\omega)e^{j\omega\tau}d\omega$

Here limits are $|\omega \pm \omega_c| \le \frac{W}{2}$

i.e. $|\omega + \omega_c| \le \frac{W}{2}$ and $|\omega - \omega_c| \le \frac{W}{2}$

$$\omega + \omega_c \le \frac{W}{2},\ \omega + \omega_c \ge \frac{-W}{2} \text{ and } \omega - \omega_c \le \frac{W}{2}, \omega - \omega_c \ge \frac{-W}{2}$$

$$\omega \le \frac{W}{2} - \omega_c,\ \omega \ge \frac{-W}{2} - \omega_c \text{ and } \omega \le \frac{W}{2} + \omega_c,\ \omega \ge \frac{-W}{2} + \omega_c$$

$$R_{XX}(\tau) = \frac{1}{2\pi}\int_{-\infty}^{\infty} \frac{N_0}{2} e^{j\omega\tau} d\omega$$

$$= \frac{1}{2\pi}\int_{-\infty}^{\infty} \frac{N_0}{2} e^{j\omega\tau} d\omega = \frac{1}{2\pi}\int_{-\omega_c - \frac{W}{2}}^{-\omega_c + \frac{W}{2}} \frac{N_0}{2} e^{j\omega\tau} d\omega + \frac{1}{2\pi}\int_{\omega_c - \frac{W}{2}}^{\omega_c + \frac{W}{2}} \frac{N_0}{2} e^{j\omega\tau} d\omega$$

$$= \frac{1}{2\pi}\frac{N_0}{2}\frac{e^{j\omega\tau}}{j\tau}\Bigg|_{-\omega_c - \frac{W}{2}}^{-\omega_c + \frac{W}{2}} + \frac{1}{\pi}\frac{N_0}{2}\frac{e^{j\omega\tau}}{j\tau}\Bigg|_{\omega_c - \frac{W}{2}}^{\omega_c + \frac{W}{2}}$$

$$= \frac{1}{2\pi}\frac{N_0}{2}\left[\frac{e^{j\left(-\omega_c + \frac{W}{2}\right)\tau}}{j\tau} - \frac{e^{j\left(-\omega_c - \frac{W}{2}\right)\tau}}{j\tau}\right] + \frac{1}{2\pi}\frac{N_0}{2}\left[\frac{e^{j\left(\omega_c + \frac{W}{2}\right)\tau}}{j\tau} - \frac{e^{j\left(\omega_c - \frac{W}{2}\right)\tau}}{j\tau}\right]$$

$$= \frac{1}{2\pi}\frac{N_0}{2} e^{-j\omega_c\tau}\left[\frac{e^{j\frac{W}{2}\tau} - e^{-j\frac{W}{2}\tau}}{j\tau}\right] + \frac{1}{2\pi}\frac{N_0}{2} e^{j\omega_c\tau}\left[\frac{e^{j\frac{W}{2}\tau} - e^{j\frac{W}{2}\tau}}{j\tau}\right]$$

$$= \frac{1}{2\pi}\frac{N_0}{2}\left[\frac{e^{j\frac{W}{2}\tau} - e^{-j\frac{W}{2}\tau}}{j\tau}\right]\left[e^{-j\omega_c\tau} + e^{j\omega_c\tau}\right]$$

$$\because \frac{e^{-j\omega_c\tau} + e^{j\omega_c\tau}}{2} = \cos\omega_c\tau \text{ and } \frac{e^{j\frac{W}{2}\tau} - e^{-j\frac{W}{2}\tau}}{2j} = \sin\frac{W}{2}\tau$$

$$= \frac{1}{\pi\tau} N_0 \sin\frac{W}{2}\tau \cos\omega_c\tau$$

$$= \frac{2N_0W}{2\pi\tau W}\sin\frac{W}{2}\tau\cos\omega_c\tau = \frac{N_0W}{2\pi}\frac{\sin\frac{W}{2}\tau}{\frac{W}{2}\tau}\cos\omega_c\tau$$

$$R_{XX}(\tau) = \frac{N_0W}{2\pi}\text{sa}\frac{W}{2}\tau\cos\omega_c\tau.$$

Example 7.6: Find the PSD for a WSS random process $X(t)$ with autocorrelation function as shown in Fig. 7.5.

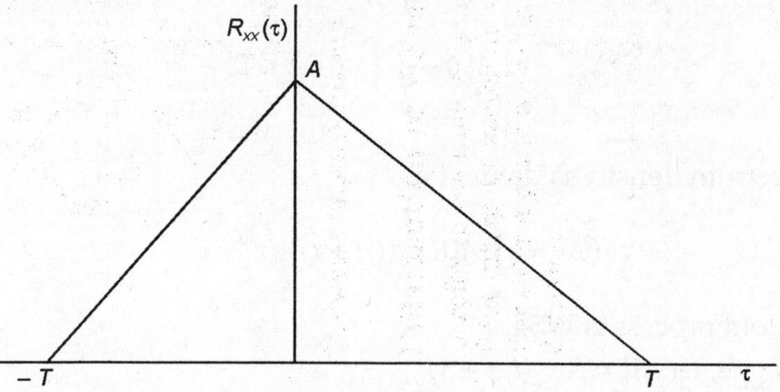

Fig. 7.5

Solution: In this problem, autocorrelation function is not given, so first we have to obtain the function of autocorrelation.

From Fig. 7.5, we know that autocorrelation is a collection of different functions.

First obtain the points.

These are $(-T, 0)$, $(0, A)$ and $(T, 0)$.

Time interval is $-T < \tau < 0$.

We have two points $(0, A)$ and $(-T, 0)$ on Y-axis $R_{XX}(\tau)$ and X-axis τ respectively.

Straight line equation using two points (x_1, y_1) and (x_2, y_2) is given by

$$y - y_1 = \frac{y_2 - y_1}{x_2 - x_1}(x - x_1)$$

Substitute the values in above equation,

$$R_{XX}(\tau) - 0 = \frac{A-0}{0-(-T)}[\tau-(-T)]$$

$$R_{XX}(\tau) = \frac{A}{T}(\tau + T)$$

$$= A\left(1+\frac{\tau}{T}\right)$$

Time interval is $0 < \tau < T$

We have two points $(0, A)$ and $(T, 0)$.

$$R_{XX}(\tau) - A = \frac{0-A}{T-0}(\tau - 0)$$

$$R_{XX}(\tau) = A - \frac{A}{T}(\tau)$$

$$= A\left(1-\frac{\tau}{T}\right)$$

From the above,

$$R_{XX}(\tau) = \begin{cases} A\left(1+\dfrac{\tau}{T}\right) & -T<\tau<0 \\ A\left(1-\dfrac{\tau}{T}\right) & 0<\tau<T \\ 0 & \text{otherwise} \end{cases}$$

Power spectrum density is defined as

$$\mathscr{S}_{XX}(\omega) = \int_{-\infty}^{\infty} A[R_{XX}(t,t+\tau)]e^{-j\omega\tau}d\tau$$

If the random process is WSS,
then $A[R_{XX}(t, t+\tau)] = R_{XX}(t, t+\tau)$

$$\mathscr{S}_{XX}(\omega) = \int_{-\infty}^{\infty} R_{XX}(t,t+\tau)e^{-j\omega\tau}d\tau$$

$$= \int_{-T}^{0} A\left(1+\frac{\tau}{T}\right)e^{-j\omega\tau}d\tau + \int_{0}^{T} A\left(1-\frac{\tau}{T}\right)e^{-j\omega\tau}d\tau$$

$$= \int_{-T}^{0} Ae^{-j\omega\tau}d\tau + \int_{-T}^{0} A\frac{\tau}{T}e^{-j\omega\tau}d\tau + \int_{0}^{T} Ae^{-j\omega\tau}d\tau - \int_{0}^{T} A\frac{\tau}{T}e^{-j\omega\tau}d\tau$$

$$= A\int_{-T}^{0} e^{-j\omega\tau}d\tau + \int_{-T}^{0} A\frac{\tau}{T}e^{-j\omega\tau}d\tau + \int_{0}^{T} Ae^{-j\omega\tau}d\tau - \int_{0}^{T} A\frac{\tau}{T}e^{-j\omega\tau}d\tau$$

$$I_1 = A\int_{-T}^{0} e^{-j\omega\tau}d\tau + \int_{-T}^{0} A\frac{\tau}{T}e^{-j\omega\tau}d\tau$$

$$= A\left.\frac{e^{-j\omega\tau}}{-j\omega}\right|_{-T}^{0} + \frac{A}{T}\left[\tau\left.\frac{e^{-j\omega\tau}}{-j\omega}\right|_{-T}^{0} - \int_{-T}^{0}\frac{1}{-j\omega}e^{-j\omega\tau}d\tau\right]$$

$$= A\left[\frac{e^{j\omega T}}{j\omega} - \frac{1}{j\omega}\right] + \frac{A}{T}\left[\left(-T\frac{e^{j\omega T}}{-j\omega}\right) - \frac{1}{(-j\omega)^2}\left.e^{-j\omega\tau}\right|_{-T}^{0}\right]$$

$$= A\left[\frac{e^{j\omega T}}{j\omega} - \frac{1}{j\omega}\right] + \frac{A}{T}\left[\left(-T\frac{e^{j\omega T}}{-j\omega}\right) + \left(\frac{1}{(-j\omega)^2}e^{j\omega T} - \frac{1}{(-j\omega)^2}\right)\right]$$

$$= A\frac{-1}{j\omega} + \frac{A}{T}\left[\left(\frac{1}{(-j\omega)^2}e^{j\omega T} - \frac{1}{(-j\omega)^2}\right)\right]$$

$$= \frac{-A}{j\omega}\left[1 + \frac{1}{(j\omega T)}\left(1-e^{j\omega T}\right)\right]$$

$$= \frac{A}{j\omega}\left[-1 - \frac{1}{(j\omega T)}\left(1-e^{j\omega T}\right)\right]$$

$$I_2 = A\int_0^T e^{-j\omega\tau}d\tau - \int_0^T A\frac{\tau}{T}e^{-j\omega\tau}d\tau$$

$$= A\frac{e^{-j\omega\tau}}{-j\omega}\bigg|_0^T - \frac{A}{T}\left[\tau\frac{e^{-j\omega\tau}}{-j\omega}\bigg|_0^T - \int_0^T\frac{1}{-j\omega}e^{-j\omega\tau}d\tau\right]$$

$$= A\left[\frac{-e^{-j\omega T}}{j\omega}+\frac{1}{j\omega}\right]-\frac{A}{T}\left[\left(T\frac{e^{-j\omega T}}{-j\omega}\right)-\frac{1}{(-j\omega)^2}e^{-j\omega\tau}\bigg|_0^T\right]$$

$$= A\left[\frac{-e^{-j\omega T}}{j\omega}+\frac{1}{j\omega}\right]-\frac{A}{T}\left[\left(T\frac{e^{-j\omega T}}{-j\omega}\right)-\left(\frac{1}{(-j\omega)^2}e^{-j\omega T}-\frac{1}{(-j\omega)^2}\right)\right]$$

$$= A\left[\frac{1}{j\omega}\right]+\frac{A}{T}\frac{1}{(-j\omega)^2}e^{-j\omega T}-\frac{A}{T}\frac{1}{(-j\omega)^2}$$

$$= \frac{A}{j\omega}\left[1+\frac{1}{(j\omega T)}\left(e^{-j\omega T}-1\right)\right]$$

$$\mathcal{S}_{XX}(\omega) = I_1 + I_2$$

$$= \frac{A}{j\omega}\left[1+\frac{1}{(j\omega T)}\left(e^{-j\omega T}-1\right)\right]+\frac{A}{j\omega}\left[-1-\frac{1}{(j\omega T)}\left(1-e^{j\omega T}\right)\right]$$

$$= \frac{A}{j\omega}\left[1+\frac{1}{(j\omega T)}\left(e^{-j\omega T}-1\right)-1-\frac{1}{(j\omega T)}\left(1-e^{j\omega T}\right)\right]$$

$$= \frac{A}{j\omega.}\left[1+\frac{1}{(j\omega T)}\left(e^{-j\omega T}-1\right)-1-\frac{1}{(j\omega T)}\left(1-e^{j\omega T}\right)\right]$$

$$= \frac{A}{j\omega}\left[\frac{1}{(j\omega T)}e^{-j\omega T}-\frac{1}{(j\omega T)}-\frac{1}{(j\omega T)}+\frac{1}{(j\omega T)}e^{j\omega T}\right]$$

$$= \frac{A}{j\omega}\left[\frac{1}{(j\omega T)}e^{j\omega T}+\frac{1}{(j\omega T)}e^{-j\omega T}-\frac{2}{(j\omega T)}\right]$$

$$= \frac{A}{j\omega}\left[\frac{1}{(j\omega T)}2\cos\omega T-\frac{2}{(j\omega T)}\right]$$

$$= \frac{A}{j\omega}\frac{2}{(j\omega T)}[\cos\omega T-1]$$

$$= \frac{A}{j\omega}\frac{2}{(j\omega T)}\left[-2\sin^2(\omega T/2)\right]$$

$$= \frac{A}{j\omega}\frac{2}{(j\omega T)}\left[-2\sin^2(\omega T/2)\right]$$

$$= \frac{-A}{j\omega}\frac{2}{(j\omega T)}\left[2\sin^2(\omega T/2)\right]$$

$$= AT\frac{\sin^2(\omega T/2)}{(\omega T/2)^2}$$

$$= AT\, Sa^2(\omega T/2)$$

The autocorrelation and PSD functions are shown in Fig. 7.6.

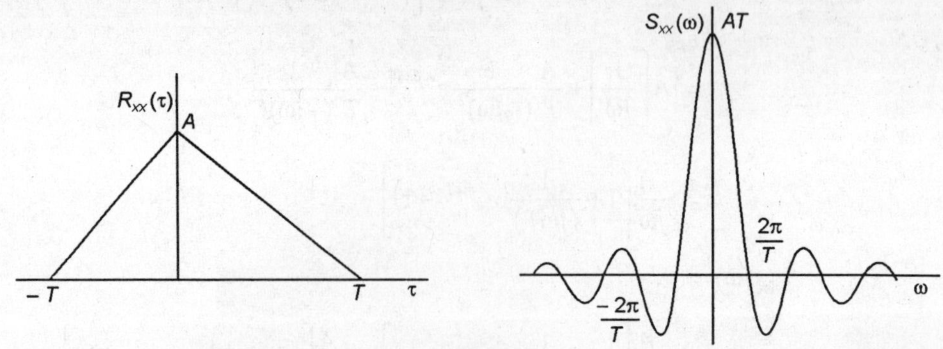

Fig. 7.6

Example 7.7: Suppose $X(t) = A\cos(\omega_c t + \theta)$ where A is a constant bias and θ is a random variable distributed uniformly on the interval $\left[0, \frac{\pi}{2}\right]$. Determine the average power in time domain as well as frequency domain.

Solution: Average power is

$$P_{XX} = \lim_{T\to\infty}\frac{1}{2T}\int_{-T}^{T} E[x^2(t)]dt = \frac{1}{2\pi}\int_{-\infty}^{\infty} \mathcal{S}_{XX}(\omega)d\omega$$

$$P_{XX} = \lim_{T\to\infty}\frac{1}{2T}\int_{-T}^{T} E[x^2(t)]dt$$

We know that $\quad E[X(t)] = \int_{-\infty}^{\infty} xf(\theta)d\theta.$

Here θ is uniformly distributed on the interval $\left[0, \frac{\pi}{2}\right]$, the density function is

$$f(\theta) = \frac{1}{\frac{\pi}{2} - 0} = \frac{2}{\pi}$$

$$E[x^2(t)] = E[A^2 \cos^2(\omega_c t + \theta)]$$

$$= \int_{-\infty}^{\infty} A^2 \cos^2(\omega_c t + \theta) f(\theta) d\theta$$

$$= A^2 \int_0^{\pi/2} \cos^2(\omega_c t + \theta) \frac{2}{\pi} d\theta$$

$$= \frac{A^2}{\pi} \int_0^{\pi/2} \cos(2\omega_c t + 2\theta) d\theta + \frac{A^2}{\pi} \int_0^{\pi/2} d\theta$$

$$= \frac{A^2}{\pi} \frac{\sin(2\omega_c t + 2\theta)}{2} \bigg|_0^{\pi/2} + \frac{A^2}{\pi} \int_0^{\pi/2} d\theta$$

$$= \frac{A^2}{\pi}\left(-\frac{\sin(2\omega_c t)}{2} - \frac{\sin(2\omega_c t)}{2}\right) + \frac{A^2}{2}$$

$$= \frac{A^2}{\pi}[-\sin(2\omega_c t)] + \frac{A^2}{2}$$

$$P_{XX} = \lim_{T\to\infty} \frac{1}{2T} \int_{-T}^{T} E[x^2(t)] dt$$

$$P_{XX} = \lim_{T\to\infty} \frac{1}{2T} \int_{-T}^{T} \left(-\frac{A^2}{\pi} \sin(2\omega_c t) + \frac{A^2}{2}\right) dt$$

$$= \lim_{T\to\infty} \frac{1}{2T} \int_{-T}^{T} -\frac{A^2}{\pi} \sin(2\omega_c t) dt + \lim_{T\to\infty} \frac{1}{2T} \int_{-T}^{T} \frac{A^2}{2} dt$$

$$= -\frac{A^2}{\pi} \lim_{T\to\infty} \frac{1}{2T} \frac{\cos(2\omega_c t)}{2\omega_c} \bigg|_{-T}^{T} + \lim_{T\to\infty} \frac{1}{2T} \frac{A^2}{2} t \Big|_{-T}^{T}$$

$$= -\frac{A^2}{\pi} \lim_{T\to\infty} \frac{1}{2T}\left(\frac{\cos(2\omega_c T)}{2\omega_c} - \frac{\cos(-2\omega_c T)}{2\omega_c}\right) + \lim_{T\to\infty} \frac{1}{2T} \frac{A^2}{2} 2T$$

$$P_{XX} = \lim_{T\to\infty} \frac{A^2}{2} = \frac{A^2}{2} \qquad \text{...(7.66)}$$

$$R_{XX}(\tau) = E[X(t)X(t+\tau)]$$

$$= E\left[\{A\cos(\omega_c t + \theta)\}\{A\cos(\omega_c t + \omega_c \tau + \theta)\}\right]$$

$$= E\left[A^2 \cos(\omega_c t + \theta)\cos(\omega_c t + \omega_c \tau + \theta)\right]$$

$$R_{XX}(\tau) = E[X(t)X(t+\tau)] = \int_{-\infty}^{\infty} X(t)X(t+\tau) f(\theta) d\theta$$

$$= \int_{-\infty}^{\infty} \left[A^2 \cos(\omega_c t + \theta) \cos(\omega_c t + \omega_c \tau + \theta)\right] f(\theta) d\theta$$

$$= \int_{0}^{\pi/2} \left[A^2 \cos(\omega_c t + \theta) \cos(\omega_c t + \omega_c \tau + \theta)\right] f(\theta) d\theta$$

$$= \frac{A^2}{2} \int_{0}^{\pi/2} [\cos(2\omega_c t + 2\theta + \omega_c \tau) + \cos(\omega_c \tau)] \frac{2}{\pi} d\theta$$

$$= \frac{A^2}{2} \int_{0}^{\pi/2} \cos(2\omega_c t + 2\theta + \omega_c \tau) \frac{2}{\pi} d\theta + \frac{A^2}{2} \int_{0}^{\pi/2} \cos(\omega_c \tau) \frac{2}{\pi} d\theta$$

$$= \frac{A^2}{\pi} \int_{0}^{\pi/2} \cos(2\omega_c t + 2\theta + \omega_c \tau) d\theta + \frac{A^2}{\pi} \cos(\omega_c \tau) \theta \Big|_2^{\pi/2}$$

$$= \frac{A^2}{\pi} \frac{\sin(2\omega_c t + 2\theta + \omega_c \tau)}{2} \Bigg|_0^{\pi/2} + \frac{A^2}{\pi} \cos(\omega_c \tau) \frac{\pi}{2}$$

$$= \frac{A^2}{\pi} \left[-\frac{\sin(2\omega_c t + \omega_c \tau)}{2} - \frac{\sin(2\omega_c t + \omega_c \tau)}{2} \right] + \frac{A^2}{2} \cos(\omega_c \tau)$$

$$= \frac{A^2}{\pi} [-\sin(2\omega_c t + \omega_c \tau)] + \frac{A^2}{2} \cos(\omega_c \tau)]$$

$$\mathcal{S}_{XX}(\omega) = \int_{-\infty}^{\infty} A[R_{XX}(t, t+\tau)] e^{-j\omega\tau} d\tau$$

$$A[R_{XX}(t, t+\tau)] = \lim_{T\to\infty} \frac{1}{2T} \int_{-T}^{T} R_{XX}(t, t+\tau) dt$$

$$= \lim_{T\to\infty} \frac{1}{2T} \int_{-T}^{T} \frac{A^2}{\pi} [-\sin(2\omega_c t + \omega_c \tau)] dt + \lim_{T\to\infty} \frac{1}{2T} \int_{-T}^{T} \frac{A^2}{2} \cos(\omega_c \tau) dt$$

$$= \frac{A^2}{\pi} \lim_{T\to\infty} \frac{1}{2T} \frac{\cos(2\omega_c t + \omega_c \tau)}{2\omega_c} \Bigg|_{-T}^{T} + \frac{A^2}{2} \cos(\omega_c \tau) \lim_{T\to\infty} \frac{1}{2T} t \Big|_{-T}^{T}$$

$$= \frac{A^2}{\pi} \lim_{T\to\infty} \frac{1}{2T} \left(\frac{\cos(2\omega_c T + \omega_c \tau)}{2\omega_c} - \frac{\cos(-2\omega_c T + \omega_c \tau)}{2\omega_c} \right)$$

$$+ \frac{A^2}{2} \cos(\omega_c \tau) \lim_{T\to\infty} \frac{1}{2T} 2T$$

$$= \frac{A^2}{2} \cos(\omega_c \tau) \lim_{T\to\infty} 1$$

$$A[R_{XX}(t, t+\tau)] = \frac{A^2}{2}\cos(\omega_c\tau)$$

$$\mathscr{S}_{XX}(\omega) = \int_{-\infty}^{\infty} \frac{A^2}{2}\cos(\omega_c\tau)e^{-j\omega\tau}d\tau = \frac{A^2}{2}\int_{-\infty}^{\infty}\left(\frac{e^{j\omega_c\tau} - e^{-j\omega_c\tau}}{2}\right)e^{-j\omega\tau}d\tau$$

$$= \frac{A^2}{2}\int_{-\infty}^{\infty}\frac{e^{-j(\omega-\omega_c)\tau}}{2}d\tau - \frac{A^2}{2}\int_{-\infty}^{\infty}\frac{e^{-j(\omega+\omega_c)\tau}}{2}d\tau$$

$$= \frac{A^2}{4}\int_{-\infty}^{\infty} e^{-j(\omega-\omega_c)\tau}d\tau - \frac{A^2}{4}\int_{-\infty}^{\infty} e^{-j(\omega+\omega_c)\tau}d\tau \quad [\because 1 \leftrightarrow 2\pi\delta(\omega)]$$

$$= \frac{A^2\pi}{2}[\delta(\omega-\omega_c) + \delta(\omega+\omega_c)]$$

$$P_{XX} = \frac{1}{2\pi}\int_{-\infty}^{\infty}\mathscr{S}_{XX}(\omega)d\omega$$

$$= \frac{1}{2\pi}\int_{-\infty}^{\infty}\frac{A^2\pi}{2}[\delta(\omega-\omega_c) + \delta(\omega+\omega_c)]d\omega$$

$$= \frac{1}{2\pi}\frac{A^2\pi}{2}\int_{-\infty}^{\infty}[\delta(\omega-\omega_c) + \delta(\omega+\omega_c)]d\omega$$

$$= \frac{A^2}{4}2 \qquad \because \int_{-\infty}^{\infty}\delta(\omega)d\omega = 1$$

$$P_{XX} = \frac{A^2}{2} \qquad ...(7.67)$$

From Eqs. (7.66) and (7.67), we get

$$\therefore \qquad P_{XX} = \lim_{T\to\infty}\frac{1}{2T}\int_{-T}^{T}E[x^2(t)]dt = \frac{1}{2\pi}\int_{-\infty}^{\infty}\mathscr{S}_{XX}(\omega)d\omega$$

EXERCISE

1. The PSD of random process is given by

$$\mathscr{S}_{WW}(\omega) = \begin{cases} \pi & |\omega| \le 1 \\ 0 & \text{else} \end{cases}$$

Find its autocorrelation function.

2. A WSS random process $X(t)$ has

$$R_{XX}(\tau) = A_0 \begin{cases} \left[1 - \frac{|\tau|}{\tau}\right] & -\tau \le t \le \tau \\ 0 & \text{elsewhere} \end{cases}$$

Find power density spectrum.

3. Consider Fig. 7.7, where $X(t)$, $Y(t)$ are random processes and $X(t)$ is WSS. Find the relation between $\mathcal{S}_{YY}(\omega)$ and $\mathcal{S}_{XX}(\omega)$.

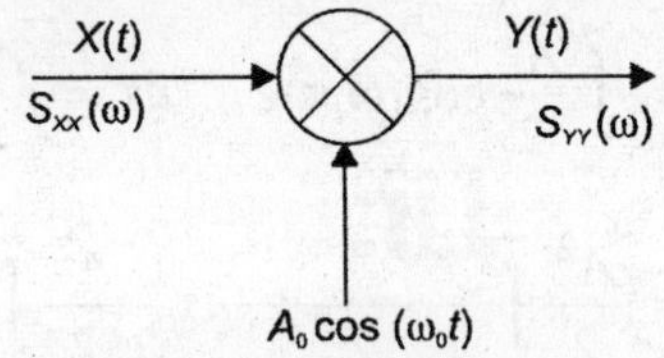

Fig. 7.7

4. A WSS noise process $N(t)$ has an autocorrelation function $R_{NN}(\tau) = Pe^{-3|\tau|}$ where P is a constant. Find and sketch its power spectrum.

5. If $R_{XX}(\tau) = ae^{|b\tau|}R(\tau) = ae^{|by|}$, find the spectral density function, where a and b are constants.

6. The power spectral density of a random process $X(t)$ is given by

$$\mathcal{S}_{XX}(\omega) = \begin{cases} \pi & |\omega| \le 1 \\ 0 & \text{else} \end{cases}$$. Find its autocorrelation function.

7. If the PSD of a WSS process is given by

$$\mathcal{S}_{XX}(\omega) = \begin{cases} \frac{b}{a}(a - |\omega|) & |\omega| \le a \\ 0 & |\omega| > a \end{cases}$$, find the autocorrelation function of the process.

8. Consider a random process $X(t) = \cos(\omega t + \theta)$ where ω is a real constant and θ is a uniform random variable in $(0, \pi/2)$. Show that $X(t)$ is not a WSS process. Also find the average power in the process

9. If the PSD of $X(t)$ is $\mathcal{S}_{XX}(\omega)$, find the PSD of $\frac{dx(t)}{dt}$.

10. If $R(\tau) = ae^{|by|}$, find the spectral density function, where a and b are constants.

11. Find the ACF of the following PSDs:

i. $\mathcal{S}_{XX}(\omega) = \dfrac{157 + 12\omega^2}{(16 + \omega^2)(9 + \omega^2)}$ ii. $\mathcal{S}_{XX}(\omega) = \dfrac{8}{(9 + \omega^2)^2}$.

12. Determine which of the following functions are valid power density spectrums and why?

i. $\mathcal{S}_{XX}(\omega) = \dfrac{\cos(8\omega)}{2 + \omega^4}$ ii. $\mathcal{S}_{XX}(\omega) = e^{-(\omega-1)^2}$.

13. If the autocorrelation function of a WSS process is $R(\tau) = Ke^{-K|\tau|}$, find its PSD.

14. Determine which of the following functions are valid power density spectrums and why?

i. $\mathcal{S}_{XX}(\omega) = \dfrac{\omega^2}{\omega^6 + 3\omega^2 + 3}$ ii. $\mathcal{S}_{XX}(\omega) = -\dfrac{\omega^2}{\omega^2 + 16}$.

8

Linear Systems with Random Inputs

8.0 INTRODUCTION

In many applications, physical systems are modeled as linear time invariant (LTI) systems. The dynamic behavior of an LTI system to deterministic inputs is described by linear differential equations. We are familiar with time and transform domain (such as Laplace and Fourier transforms) techniques to solve these differential equations. In this chapter, we develop techniques of describing the response of a linear system when the applied waveform is random. We begin by discussing some basic aspects of linear system in the following section. The purpose of this chapter is analysis of the response of a system and finding an LTI system that can optimally estimate an unobserved random process from an observed process, i.e. the observed random process is statistically related to the unobserved random process. For example, we may have to find LTI system (also called a filter) to estimate the signal from the noisy observations.

8.1 LINEAR SYSTEMS WITH DETERMINISTIC INPUTS

A system is modeled by a transformation $\mathcal{L}$ that maps an input signal $x(t)$ to an output signal $y(t)$ as shown in Fig. 8.1. We can thus write,

$$y(t) = \mathcal{L}[x(t)] \qquad \text{...(8.1)}$$

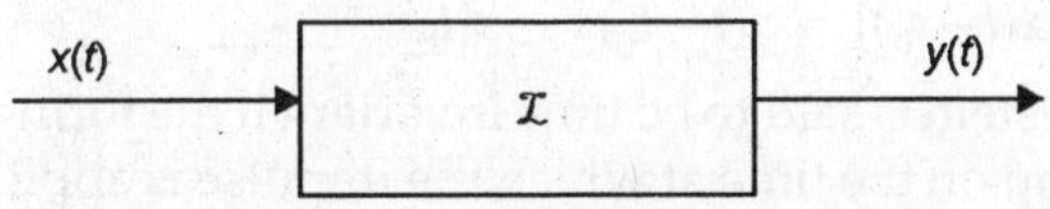

Fig. 8.1: Block diagram of a linear system

8.1.1 Linear System

The system is called linear if its response to a sum of inputs $x_n(t)$, $n = 1, 2,...,N$ is equal to the sum of responses taken separately. Thus, if $x_n(t)$ causes a response $y_n(t)$, $n = 1, 2,...,N$, then for a linear system

$$y(t) = \mathcal{L}\left[\sum_{n=1}^{N} \alpha_n x_n(t)\right] = \sum_{n=1}^{N} \alpha_n \mathcal{L}[x_n(t)] = \sum_{n=1}^{N} \alpha_n y_n(t) \qquad \text{..(8.2)}$$

must hold, where α_n's are arbitrary constants and N may be infinite.

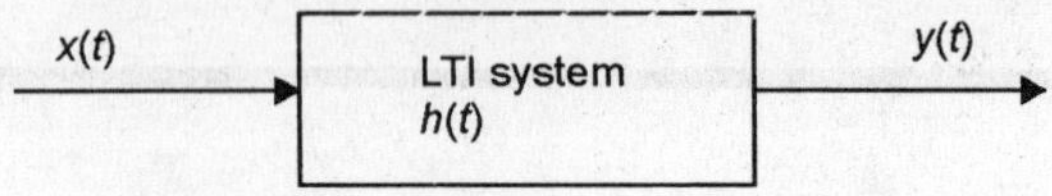

Fig. 8.2: A linear, time invariant system

We know that the property of impulse function is

$$\int_{-\infty}^{\infty} \phi(x)\delta(x_0 - x)dx = \phi(x_0)$$

Substituting $\phi(x) = x(\tau)$, $x_0 = t$ and $x = \tau$ in the above equation, we get

$$x(t) = \int_{-\infty}^{\infty} x(\tau)\delta(t-\tau)d\tau \qquad ...(8.3)$$

By substituting Eq. (8.3) in Eq. (8.1), we get

$$y(t) = \mathcal{L}[x(t)] = \mathcal{L}\left[\int_{-\infty}^{\infty} x(\tau)\delta(t-\tau)d\tau\right] = \int_{-\infty}^{\infty} x(\tau)\mathcal{L}[\delta(t-\tau)]d\tau \qquad ...(8.4)$$

We now define a new function $h(t, \tau)$ as the impulse response of the linear system, i.e.

$$\mathcal{L}[\delta(t-\tau)] = h(t, \tau) \qquad ...(8.5)$$

Equation (8.4) can be rewritten as

$$y(t) = \int_{-\infty}^{\infty} x(\tau)h(t,\tau)d\tau \qquad ...(8.6)$$

which shows that the response of a general linear system is completely determined by its impulse response.

8.1.2 Linear Time Invariant System

Consider a linear system with $y(t) = \mathcal{L}[x(t)]$. The system is called time invariant if

$$\mathcal{L}[x(t-t_0)] = y(t-t_0) \qquad \forall t_0 \qquad ...(8.7)$$

A general linear system is said to be time invariant if the form of its impulse response $h(t, \xi)$ does not depend on the time at which the impulse is applied. Thus, if an impulse $\delta(t)$, occurring at $t = 0$, causes the response $h(t)$, and then an impulse $\delta(t - \xi)$, occurring at $t = \xi$, must cause the response $h(t - \xi)$ if the system is time invariant. This fact means that

$$h(t, \tau) = h(t - \tau) \qquad ...(8.8)$$

From above equation,

$$y(t) = \int_{-\infty}^{\infty} x(\tau)h(t-\tau)d\tau \qquad ...(8.9)$$

Equation (8.9) is known as the *convolution integral* of $x(t)$ and $h(t)$; it is also written in the short form as

$$y(t) = x(t) * h(t) \qquad ...(8.10)$$

By changing the variables, Eq. (8.9) can be written as

$$y(t) = \int_{-\infty}^{\infty} h(\tau)x(t-\tau)d\tau. \qquad ...(8.11)$$

8.1.3 Linear Time Invariant System: Transfer Function

By Fourier transformation of $y(t)$, we may derive an equivalent characterization in the frequency domain. Hence, if $X(\omega)$, $Y(\omega)$ and $H(\omega)$ are the respective Fourier transform of $x(t)$, $y(t)$ and $h(t)$, then

$$Y(\omega) = \int_{-\infty}^{\infty} y(t)e^{-j\omega t}dt \qquad ...(8.12)$$

But
$$y(t) = \int_{-\infty}^{\infty} x(\tau)h(t-\tau)d\tau$$

Substituting $y(t)$ in Eq. (8.12), we get

$$Y(\omega) = \int_{-\infty}^{\infty}\left[\int_{-\infty}^{\infty} x(\tau)h(t-\tau)d\tau\right]e^{-j\omega t}dt$$

$$= \int_{-\infty}^{\infty} x(\tau)\left[\int_{-\infty}^{\infty} h(t-\tau)e^{-j\omega(t-\tau)}dt\right]e^{-j\omega\tau}d\tau$$

$$= \int_{-\infty}^{\infty} x(\tau)H(\omega)e^{-j\omega\tau}d\tau \qquad \because \int_{-\infty}^{\infty} h(t)e^{-j\omega t}dt = H(\omega)$$

$$Y(\omega) = X(\omega)\,H(\omega) \qquad ...(8.13)$$

The function $H(\omega)$ is called the *transfer function* of the system.

From Eq. (8.13), the Fourier transform of the response of any linear time invariant system is equal to the product of the transform of input signal and the transfer function of the network impulse response.

The transfer function is

$$H(\omega) = \frac{Y(\omega)}{X(\omega)}. \qquad ...(8.14)$$

8.2 LINEAR SYSTEMS WITH RANDOM INPUTS

Let the input be random process $X(t)$ which has known properties such as its first and second moments, possibly its PDF, and whether or not it is uncorrelated. Typically, the goal in the analysis or design of a system is to modify $X(t)$ in order to generate an output $Y(t)$ that has some desired characteristics. For example, $X(t)$ might contain a communication signal embedded in additive noise, and the goal of the system is to filter $X(t)$ such that $Y(t)$ has reduced noise, and without adversely affecting the underlying signal of interest.

If the possible inputs to the system are $x(t)$, i.e. sample functions of a stochastic process $X(t)$, then the outputs, $y(t)$ are sample functions of another stochastic process, $Y(t)$. Because $y(t)$ is the convolution of $x(t)$ and $h(t)$, we adopt the following notation for the relationship of $Y(t)$ to $X(t)$:

$$Y(t) = \int_{-\infty}^{\infty} X(\tau)h(t-\tau)d\xi = \int_{-\infty}^{\infty} h(\tau)X(t-\tau)d\tau \qquad \text{...(8.15)}$$

8.2.1 Mean Value of System Response

The expected value of $Y(t)$ is

$$E[Y(t)] = E\left[\int_{-\infty}^{\infty} X(\tau)h(t-\tau)d\tau\right] = E\left[\int_{-\infty}^{\infty} h(\tau)X(t-\tau)d\tau\right] \qquad \text{...(8.16)}$$

$$E[Y(t)] = \int_{-\infty}^{\infty} E[X(t-\tau)]h(\tau)d\tau \qquad \text{...(8.17)}$$

From Eq. (8.17), the expected value of $Y(t)$ is the convolution of $h(t)$ and $E[X(t)]$. If $X(t)$ is a WSS random process,

$$E[Y(t)] = E[X(t-\tau)]\int_{-\infty}^{\infty} h(\tau)d\tau = \mu_X \int_{-\infty}^{\infty} h(\tau)d\tau \qquad \text{...(8.18)}$$

$$\because \quad H(\omega)\big|_{\omega=0} = \left.\int_{-\infty}^{\infty} h(\tau)e^{-j\omega\tau}d\tau\right|_{\omega=0} \Rightarrow H(0) = \int_{-\infty}^{\infty} h(\tau)d\tau$$

$$\mu_Y = \mu_X \int_{-\infty}^{\infty} h(\tau)d\tau = \mu_X H(0), \text{ which is a constant value.}$$

Hence if $X(t)$ is a WSS random process, the system response $Y(t)$ is also WSS random process.

8.2.2 Mean Square Value of System Response

The mean square value of $Y(t)$ is

$$E\left[Y^2(t)\right] = E\left[\{h(t)*X(t)\}^2\right] = E\left[\{h(t)*X(t)\}\{h(t)*X(t)\}\right]$$

$$= E\left[\int_{-\infty}^{\infty} [X(t-\tau_1)h(\tau_1)]d\tau_1 \int_{-\infty}^{\infty} [X(t-\tau_2)h(\tau_2)]d\tau_2\right]$$

$$= E\left[\int_{-\infty}^{\infty}\int_{-\infty}^{\infty} X(t-\tau_1)X(t-\tau_2)h(\tau_1)h(\tau_2)d\tau_1 d\tau_2\right]$$

$$E\left[Y^2(t)\right] = \int_{-\infty}^{\infty}\int_{-\infty}^{\infty} E\left[X(t-\tau_1)X(t-\tau_2)\right]h(\tau_1)h(\tau_2)d\tau_1 d\tau_2 \qquad \text{...(8.19)}$$

where τ_1, τ_2 are constant values.
If $X(t)$ is a WSS random process,

$$E[Y^2(t)] = \int_{-\infty}^{\infty}\int_{-\infty}^{\infty} R_{XX}(\tau_1 - \tau_2)h(\tau_1)h(\tau_2)d\tau_1 d\tau_2 = \text{a constant} \quad ...(8.20)$$

8.2.3 Autocorrelation Function of System Response

We know that $$Y(t) = \int_{-\infty}^{\infty} X(t-\tau_1)h(\tau_1)d\tau_1$$

$$Y(t+\tau) = \int_{-\infty}^{\infty} X(t+\tau-\tau_2)h(\tau_2)d\tau_2$$

$$R_{YY}(t,\tau) = R_{YY}(\tau) = E[Y(t)Y(t+\tau)] = E\left[\int_{-\infty}^{\infty} X(t-\tau_1)h(\tau_1)d\tau_1 \int_{-\infty}^{\infty} X(t+\tau-\tau_2)h(\tau_2)d\tau_2\right]$$

$$R_{YY}(t,\tau) = R_{YY}(\tau) = \int_{-\infty}^{\infty}\int_{-\infty}^{\infty} E[X(t-\tau_1)X(t+\tau-\tau_2)]h(\tau_1)h(\tau_2)d\tau_1 d\tau_2 \quad ...(8.21)$$

We know that

$$E[X(t-\tau_1)X(t+\tau-\tau_2)] = R_{XX}(\tau_1+\tau-\tau_2)$$

$$R_{YY}(t,\tau) = R_{YY}(\tau) = \int_{-\infty}^{\infty}\int_{-\infty}^{\infty} R_{XX}(\tau_1+\tau-\tau_2)h(\tau_1)h(\tau_2)d\tau_1 d\tau_2 \quad ...(8.22)$$

If $X(t)$ is a WSS random process, it is observed that the output autocorrelation function is a function of τ only. Hence, the output random process $Y(t)$ is also WSS random process.

The relation between cross-correlation and autocorrelation of the response is

$$R_{YY}(t,\tau) = R_{YY}(\tau) = E[Y(t)Y(t+\tau)] = E\left[Y(t)\int_{-\infty}^{\infty} X(t+\tau-\tau_2)h(\tau_2)d\tau_2\right]$$

$$R_{YY}(\tau) = E\left[\int_{-\infty}^{\infty} Y(t)X(t+\tau-\tau_2)h(\tau_2)d\tau_2\right]$$

$$= \int_{-\infty}^{\infty} E[Y(t)X(t+\tau-\tau_2)]h(\tau_2)d\tau_2$$

$$R_{YY}(\tau) = \int_{-\infty}^{\infty} R_{YX}(\tau-\tau_2)h(\tau_2)d\tau_2 = R_{YX}(\tau) * h(\tau) \quad ...(8.23)$$

Similarly,

$$R_{YY}(\tau) = R_{XY}(\tau) * h(-\tau) \quad ...(8.24)$$

8.2.4 Cross-correlation Functions of Input and Output

We know that

$$Y(t) = \int_{-\infty}^{\infty} X(t-\tau_1)h(\tau_1)d\tau_1$$

$$Y(t+\tau) = \int_{-\infty}^{\infty} X(t+\tau-\tau_2)h(\tau_2)d\tau_2$$

$$R_{XY}(t,\tau) = R_{XY}(\tau) = E[X(t)Y(t+\tau)] = E\left[X(t)\int_{-\infty}^{\infty} X(t+\tau-\tau_2)h(\tau_2)d\tau_2\right]$$

$$= \int_{-\infty}^{\infty} E\left[X(t)X(t+\tau-\tau_2)\right]h(\tau_2)d\tau_2$$

$$R_{XY}(t,\tau) = R_{XY}(\tau) = \int_{-\infty}^{\infty} R_{XX}(\tau-\tau_2)h(\tau_2)d\tau_2 \quad \text{...(8.25)}$$

which is the convolution of $R_{XX}(\tau)$ and $h(\tau)$. Therefore, $R_{XY}(\tau) = R_{XX}(\tau) * h(\tau)$. Similarly, we can show that $R_{YX}(\tau) = R_{XX}(\tau) * h(-\tau)$.

This shows that $X(t)$ and $Y(t)$ are jointly WSS.

From Eq. (8.22),

$$R_{YY}(t,\tau) = R_{YY}(\tau) = \int_{-\infty}^{\infty}\underbrace{\int_{-\infty}^{\infty} R_{XX}(\tau_1+\tau-\tau_2)h(\tau_2)d\tau_2}_{R_{XY}(\tau_1+\tau)}\, h(\tau_1)d\tau_1$$

$$R_{YY}(t,\tau) = R_{YY}(\tau) = \int_{-\infty}^{\infty} R_{XY}(\tau_1+\tau)h(\tau_1)d\tau_1 \quad \text{...(8.26)}$$

Example 8.2.1: A linear system has an impulse response of $h(t) = 10[u(t) - u(t-1)]$. A white noise process having a two-sided spectral density $2V^2/\text{Hz}$ is applied to the input of the system. Find:

(a) The mean value of the output of the system.

(b) The mean square value of the output.

(c) The variance of the output.

Solution: We know that the mean value of white noise process is zero.

Hence the mean of the response is $\mu_Y = \mu_X H(0) = 0$

The mean square value of the output is

$$E[Y^2] = \int_{-\infty}^{\infty}\int_{-\infty}^{\infty} R_{XX}(\tau_1-\tau_2)h(\tau_1)h(\tau_2)d\tau_1 d\tau_2$$

$$= 2\int_0^1\int_0 (10)(10)d\tau_1 d\tau_2 = 200$$

The variance of the output is

$$\sigma_Y^2 = E[Y^2] - \mu_Y^2 = 200 - 0 = 200.$$

8.3 SPECTRAL CHARACTERISTICS OF SYSTEM RESPONSE

Spectral characteristics mean converting time domain signal into frequency domain. Consider a random process $X(t)$ which is applied through an LTI system and the output response of the system is $Y(t)$. It is noted that if $X(t)$ is a WSS random process with autocorrelation function $R_{XX}(\tau)$, $Y(t)$ is also a WSS and the processes $X(t)$ and $Y(t)$ are jointly WSS. The power spectral characteristics of the output $Y(t)$ can be obtained by Fourier transform of output autocorrelation function.

8.3.1 Power Density Spectrum of Response

Consider that a random process $X(t)$ is applied to an LTI system with transfer function $H(\omega)$.

Let the output response be $Y(t)$ and the power spectrum of the $X(t)$ be $S_{XX}(\omega)$. From Eq. (8.22), the autocorrelation of output response is

$$R_{YY}(t, \tau) = R_{YY}(\tau) = \int_{-\infty}^{\infty}\int_{-\infty}^{\infty} R_{XX}(\tau_1 + \tau - \tau_2) h(\tau_1) h(\tau_2) d\tau_1 d\tau_2$$

Apply Fourier transform on both sides

$$F[R_{YY}(\tau)] = F\left(\int_{-\infty}^{\infty}\int_{-\infty}^{\infty} R_{XX}(\tau_1 + \tau - \tau_2) h(\tau_1) h(\tau_2) d\tau_1 d\tau_2\right)$$

$$S_{YY}(\omega) = \int_{-\infty}^{\infty}\int_{-\infty}^{\infty}\int_{-\infty}^{\infty} R_{XX}(\tau_1 + \tau - \tau_2) h(\tau_1) h(\tau_2) e^{-j\omega\tau} d\tau_1 d\tau_2 d\tau$$

$$S_{YY}(\omega) = \int_{-\infty}^{\infty} h(\tau_2) \int_{-\infty}^{\infty} h(\tau_1) \int_{-\infty}^{\infty} R_{XX}(\tau_1 + \tau - \tau_2) e^{-j\omega(\tau+\tau_1-\tau_2)} e^{j\omega\tau_1} e^{-j\omega\tau_2} d\tau d\tau_1 d\tau_2$$

$$S_{YY}(\omega) = S_{XX}(\omega) \int_{-\infty}^{\infty} h(\tau_2) e^{-j\omega\tau_2} d\tau_2 \int_{-\infty}^{\infty} h(\tau_1) e^{j\omega\tau_1} d\tau_1 = S_{XX}(\omega) H^*(\omega) H(\omega)$$

$$S_{YY}(\omega) = S_{XX}(\omega) |H(\omega)|^2 \qquad \text{...(8.27)}$$

8.3.2 Cross Power Density Spectrums of Input and Output

Consider that a random process $X(t)$ is applied to an LTI system with transfer function $H(\omega)$.

Let the output response be $Y(t)$ and the power spectrum of the $X(t)$ be $S_{XX}(\omega)$. From Eq. (8.23), the autocorrelation of output response is

$$R_{XY}(t, \tau) = R_{XY}(\tau) = \int_{-\infty}^{\infty} R_{XX}(\tau - \tau_2) h(\tau_2) d\tau_2$$

Apply Fourier transform on both sides,

$$F[R_{XY}(\tau)] = F\left(\int_{-\infty}^{\infty} R_{XX}(\tau-\tau_2)h(\tau_2)d\tau_2\right)$$

$$S_{XY}(\omega) = \int_{-\infty}^{\infty}\int_{-\infty}^{\infty} R_{XX}(\tau-\tau_2)h(\tau_2)e^{-j\omega\tau}d\tau d\tau_2$$

$$= \int_{-\infty}^{\infty}\int_{-\infty}^{\infty} R_{XX}(\tau-\tau_2)h(\tau_2)e^{-j\omega(\tau-\tau_2)}e^{-j\omega\tau_2}d\tau d\tau_2$$

$$S_{XY}(\omega) = S_{XX}(\omega)\int_{-\infty}^{\infty} h(\tau_2)e^{-j\omega\tau_2}d\tau_2 = S_{XX}(\omega)H(\omega)$$

$$S_{XY}(\omega) = S_{XX}(\omega)H(\omega) \qquad ...(8.28)$$

Similarly, $S_{YX}(\omega) = S_{XX}(\omega)H(-\omega)$. ...(8.29)

Example 8.3.1: A white-noise process $X(t)$ with autocorrelation function $R_{XX}(\tau) = \sigma^2\delta(\tau)$ is passed through a linear system whose impulse response $h(t)$ is given by $h(t) = e^{-t}u(t)$. Find the mean value μ_Y of the output, the cross-correlation function $R_{XY}(\tau)$, the output autocorrelation function $R_{YY}(\tau)$, and the variance of the output σ_Y^2.

Solution: The transfer function $H(\omega) = \dfrac{1}{1+j\omega}$ and $|H(\omega)|^2 = \dfrac{1}{1+\omega^2}$

The mean $\mu_Y = \mu_X H(0) = 0 \quad \because \mu_X = 0$

The cross power spectral density function is

$$S_{XY}(\omega) = S_{XX}(\omega)H(\omega) = \sigma^2\frac{1}{1+j\omega}$$

The output power spectral density function is

$$S_{YY}(\omega) = S_{XX}(\omega)|H(\omega)|^2 = \sigma^2\frac{1}{1+\omega^2}$$

The cross-correlation is

$$R_{XY}(\tau) = IFT[S_{XY}(\omega)] = \sigma^2 e^{-\tau}u(\tau)$$

The cross-correlation of output is

$$R_{YY}(\tau) = IFT[S_{YY}(\omega)] = \frac{\sigma^2}{2}e^{-|\tau|}.$$

8.4 TYPES OF RANDOM PROCESS

Depending upon frequency components, random processes can be characterized into different types. Some are as follows:

i. Low pass random process
ii. Band pass random process

iii. Band limited random process
iv. Narrow band random process

8.4.1 Low Pass

If power spectrum density of random process is significant in the vicinity of $f = 0$ (at low frequencies) and very small (approaching 0) at high frequencies, the random process is called low pass random process. In other words, a low pass random process has most of its power contracted at low frequencies.

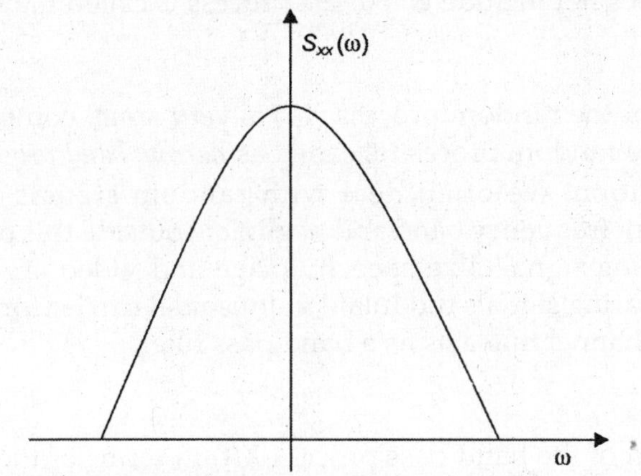

Fig. 8.3: Low pass random process

8.4.2 Band Pass

A random process $X(t)$ is called a *band pass process* if its power spectrum $S_{XX}(\omega)$ is zero outside a band $\omega_c - \frac{B}{2} \le |\omega| \le \omega_c + \frac{B}{2}$. B is called the bandwidth and $\omega_c \neq 0$ is called the centre frequency of the band pass process $X(t)$.

Band pass random processes take the form of band pass signals, and can be written in terms of phase and quadrature components as follows:

$$X(t) = X_I(t)\cos(\omega_c t + \theta) + X_Q(t)\sin(\omega_c t + \theta) \qquad ...(8.30)$$

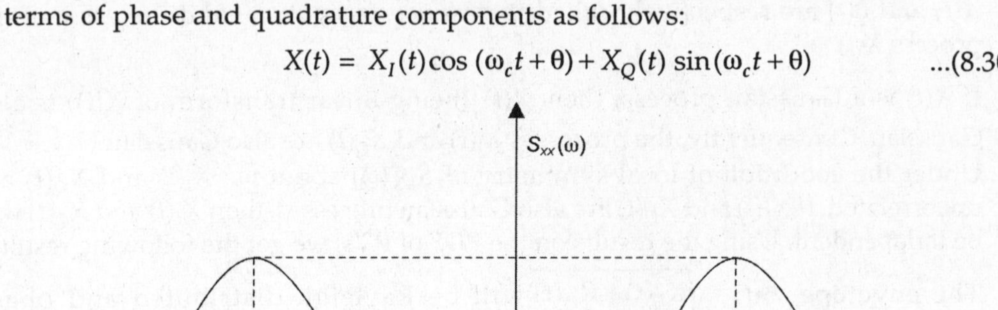

Fig. 8.4: Band pass random process

For example, modulated signals with carrier frequency ω_0 and bandwidth W are band pass random processes. The noise transmitting over a communication channel can be modeled as a band pass process.

8.4.3 Band Limited

A low pass random process is said to be *band limited random process,* if the power spectral density $S_{XX}(\omega)$ of a low pass random process $X(t)$ is zero outside the band $|\omega| \geq B$, or if the power spectrum of a band pass process is zero outside some frequency band of bandwidth W that does not include $\omega = 0$, the process is called band limited process.

8.4.4 Narrow Band

If the bandwidth W of the random process $X(t)$ is very small compared to the centre frequency ω_c, then the random process is called as *narrow band process.*

In telecommunication, we often deal with random signals which have PSD concentrated in a small frequency band and negligible outside this band. For example, the information bearing signals like speech, image and video are low pass signals. These information bearing signals modulate a sinusoidal carrier for transmitting over the communication channel that acts as a band pass filter.

Properties

i. The representation of the band pass process $X(t)$ in terms of the in-phase and the quadrature phase components is not unique. By selecting different ω_c, we can have different representations.

ii. The band pass process $X(t)$ can be represented as

$$X(t) = A(t)\cos(\omega_c t + \theta) \quad \text{...(8.31)}$$

where

$$A(t) = \sqrt{X_I^2(t) + X_Q^2(t)} \quad \text{...(8.32)}$$

and

$$\theta(t) = \tan^{-1}\left(\frac{X_Q(t)}{X_I(t)}\right) \quad \text{...(8.33)}$$

$A(t)$ and $\theta(t)$ are respectively called the *envelope(magnitude)* and the *phase* of the process $X(t)$.

iii. If $X(t)$ is a Gaussian process, then $\hat{X}(t)$ (being linear transform of $X(t)$) is also Gaussian. Consequently, the processes $X_I(t)$ and $X_Q(t)$ are also Gaussian.

iv. Under the condition of local symmetry of $S_{XX}(\omega)$ about ω_c, $X_I(t)$ and $X_Q(t)$ are uncorrelated. If $X_I(t)$ and $X_Q(t)$ are also Gaussian processes, then $X_I(t)$ and $X_Q(t)$ will be independent. Using the results on the PDF of RVs, we get the following result.

The envelope $A(t) = \sqrt{X_I^2(t) + X_Q^2(t)}$ will be Rayleigh distributed and phase $\theta(t) = \tan^{-1}\left(\frac{X_Q(t)}{X_I(t)}\right)$ is uniformly distributed over $[0, 2\pi]$. Thus,

$$f_{A(t)}(x) = \begin{cases} \dfrac{x}{R_{X_I}(0)} e^{\frac{-x^2}{R_{X_I}(0)}} & \text{for } x > 0 \\ 0 & \text{otherwise} \end{cases} \quad \text{...(8.34)}$$

8.5 MODELING OF NOISE SOURCES

A random fluctuation in an electrical (voltage or current) signal causes noise. It exists in all electronic systems. Noise added by circuit elements can conceal or obscure low-level signals, adding impairments to voice or video reception, uncertainty to bit detection in digital systems. While noise is generally unwanted, it can serve a useful purpose in some applications, such as random number generation or dithering.

8.5.1 Noise in Communication Systems

White Noise

One of the very important random processes is the *white noise* process. Noises in many practical situations are approximated by the white noise process. Most importantly, the white noise plays an important role in modeling of WSS signals.

A white noise process $W(t)$ is a random process that has constant power spectral density at all frequencies. Thus, the PSD is defined as

$$\mathcal{S}_{WW}(\omega) = \frac{N_0}{2} \quad \text{for } -\infty < \omega < \infty \qquad \text{...(8.35)}$$

where N_0 is a real constant and called the *intensity* of the white noise and the factor $\frac{1}{2}$ indicates that half the power is associated with positive frequencies and half with negative frequencies. Corresponding autocorrelation function is given by

$$R_{WW}(\tau) = \frac{N_0}{2}\delta(\tau) \qquad \text{...(8.36)}$$

where $\delta(\tau)$ is the Dirac delta.

The average power of white noise is

$$P_{avg} = \frac{1}{2\pi}\int_{-\infty}^{\infty} \frac{N_0}{2} d\omega \to \infty \qquad \text{...(8.37)}$$

The autocorrelation function and the PSD of a white noise process is shown in Fig. 8.5.

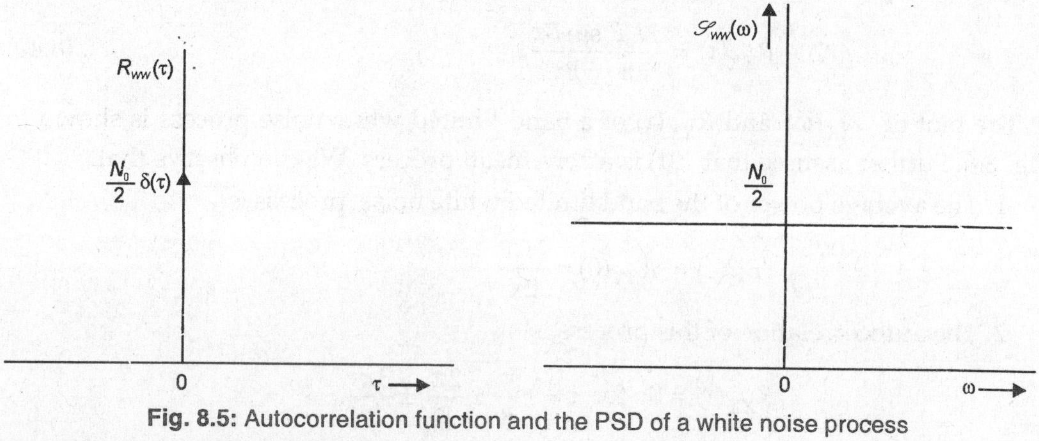

Fig. 8.5: Autocorrelation function and the PSD of a white noise process

Properties

1. The term *white noise* is analogous to *white light* which contains all visible light frequencies.
2. White noise is generally assumed to be *zero mean* .
3. A white noise process is unpredictable as the noise samples at different instants of time are uncorrelated:

$$C_{WW}(t_i, t_j) = 0 \qquad \text{for } t_i \neq t_j \qquad ...(8.38)$$

   Thus, the samples of a white noise process are uncorrelated no matter how closely the samples are placed. Thus, white noise has an infinite variance, $\sigma_W^2 \to \infty$.
4. White noise is a mathematical abstraction; it cannot be physically realized since it has infinite average power.
5. If the system bandwidth (BW) is sufficiently narrower than the noise BW and noise PSD is flat, we can model it as a white noise process. Thermal noise, which is the noise generated in resistors due to random motion electrons, is well modeled as white noise, since they have very flat PSD over very wide band (GHz).
6. A white noise process can have any probability density function. Particularly, if the white noise process $W(t)$ is a Gaussian random process, then $W(t)$ is called a white Gaussian random process.
7. A white noise process is called a *strict sense white noise process* if the noise samples at distinct instants of time are independent. A white Gaussian noise process is a strict sense white noise process. Such a process represents a pure random process, because its samples at arbitrarily close intervals will also be independent.

Band Limited White Noise

A white noise process is said to be band limited white noise, if its PSD is non-zero constant over a finite frequency band and zero for all other frequency band. Thus, the WSS process $X(t)$ is band limited white noise if

$$\mathcal{S}_{XX}(\omega) = \frac{N_0}{2} \qquad \text{for } -B < \omega < B \qquad ...(8.39)$$

For example, thermal noise which has constant PSD up to very high frequency is better modeled by a band limited white noise process.

The corresponding autocorrelation function $R_{XX}(\tau)$ is given by

$$R_{XX}(\tau) = \frac{N_0 B}{2\pi}\frac{\sin B\tau}{B\tau} \qquad ...(8.40)$$

The plot of $\mathcal{S}_{XX}(\omega)$ and $R_{XX}(\tau)$ of a band limited white noise process is shown in Fig. 8.6. Further assume that $X(t)$ is a zero mean process. We can observe that:

1. The average power of the band limited white noise process is

$$E[X^2] = R_{XX}(0) = \frac{N_0 B}{2\pi}$$

2. The autocorrelation of this process is

$$R_{XX}(\tau) = 0 \text{ for } \tau = \pm\frac{\pi}{B}, \pm\frac{2\pi}{B}, \pm\frac{3\pi}{B}, ...$$

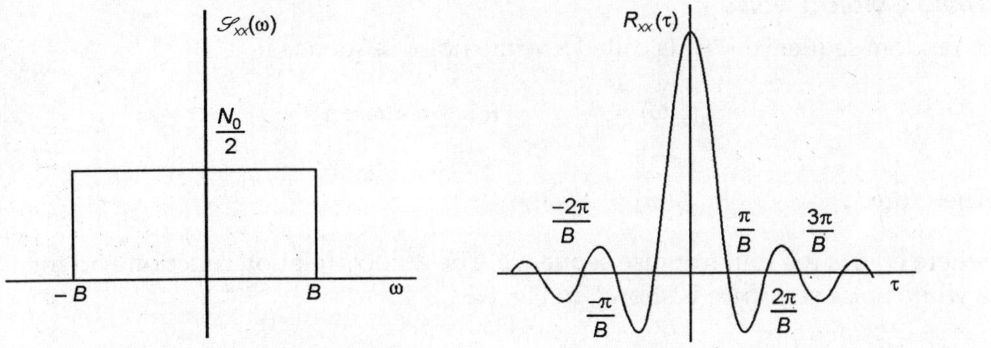

Fig. 8.6: $\mathcal{S}_{XX}(\omega)$ and $R_{XX}(\tau)$ of a band limited white noise process

This means that $X(t)$ and $X\left(t+\dfrac{n\pi}{B}\right)$ are uncorrelated for any value of nonzero integer n. Thus, we can get uncorrelated samples by sampling a band limited white noise process at a uniform sampling interval of $\dfrac{\pi}{B}$.

A band limited white noise process may also be a band pass process with PSD as shown in Fig. 8.7 and given by

$$\mathcal{S}_{XX}(\omega) = \begin{cases} \dfrac{N_0}{2} & |\omega - \omega_c| < \dfrac{B}{2} \\ 0 & \text{otherwise} \end{cases} \qquad \text{...(8.41)}$$

The corresponding autocorrelation function is given by

$$R_{XX}(\tau) = \frac{N_0 B}{2\pi} sa \frac{B}{2}\tau \cos \omega_c \tau \qquad \text{...(8.42)}$$

Colored Noise

A noise process which is not white is called *colored noise*. Thus, the noise process $X(t)$ with $R_{XX}(\tau) = a^2 e^{-b|\tau|}$ for $b > 0$ and PSD $\mathcal{S}_{XX}(\omega) = \dfrac{2a^2 b}{b^2 + \omega^2}$ is an example of colored noise.

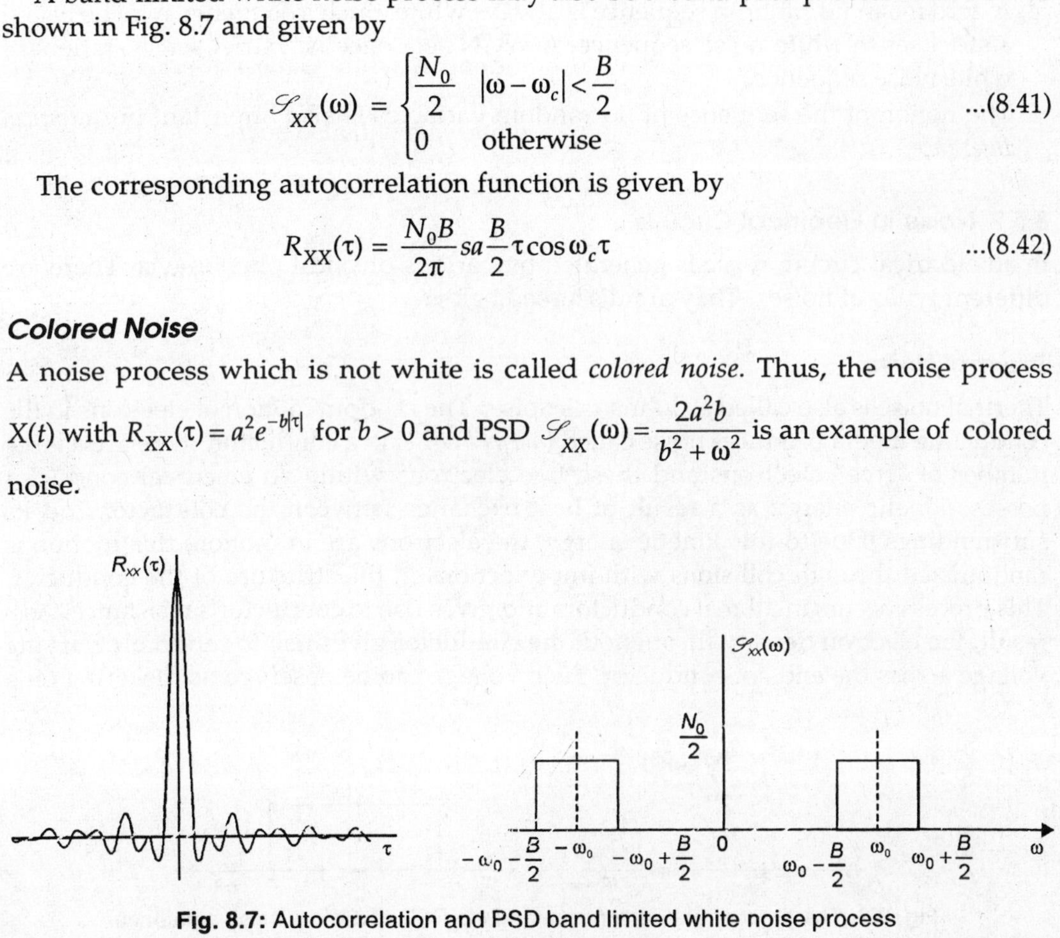

Fig. 8.7: Autocorrelation and PSD band limited white noise process

White Colored Noise

A random sequence $W(t)$ is called a white noise sequence if

$$\mathcal{S}_{WW}(\omega) = \frac{N}{2} \qquad \text{for } -\pi < \omega < \pi \qquad ...(8.43)$$

Therefore

$$R_{WW}(m) = \frac{N_0}{2}\delta(m) \qquad ...(8.44)$$

where $\delta(m)$ is the unit impulse sequence. The autocorrelation function and the PSD of a white noise sequence is shown in Fig. 8.8.

Properties

1. The average power of the white noise sequence is $E[X^2(n)] = \frac{1}{2\pi}\frac{N}{2} \times 2\pi = \frac{N}{2}$.
2. The average power of the white noise sequence is finite and uniformly distributed over all frequencies.
3. If the white noise sequence $W(t)$ is a Gaussian sequence, then $W(t)$ is called a *white Gaussian noise* (WGN) sequence.
4. A zero mean iid random sequence is always white. Such a sequence may be called a strict sense white noise sequence. A WGN sequence is a strict sense stationary white noise sequence.
5. The notion of the sequence of iid random variables is also important in *statistical inference.*

8.5.2 Noise in Electrical Circuits

In an electrical circuit, noise is generated by various physical phenomena. There are different types of noises. They are discussed below.

Thermal Noise

Thermal noise is also called as Johnson's noise. The random motion of electrons in the conducting media produces noise called *thermal noise*. In a conductor, there are a large number of "free" electrons and these free electrons within an electrical conductor possess kinetic energy as a result of heat exchange between the conductor and its surroundings. Due to this kinetic energy, the electrons are in motion; this motion is randomized through collisions with imperfections in the structure of the conductor. This process occurs in all real conductors and gives rise to conductor's resistance. As a result, the electron density throughout the conductor gives rise to randomly varying voltage across the ends of conductor. Such voltage can be observed as flickering on a

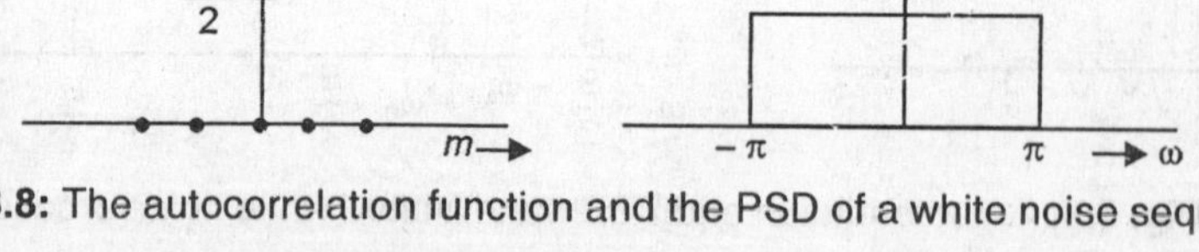

Fig. 8.8: The autocorrelation function and the PSD of a white noise sequence

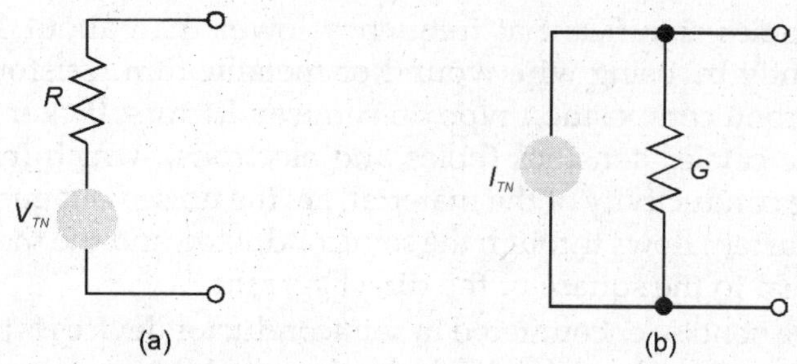

Fig. 8.9: Models of a noisy resistor: (a) Thevenin equivalent circuit (b) Norton equivalent circuit

very sensitive voltmeter. The average or mean noise voltage across the conductor is zero, but the root mean square value is finite and can be measured. The mean square value of the noise voltage is proportional to the resistance of the conductor, to its absolute temperature and to the frequency bandwidth of the device measuring the noise.

For a given resistor with resistance *R* ohms, the mean square value of the thermal noise voltage measured across the terminals of the resistor is

$$E[V_{\text{thermal}}^2] = 4KTBR \qquad \text{...(8.45)}$$

where *B* is the bandwidth (in hertz) over which the noise voltage is measured, *K* is Boltzmann's constant, 1.38×10^{-23} W/Hz/K, and *T* is temperature in K.

The mean square value of the noise current generator is

$$E[I_{\text{thermal}}^2] = \frac{E[V_{\text{thermal}}^2]}{R^2} = \frac{4KTBR}{R^2} = 4KTBG \qquad \text{...(8.46)}$$

Shot Noise

While it is often appropriate to consider electrical current as a continuous quantity, it must be kept in mind that it is composed of discrete electrons, each with a fixed charge. Current is quantized. Rather than measuring a continuous flow, the measurement of current is composed of sensing the effects of individual electrons arriving at a detector with a uniform distribution over time. The effect of this variation in time of arrival is called shot noise or corpuscular noise. Shot noise is produced by random fluctuations of current flow in electronic devices. It is present in both vacuum tubes and semiconductor devices. In vacuum tubes, shot noise arises due to random emission of electrons from the cathode. In semiconductor devices, this effect arises due to random diffusion of minority carriers and the random generation and recombination of hole–electron pairs.

$$E[I_{\text{shot}}^2] = 2IqB\,\text{amp}^2 \qquad \text{...(8.47)}$$

where *q* is the electron charge equal to 1.6×10^{-19} coulombs, *i* is the mean value of the current in amperes, and *B* is the bandwidth of the measuring instrument in hertz.

Besides thermal and shot noise, transistors exhibit a low frequency noise phenomenon known as flicker noise. The mean square value of flicke. noise is inversely proportional to frequency; hence it is also referred to as "one-over-*f*" noise. The cause of flicker noise is not well understood and is recognizable by its frequency dependence.

Flicker noise becomes significant at frequency lower than about 100 Hz. It can be reduced significantly by using wire wound or metallic film resistors rather than the more common carbon composition type. In semiconductors, flicker noise arises from fluctuations in the carrier densities (holes and electrons), which in turn give rise to fluctuations in the conductivity of the material, i.e. the noise voltage will be developed whenever direct current flows through the semiconductor, and the mean square voltage will be proportional to the square of the direct current.

Yet another type of noise encountered in semiconductor devices is burst noise, whose mean square value falls off as $1/f^2$. Flicker noise and burst noise are both nonwhite, with their degrading effects being observed at low frequencies. Ordinarily, they can be ignored above a few kilohertz. On the other hand, thermal noise and shot noise are both white for all practical purposes; hence their degrading influence on the operation of a communication system extends right across the complete frequency band of interest.

8.5.3 Noise Measurement

The noise of a system can be defined by three ways. They are as follows:

1. Noise factor
2. Noise figure
3. Noise temperature

Noise Factor

The noise factor of a two-port device is defined as the ratio of the total available output noise power (due to the device and the source) per unit bandwidth to the portion thereof due to the source.

It is also defined as the ratio of the output noise power to the input noise power.

$$N_{\text{factor}} = \frac{\text{Output noise power}}{\text{Input noise power}} = \left.\frac{P_{NO}}{P_{NI}}\right|_{T=290\text{K}} \qquad \text{...(8.48)}$$

It is also defined as the ratio of signal-to-noise ratio at the input to the signal-to-noise ratio at the output.

$$N_{\text{factor}} = F = \frac{\text{input signal–noise ratio}}{\text{output signal–noise ratio}} = \frac{(SNR)_{\text{input}}}{(SNR)_{\text{output}}} \qquad \text{...(8.49)}$$

The input noise power P_{NI} is defined as the product of the source noise at standard temperature (T_0) and the amplifier gain (G):

$$P_{NI} = GKBT_0 \qquad \text{...(8.50)}$$

where B is the bandwidth, K is Boltzmann's constant and $T_0 = 290\,\text{K}$.

$$F = \frac{P_{NO}}{P_{NI}} = \frac{P_{NO}}{GKBT_0}$$

$$F = \frac{GKBT_0 + N_A^2}{GKBT_0} = 1 + \frac{N_A^2}{GKBT_0} \qquad \text{...(8.51)}$$

where N_A is the noise voltage added by the DUT (device under test).

$$F = \frac{\mathcal{S}_{\text{N output}}(\omega)}{G(\omega)\mathcal{S}_{\text{N input}}(\omega)} \quad \text{...(8.52)}$$

where $\mathcal{S}_{\text{N output}}(\omega)$ and $\mathcal{S}_{\text{N input}}(\omega)$ are power spectral density of output and input respectively.

$$S_{\text{N output}}(\omega) \text{ in dB} = F \text{ in dB} + S_{\text{N input}}(\omega) \text{ in dB} + G \text{ in dB} \quad \text{...(8.53)}$$

Noise Figure

Expressing noise factor in decibels is called *noise figure.*

$$N_F = 10 \log F \text{ in dB} \quad \text{...(8.54)}$$

It is used to measure the quality of amplifier. Thus, it is a figure of merit.

Noise Temperature

A disadvantage of the noise figure is that when it is used to compare low noise devices, the values are all close to unity, which makes the comparison difficult. In such cases, it is preferable to use the equivalent noise temperature.

Effective noise temperature refers to the temperature that a matched input resistance would require to exhibit the same added noise.

$$N_A^2 = GKBT_e \quad \text{...(8.55)}$$

$$F = 1 + \frac{N_A^2}{GKBT_0} = 1 + \frac{T_e}{T_0}$$

$$T_e = T_0(F-1) \quad \text{...(8.56)}$$

where T_0 is reference temperature, usually 290 K.

8.5.4 Noise Figure of Cascaded Stages

Consider a two-port network of K-stages. Assume that the thermal noise is present at the input of the first stage. This thermal noise is amplified by the first stage gain G_1. The total noise power at the output of the first stage is the sum of the amplified thermal noise and additive noise produced by the first stage amplifier.

$$|N_1|^1 = G_1 T_0 KB + |N_{A1}|^2 \quad \text{...(8.57)}$$

The input of the second stage is the output of the first stage. Hence, the total noise power at the output of the second stage is

$$|N_2|^2 = G_2|N_1|^2 + |N_{A2}|^2 = G_2[G_1 T_0 KB + |N_{A1}|^2] + |N_{A2}|^2$$

$$|N_2|^2 = G_2 G_1 T_0 KB + G_2 |N_{A1}|^2 + |N_{A2}|^2 \quad \text{...(8.58)}$$

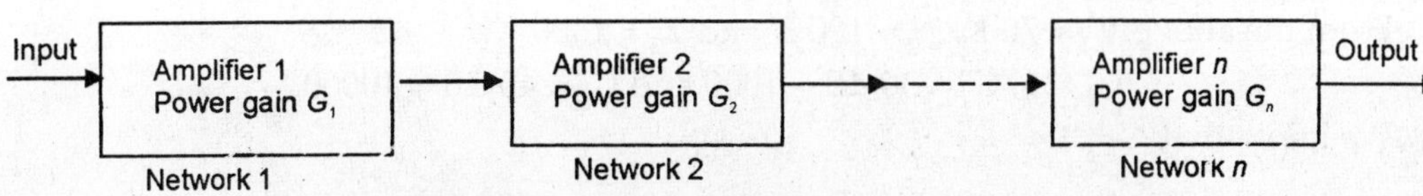

Fig. 8.10: Cascaded n-stage noisy network

Similarly for K^{th} stage, the total noise power is

$$|N_K|^2 = G_K...G_2G_1T_0KB + G_K...G_3G_2|N_{A1}|^2 + G_K...G_4G_3|N_{A2}|^2 + \cdots + |N_{AK}|^2$$

$$F_K = \frac{|N_K|^2}{GKBT_0} = \frac{G_K...G_2G_1T_0KB + G_K...G_3G_2|N_{A1}|^2 + G_K...G_4G_3|N_{A2}|^2 + \cdots + |N_{AK}|^2}{GKBT_0}$$

$$F_K = 1 + \frac{G_K...G_3G_2|N_{A1}|^2}{GKBT_0} + \frac{G_K...G_4G_3|N_{A2}|^2}{GKBT_0} + \frac{|N_{AK}|^2}{GKBT_0}$$

$$F_K = 1 + \frac{|N_{A1}|^2}{G_1KBT_0} + \frac{|N_{A2}|^2}{G_1G_2KBT_0} + \frac{|N_{A3}|^2}{G_1G_2G_3KBT_0} + \cdots + \frac{|N_{AK}|^2}{G_1G_2...G_KKBT_0}$$

$$F_K = F_1 + \frac{1}{G_1}\left(1 + \frac{|N_{A2}|^2}{G_2KBT_0} - 1\right) + \frac{1}{G_1G_2}\left(1 + \frac{|N_{A3}|^2}{G_3KBT_0} - 1\right)$$

$$+ \cdots + \frac{1}{G_1G_2...G_{K-1}}\left(1 + \frac{|N_{AK}|^2}{G_KKBT_0} - 1\right)$$

$$F_K = F_1 + \frac{F_2 - 1}{G_1} + \frac{F_3 - 1}{G_1G_2} + \cdots + \frac{F_K - 1}{G_1G_2...G_{K-1}} \qquad ...(8.59)$$

Example 8.5.1: Calculate the thermal noise power available from any resistor at room temperature (290 K) for a bandwidth of 2 MHz. Also calculate the corresponding noise voltage, given that $R = 100\ \Omega$.

Solution: For a 2 MHz bandwidth, the noise power is

$$P_n = KTB = 1.38 \times 10^{-23} \times 290 \times 2 \times 10^6 = 8 \times 10^{-15}\ \text{W}$$

$$E_n = \sqrt{4RKTB} = \sqrt{4 \times 100 \times 1.38 \times 10^{-23} \times 290 \times 2 \times 10^6}$$

$$= \sqrt{3.2 \times 10^{-12}} = 1.78\ \mu\text{V}.$$

EXAMPLES

Example 8.1: A satellite receiver consists of an antenna that has a noise temperature of 70 K. This is followed by a low noise amplifier with 30 dB of gain and noise temperature of 100 K. The low noise amplifier, in turn, is followed by a second stage amplifier with a gain of 40 dB and noise temperature of 1400 K. What is the system noise temperature?

Solution: Since the antenna does not provide any electrical gain to the received signal, $G_1 = 1$.

From the above, $T_1 = 70$ K, $T_2 = 100$ K and $T_3 = 1400$ K

$G_1 = 1$, $G_2 = 30$ dB $= 1000$ and $G_3 = 40$ dB $= 10000$.

We know that

$$T_e = T_1 + \frac{T_2}{G_1} + \frac{T_3}{G_1G_2} = 70 + \frac{100}{1} + \frac{1400}{1 \times 1000} = 171.4\text{K}$$

Example 8.2: The noise figure of an amplifier is 10 dB. Calculate the output signal to noise ratio when the input signal to noise ratio is 25 dB.

Solution: From the definition of noise factor,

$$(SNR_{output})_{dB} = (SNR_{input})_{dB} - F(\text{in dB}) = 25 - 10 = 15 \text{ dB}$$

Example 8.3: An amplifier has a noise figure of 20 dB and gain of 10 dB. Calculate equivalent amplifier input noise for a bandwidth of 5 MHz.

Solution: From the above data, noise figure F = 20 dB = 100 and B = 5 MHz.

We know that $$F = 1 + \frac{N_A^2}{GKBT_0}$$

$$N_A^2 = (F-1)GKBT_0$$

$$N_A^2 = (100-1) \times 10 \times 1.38 \times 10^{-23} \times 5 \times 10^6 \times 290 = 1.98 \times 10^{-11} \text{ W.}$$

Example 8.4: A mixer stage has a noise figure of 30 dB and this is preceded by an amplifier that has a noise figure of 10 dB and an available power gain of 20 dB. Calculate overall noise figure referred to the input.

Solution: It is necessary to convert all decibel values to the equivalent power ratios.

From the given data, F_2 = 20 dB = 100, F_1 = 10 dB = 1 and G_1 = 20 dB = 100

$$F = F_1 + \frac{F_2 - 1}{G_1} = 10 + \frac{100-1}{100} = 10.99$$

$$F(\text{in dB}) = 10.4099 \text{ dB.}$$

Example 8.5: Calculate the shot noise component of the current present on the direct current of 2 mA flowing across a semiconductor junction, given that the effective noise bandwidth is 0.5 MHz.

Solution: $$E[I_{shot}^2] = 2IqB$$

$$E[I_{shot}^2] = 2 \times 2 \times 10^{-3} \times 1.6 \times 10^{-19} \times 0.5 \times 10^6 = 3.2 \times 10^{-16} \text{ amp}^2 = 17.88 \text{ nA.}$$

Example 8.6: A receiver has a noise figure of 15 dB and it is fed by a low noise amplifier that has gain of 50 dB and a noise temperature of 100 K. Calculate the noise temperature of the receiver and the overall noise temperature of the receiving system.

Solution: From the given data, F = 15 dB = 31.6 and the gain G = 50 dB = 100000.

We know that $$T_e = T_0(F-1)$$

$$T_e = (31.6 - 1) \times 290 = 8874 \text{ K}$$

The overall noise temperature is

$$T_e = T_1 + \frac{T_2}{G_1} = 100 + \frac{8874}{100000} = 100.08 \text{ K.}$$

Example 8.7: Find the power spectral density and average power of the response of the system shown in Fig. 8.11 when the input $X(t)$ is white noise.

Solution: From the RL circuit, the transfer function is given by

$$H(\omega) = \frac{V_o(\omega)}{V_i(\omega)} = \frac{R}{R + j\omega L} = \frac{1}{1 + \frac{j\omega L}{R}}$$

$$|H(\omega)| = \frac{1}{1 + \left(\frac{\omega L}{R}\right)^2}$$

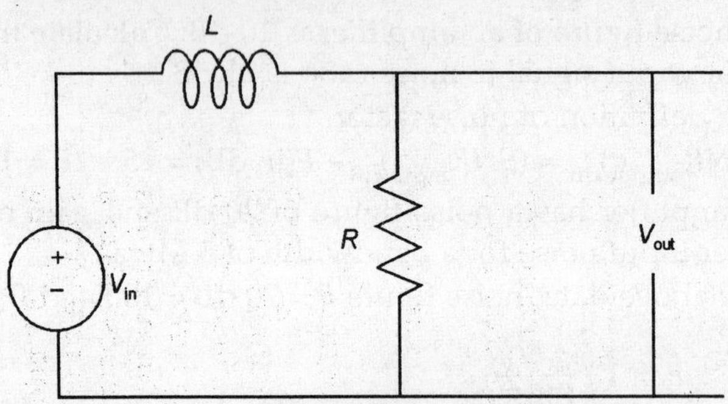

Fig. 8.11

We know that the power spectral density of response is

$$\mathcal{S}_{YY}(\omega) = \mathcal{S}_{XX}(\omega)|H(\omega)|^2$$

$$\mathcal{S}_{YY}(\omega) = \frac{N_0}{2}|H(\omega)|^2 = \frac{N_0}{2}\left[\frac{1}{1+\left(\frac{\omega L}{R}\right)^2}\right]$$

The average power of the response is given by

$$P_{YY} = \frac{1}{2\pi}\int_{-\infty}^{\infty} \mathcal{S}_{YY}(\omega)\ d\omega = R_{YY}(0) = \int_{-\infty}^{\infty} R_{XX}(0)h^2(t)dt$$

$$P_{YY} = \frac{1}{2\pi}\int_{-\infty}^{\infty} \mathcal{S}_{YY}(\omega)\,d\omega = \frac{N_0}{4\pi}\int_0^{\infty}\frac{1}{1+(\omega L/R)^2}\,d\omega = \frac{N_0 R}{4L}$$

Another approach is as follows:

$$H(\omega) = \frac{1}{1+\frac{j\omega L}{R}}$$

$$h(t) = \frac{1}{1+\frac{j\omega L}{R}} = \frac{R}{L}e^{-\frac{R}{L}t}u(t)$$

The average power of the response is given by

$$P_{YY} = \overline{Y^2} = \frac{N_0}{2}\int_0^{\infty} h^2(t)dt = \frac{N_0}{2}\left(\frac{R}{L}\right)^2\int_0^{\infty} e^{-\frac{2R}{L}t}dt = \frac{N_0}{2}\left(\frac{R}{L}\right)^2 \left.\frac{e^{-\frac{2R}{L}t}}{-\frac{2R}{L}}\right|_0^{\infty} = \frac{N_0}{4}\left(\frac{R}{L}\right).$$

Example 8.8: A white Gaussian noise $X(t)$ with autocorrelation $R_{XX}(\tau) = \frac{N_0}{2}\delta(\tau)$ is given as input to a filter with transfer function

$$h(t) = \begin{cases}\frac{1}{T} & 0 \le t \le T \\ 0 & \text{otherwise}\end{cases}$$

Find: (a) Expected value of the response $Y(t)$.
(b) Cross-correlation of input–output.
(c) Autocorrelation of the response.

Solution: Since the input is white noise, the expected value of white noise is zero.

The expected value of the response is given by

$$E[Y(t)] = \mu_X H(0) = 0$$

The cross-correlation is defined as

$$R_{XY}(t,\tau) = R_{XY}(\tau) = \int_{-\infty}^{\infty} R_{XX}(\tau-\tau_2)h(\tau_2)d\tau_2$$

$$R_{XY}(\tau) = \frac{N_0}{2T}\int_0^T \delta(\tau-\tau_2)d\tau_2 = \begin{cases} \frac{N_0}{2T} & 0 \le \tau \le T \\ 0 & \text{otherwise} \end{cases}$$

The autocorrelation is defined as

$$R_{YY}(\tau) = R_{XY}(\tau) * h(-\tau)$$

$$R_{YY}(\tau) = \int_{-\infty}^{\infty} R_{XY}(\tau-\tau_2)h(-\tau_2)d\tau_2$$

$$= \int_0^T R_{XY}(\tau-\tau_2)h(-\tau_2)d\tau_2 = \frac{1}{T}\int_{-T}^{0} R_{XY}(\tau-\tau_2)d\tau_2$$

where $$R_{XY}(\tau) = \begin{cases} \frac{N_0}{2T} & 0 \le \tau-\tau_2 \le T \\ 0 & \text{otherwise} \end{cases} = \begin{cases} \frac{N_0}{2T} & \tau-T \le \tau_2 \le \tau \\ 0 & \text{otherwise} \end{cases}$$

To find $R_{YY}(\tau)$, there are three cases:

Case 1: $\tau < T$

In this case $R_{XY}(\tau) = 0$

Since $R_{XY}(\tau) = 0$, the integral value is zero, i.e. $R_{YY}(\tau) = 0$

Case 2: $-T \le \tau \le 0$

The common area between $R_{XY}(\tau)$ and $h(-\tau)$ is

$$R_{YY}(\tau) = \frac{1}{T}\int_{-T}^{\tau} R_{XY}(\tau-\tau_2)d\tau_2 = \frac{1}{T}\int_{-T}^{\tau}\frac{N_0}{2T}d\tau_2 = \frac{N_0}{2T^2}\tau_2\Big|_{-T}^{\tau} = \frac{N_0(\tau+T)}{2T^2}$$

Case 3: $0 \le t \le T$

The common area between $R_{XY}(\tau)$ and $h(-\tau)$ is

$$R_{YY}(\tau) = \frac{1}{T}\int_{\tau-T}^{0} R_{XY}(\tau-\tau_2)d\tau_2 = \frac{1}{T}\int_{\tau-T}^{0}\frac{N_0}{2T}d\tau_2 = \frac{N_0}{2T^2}\tau_2\Big|_{\tau-T}^{0} = \frac{N_0(T-\tau)}{2T^2}$$

Hence $$R_{YY}(\tau) = \begin{cases} \frac{N_0(T-|\tau|)}{2T^2} & |\tau| \le T \\ 0 & \text{otherwise} \end{cases}$$

Example 8.9: A stationary random process $X(t)$ having an autocorrelation function $R_{XX}(\tau) = 2e^{-4|\tau|}$ is applied to the network shown in Fig. 8.12. Find:

(a) $S_{XX}(\omega)$ (b) $|H(\omega)|^2$ (c) $S_{YY}(\omega)$

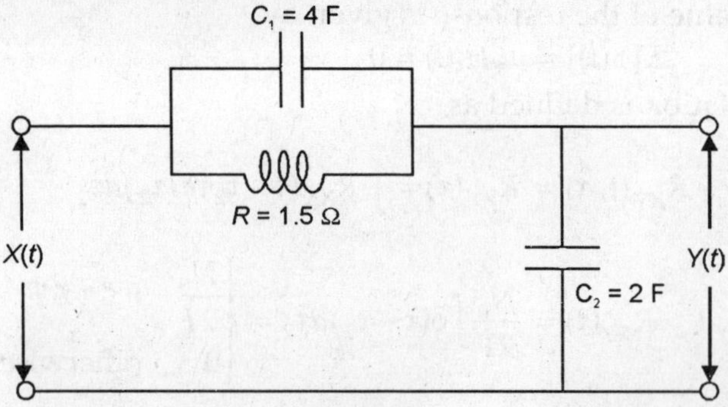

Fig. 8.12

Solution: $S_{XX}(\omega) = \int_{-\infty}^{\infty} 2e^{-4|\tau|}e^{-j\omega\tau}d\tau = \int_{-\infty}^{0} 2e^{4\tau}e^{-j\omega\tau}d\tau + \int_{0}^{\infty} 2e^{-4\tau}e^{-j\omega\tau}d\tau$

$$= \int_{-\infty}^{0} 2e^{(4-j\omega)\tau}d\tau + \int_{0}^{\infty} 2e^{-(j\omega+4)\tau}d\tau = 2\frac{e^{(4-j\omega)\tau}}{(4-j\omega)}\Bigg|_{-\infty}^{0} + 2\frac{e^{-(j\omega+4)\tau}}{-(j\omega+4)}\Bigg|_{0}^{\infty}$$

$$S_{XX}(\omega) = 2\frac{1}{(4-\omega j)} + 2\frac{1}{(4+j\omega)} = \frac{16}{16+\omega^2}$$

By applying Laplace transform and kvl:

$$X(s) = \frac{1.5}{1+6s}I(s) + \frac{1}{2s}I(s) = I(s)\left(\frac{1.5}{1+6s} + \frac{1}{2s}\right) = I(s)\left(\frac{9s+1}{2s(1+6s)}\right)$$

$$I(s) = \frac{2s(1+6s)}{9s+1}X(s)$$

$$Y(s) = \frac{1}{2s}I(s) = \left(\frac{1}{2s}\right)\frac{2s(1+6s)}{9s+1}X(s) = \frac{(1+6s)}{9s+1}X(s)$$

$$\frac{Y(s)}{X(s)} = H(s) = \frac{1+6s}{9s+1}$$

$$H(\omega) = H(s)\big|_{s=j\omega} = \frac{1+6j\omega}{1+9j\omega}$$

$$|H(\omega)| = \frac{|1+6j\omega|}{|1+9j\omega|} = \frac{\sqrt{1+36\omega^2}}{\sqrt{1+81\omega^2}}$$

$$|H(\omega)|^2 = \frac{1+36\omega^2}{1+81\omega^2}$$

$$S_{YY}(\omega) = S_{XX}(\omega)|H(\omega)|^2 = \frac{16(1+36\omega^2)}{(16+\omega^2)(1+81\omega^2)}.$$

Example 8.10: A linear system has an impulse response of $h(t) = e^{-4t}\, u(t)$. The input to this system is a sample function from a random process having an autocorrelation function of $R_{XX}(\tau) = e^{-\tau}$. Find the value of the autocorrelation of the system for:

(a) $\tau = 0$ (b) $\tau = 0.5$ (c) $\tau = 1$.

Solution: $R_{YY}(t, \tau) = R_{YY}(\tau) = \int\limits_{-\infty}^{\infty}\int\limits_{-\infty}^{\infty} R_{XX}(\tau_1 + \tau - \tau_2) h(\tau_1) h(\tau_2) d\tau_1 d\tau_2$

$$R_{YY}(t, \tau) = R_{YY}(\tau) = \int\limits_{0}^{\infty}\int\limits_{0}^{\infty} e^{-(\tau_1+\tau-\tau_2)} e^{-4\tau_1} e^{-4\tau_2} d\tau_1 d\tau_2 = e^{-\tau}\int\limits_{0}^{\infty} e^{-3\tau_2}\left(\int\limits_{0}^{\infty} e^{-5\tau_1} d\tau_1\right) d\tau_2$$

$$= e^{-\tau}\left.\frac{e^{-3\tau_2}}{-3}\right|_0^{\infty}\left.\frac{e^{-5\tau_2}}{-5}\right|_0^{\infty} = \frac{e^{-\tau}}{15}$$

(a) $\tau = 0$, $R_{YY}(0) = \dfrac{1}{15}$

(b) $\tau = 0.5$, $R_{YY}(0.5) = \dfrac{e^{-0.5}}{15} = 0.0404$

(c) $\tau = 1$, $R_{YY}(1) = \dfrac{e^{-1}}{15} = 0.0245.$

Example 8.11: Find the equivalent noise bandwidth of an RC filter shown in Fig. 8.13.

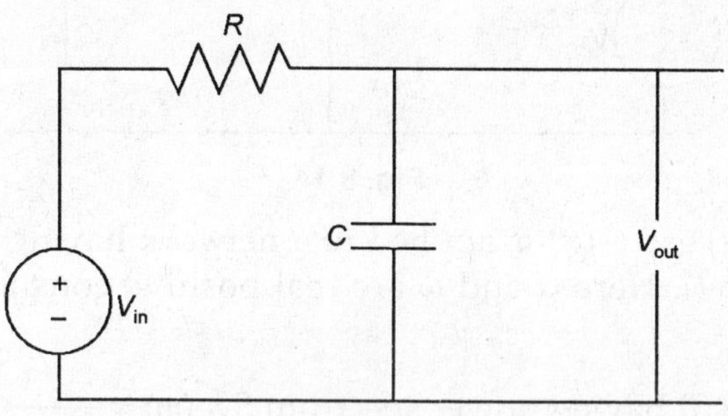

Fig. 8.13

Solution: Equivalent noise bandwidth $B_N = \dfrac{\int\limits_0^{\infty} |H(f)|^2 df}{|H(0)|^2}$

The transfer function of RC filter is given by

$$H(f) = \frac{1}{1 + j2\pi fRC}$$

$$|H(f)| = \frac{1}{\sqrt{1 + (2\pi fRC)^2}}$$

$$|H(f)|^2 = \frac{1}{1 + (2\pi fRC)^2}$$

Equivalent noise bandwidth is given by

$$B_N = \frac{\int_0^\infty |H(f)|^2\, df}{|H(0)|^2} = \frac{1}{|H(0)|^2}\int_0^\infty \frac{1}{1+(2\pi fRC)^2}\, df = 1\left(\int_0^\infty \frac{1}{1+(2\pi fRC)^2}\, df\right)$$

$$= \int_0^\infty \frac{1}{1+(2\pi fRC)^2}\, df = \frac{1}{2\pi RC}\tan^{-1}(2\pi fRC)\Big|_0^\infty = \frac{1}{2\pi RC}\left(\frac{\pi}{2}-0\right) = \frac{1}{4RC}$$

$$= \frac{1}{4RC} = \frac{\pi}{4\pi RC} = 0.5\pi f_c$$

where f_c = 3 dB is the cut-off frequency.

EXERCISE

1. Find the power spectral density and average power of the response of the system shown in Fig. 8.14. when the input $X(t)$ is white noise.

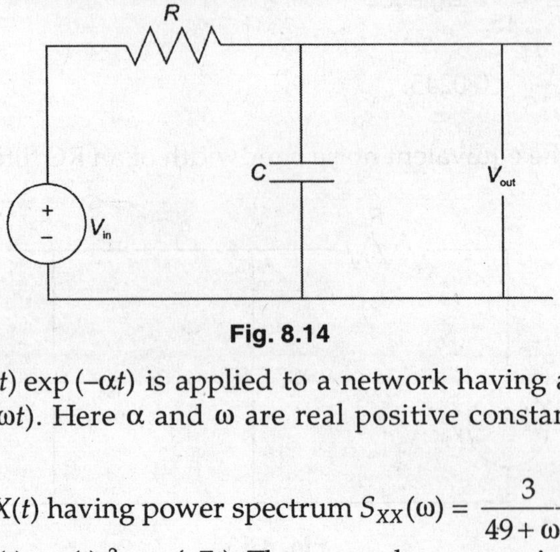

Fig. 8.14

2. A signal $x(t) = u(t)\exp(-\alpha t)$ is applied to a network having an impulse response $h(t) = \omega u(t)\exp(-\omega t)$. Here α and ω are real positive constants. Find the system response.

3. A random noise $X(t)$ having power spectrum $S_{XX}(\omega) = \dfrac{3}{49+\omega^2}$ is applied to a network for which $h(t) = u(t)t^2\exp(-7t)$. The network response is denoted by $Y(t)$.
   (a) What is the average power of $X(t)$?
   (b) Find the power spectrum of $Y(t)$.
   (c) Find average power of $Y(t)$.

4. A WSS process $X(t)$ having $R_{XX}(\tau) = Ae^{-a|\tau|}$ where A and a are real constants, is applied to input of an LTI system with $h(t) = u(t)\exp(-bt)$, where b is a real positive constant. Find the PSD of the output of system.

5. Find the overall noise figure and equivalent input noise temperature of the circuit shown in Fig. 8.15. Assume room temperature to be 27°C.

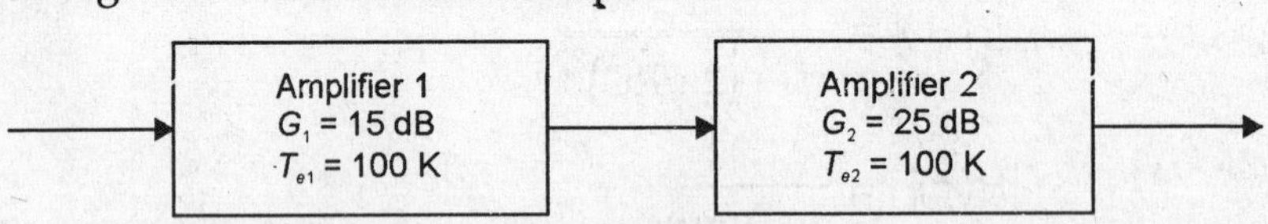

Fig. 8.15

6. White noise process with power density $\frac{N_0}{2} = 6 \times 10^{-6} \frac{\text{W}}{\text{Hz}}$ is applied to an ideal filter (gain = 1) with bandwidth W(rad/s). Find W so that the output average noise power is 15 W.

7. A linear system has a transfer function given by $H(\omega) = \frac{\omega}{\omega^2 + 10\omega + 50}$.

   Determine the power spectral density of the output when the input function is:

   (a) A stationary random process $X(t)$ with an autocorrelation function $R_{XX}(\tau) = 5e^{-|\tau|}$.

   (b) White noise that has a mean square value of 2 V^2/Hz.

8. Compute the noise spectral density (W/Hz) of:
   (a) An ideal resistor at nominal temperature of 290 K.
   (b) An amplifier with an equivalent noise temperature of 25,000 K.

9. The input $X(t)$ to a linear time invariant filter is a wide sense stationary random process with expected value $\mu_X = -5$ volts. The impulse response of the filter is

$$h(t) = \begin{cases} 1 - 10^4 t^2 & 0 \le t \le 10^{-2} \text{ sec} \\ 0 & \text{otherwise} \end{cases}$$

   What is the expected value of the output process $Y(t)$?

10. $X(t)$, the input to a linear time invariant filter is a wide sense stationary stochastic process with expected value $\mu_X = 10$ volts. The filter output $Y(t)$ is a wide sense stationary stochastic process with expected $\mu_Y = 1$ volt. The filter impulse response is

$$h(t) = \begin{cases} e^{-at} & t \ge 0 \\ 0 & \text{otherwise} \end{cases}$$

   What is the value of the time constant a?

11. A white Gaussian noise signal $W(t)$ with autocorrelation function $R_{WW}(\tau) = \eta_0 \delta(\tau)$ is passed through an LTI filter $h(t)$. Show that the response $Y(t)$ has average power

$$E[Y^2(t)] = \eta_0 \int_{-\infty}^{\infty} h^2(\tau)\, d\tau.$$

12. A wide sense stationary process $X(t)$ with autocorrelation function $R_{XX}(\tau) = 100e^{-100|\tau|}$ is the input to an RC filter with impulse response

$$h(t) = \begin{cases} e^{-t/RC} & t \ge 0 \\ 0 & \text{otherwise} \end{cases}$$

   The filter output process has average power $E[Y^2(t)] = 100$.
   (a) Find the output autocorrelation $R_{YY}(\tau)$.
   (b) What is the value of RC?

13. A wide sense stationary process $X(t)$ with power spectral density

$$S_{YY}(f) = \begin{cases} 10^{-6} & |f| \le 50 \\ 0 & \text{otherwise} \end{cases}$$

is the input to an RC filter with frequency response

$$H(f) = \frac{1}{100\pi + j2\pi f}$$

The filter output is the stochastic process $Y(t)$. Find:

(a) $E[X^2(t)]$

(b) $S_{XY}(f)$

(c) $S_{YX}(f)$

(d) $S_{YY}(f)$

(e) $E[Y^2(t)]$.

14. Let $W(t)$ denote a wide sense stationary Gaussian noise process with $\mu_W = 0$ and power spectral density $S_{WW}(f)$.

(a) What is $R_{WW}(\tau)$, the autocorrelation of $W(t)$?

(b) $W(t)$ is the input to a linear time invariant filter with impulse response

$$H(f) = \begin{cases} 1 & |f| \leq B/2 \\ 0 & \text{otherwise} \end{cases}$$

The filter output is $Y(t)$. What is the power spectral density function of $Y(t)$?

(c) What is the average power of $Y(t)$?

(d) What is the expected value of the filter output?

APPENDICES

Appendix A

Gaussian Distribution Function

The general Gaussian or normal probability density and distribution functions are:

$$f_X(x) = \frac{1}{\sqrt{2\pi\sigma^2}} e^{-(x-\mu)^2/2\sigma^2} \quad \text{...(A.1)}$$

$$F_X(x) = \frac{1}{\sqrt{2\pi\sigma^2}} \int_{-\infty}^{x} e^{-(\xi-\mu)^2/2\sigma^2} d\xi = \phi\left(\frac{x-\mu}{\sigma}\right) \quad \text{...(A.2)}$$

where $0 < \mu < \infty$, $0 < \sigma$ are constants and $\phi(\cdot)$ is the normal distribution for $\mu = 0$, $\sigma = 1$.

The normalized or standard Gaussian distribution is

$$\phi(x) = \frac{1}{\sqrt{2\pi}} \int_{-\infty}^{x} e^{-\xi^2/2} d\xi \quad \text{...(A.3)}$$

When $\mu \neq 0$, $\sigma \neq 1$, $F_X(x)$ can be found from $\phi(x)$ by using Eq. (A.2). For negative values of x,

$$\phi(-x) = 1 - \phi(x) \quad \text{...(A.4)}$$

The values of $\phi(x)$ are listed in Tables A.1 and A.2.

Table A.1: The standard normal distribution values ($x \leq 0$)

$-x$	0.00	0.01	0.02	0.03	0.04	0.05	0.06	0.07	0.08	0.09
0.0	0.50000	0.49601	0.49202	0.48803	0.48405	0.48006	0.47608	0.47210	0.46812	0.46414
0.1	0.46017	0.45621	0.45224	0.44828	0.44433	0.44038	0.43644	0.43251	0.42858	0.42466
0.2	0.42074	0.41683	0.41294	0.40905	0.40517	0.40129	0.39743	0.39358	0.38974	0.38591
0.3	0.38209	0.37828	0.37448	0.37070	0.36693	0.36317	0.35942	0.35569	0.35197	0.34827
0.4	0.34458	0.34090	0.33724	0.33360	0.32997	0.32636	0.32276	0.31918	0.31561	0.31207
0.5	0.30854	0.30503	0.30153	0.29806	0.29460	0.29116	0.28774	0.28434	0.28096	0.27760
0.6	0.27425	0.27093	0.26763	0.26435	0.26109	0.25785	0.25463	0.25143	0.24825	0.24510
0.7	0.24196	0.23885	0.23576	0.23270	0.22965	0.22663	0.22363	0.22065	0.21770	0.21476
0.8	0.21186	0.20897	0.20611	0.20327	0.20045	0.19766	0.19489	0.19215	0.18943	0.18673
0.9	0.18406	0.18141	0.17879	0.17619	0.17361	0.17106	0.16853	0.16602	0.16354	0.16109
1.0	0.15866	0.15625	0.15386	0.15151	0.14917	0.14686	0.14457	0.14231	0.14007	0.13786

Contd.

Table A.1: The standard normal distribution values ($x \leq 0$) *(Contd.)*

$-x$	0.00	0.01	0.02	0.03	0.04	0.05	0.06	0.07	0.08	0.09
1.1	0.13567	0.13350	0.13136	0.12924	0.12714	0.12507	0.12302	0.12100	0.11900	0.11702
1.2	0.11507	0.11314	0.11123	0.10935	0.10749	0.10565	0.10384	0.10204	0.10027	0.09853
1.3	0.09680	0.09510	0.09342	0.09176	0.09012	0.08851	0.08692	0.08534	0.08379	0.08226
1.4	0.08076	0.07927	0.07780	0.07636	0.07493	0.07353	0.07215	0.07078	0.06944	0.06811
1.5	0.06681	0.06552	0.06426	0.06301	0.06178	0.06057	0.05938	0.05821	0.05705	0.05592
1.6	0.05480	0.05370	0.05262	0.05155	0.05050	0.04947	0.04846	0.04746	0.04648	0.04551
1.7	0.04457	0.04363	0.04272	0.04182	0.04093	0.04006	0.03920	0.03836	0.03754	0.03673
1.8	0.03593	0.03515	0.03438	0.03363	0.03288	0.03216	0.03144	0.03074	0.03005	0.02938
1.9	0.02872	0.02807	0.02743	0.02680	0.02619	0.02559	0.02500	0.02442	0.02385	0.02330
2.0	0.02275	0.02222	0.02169	0.02118	0.02068	0.02018	0.01970	0.01923	0.01876	0.01831
2.1	0.01786	0.01743	0.01700	0.01659	0.01618	0.01578	0.01539	0.01500	0.01463	0.01426
2.2	0.01390	0.01355	0.01321	0.01287	0.01255	0.01222	0.01191	0.01160	0.01130	0.01101
2.3	0.01072	0.01044	0.01017	0.00990	0.00964	0.00939	0.00914	0.00889	0.00866	0.00842
2.4	0.00820	0.00798	0.00776	0.00755	0.00734	0.00714	0.00695	0.00676	0.00657	0.00639
2.5	0.00621	0.00604	0.00587	0.00570	0.00554	0.00539	0.00523	0.00509	0.00494	0.00480
2.6	0.00466	0.00453	0.00440	0.00427	0.00415	0.00403	0.00391	0.00379	0.00368	0.00357
2.7	0.00347	0.00336	0.00326	0.00317	0.00307	0.00298	0.00289	0.00280	0.00272	0.00264
2.8	0.00256	0.00248	0.00240	0.00233	0.00226	0.00219	0.00212	0.00205	0.00199	0.00193
2.9	0.00187	0.00181	0.00175	0.00170	0.00164	0.00159	0.00154	0.00149	0.00144	0.00140
3.0	0.00135	0.00131	0.00126	0.00122	0.00118	0.00114	0.00111	0.00107	0.00104	0.00100
3.1	0.00097	0.00094	0.00090	0.00087	0.00085	0.00082	0.00079	0.00076	0.00074	0.00071
3.2	0.00069	0.00066	0.00064	0.00062	0.00060	0.00058	0.00056	0.00054	0.00052	0.00050
3.3	0.00048	0.00047	0.00045	0.00043	0.00042	0.00040	0.00039	0.00038	0.00036	0.00035
3.4	0.00034	0.00033	0.00031	0.00030	0.00029	0.00028	0.00027	0.00026	0.00025	0.00024
3.5	0.00023	0.00022	0.00022	0.00021	0.00020	0.00019	0.00019	0.00018	0.00017	0.00017

Table A.2: The standard normal distribution values ($x \geq 0$)

x	0.00	0.01	0.02	0.03	0.04	0.05	0.06	0.07	0.08	0.09
0.0	0.50000	0.50399	0.50798	0.51197	0.51595	0.51994	0.52392	0.52790	0.53188	0.53586
0.1	0.53983	0.54380	0.54776	0.55172	0.55567	0.55962	0.56356	0.56749	0.57142	0.57535
0.2	0.57926	0.58317	0.58706	0.59095	0.59483	0.59871	0.60257	0.60642	0.61026	0.61409
0.3	0.61791	0.62172	0.62552	0.62930	0.63307	0.63683	0.64058	0.64431	0.64803	0.65173
0.4	0.65542	0.65910	0.66276	0.66640	0.67003	0.67364	0.67724	0.68082	0.68439	0.68793
0.5	0.69146	0.69497	0.69847	0.70194	0.70540	0.70884	0.71226	0.71566	0.71904	0.72240
0.6	0.72575	0.72907	0.73237	0.73565	0.73891	0.74215	0.74537	0.74857	0.75175	0.75490
0.7	0.75804	0.76115	0.76424	0.76730	0.77035	0.77337	0.77637	0.77935	0.78230	0.78524
0.8	0.78814	0.79103	0.79389	0.79673	0.79955	0.80234	0.80511	0.80785	0.81057	0.81327
0.9	0.81594	0.81859	0.82121	0.82381	0.82639	0.82894	0.83147	0.83398	0.83646	0.83891
1.0	0.84134	0.84375	0.84614	0.84849	0.85083	0.85314	0.85543	0.85769	0.85993	0.86214
1.1	0.86433	0.86650	0.86864	0.87076	0.87286	0.87493	0.87698	0.87900	0.88100	0.88298
1.2	0.88493	0.88686	0.88877	0.89065	0.89251	0.89435	0.89617	0.89796	0.89973	0.90147
1.3	0.90320	0.90490	0.90658	0.90824	0.90988	0.91149	0.91308	0.91466	0.91621	0.91774
1.4	0.91924	0.92073	0.92220	0.92364	0.92507	0.92647	0.92785	0.92922	0.93056	0.93189
1.5	0.93319	0.93448	0.93574	0.93699	0.93822	0.93943	0.94062	0.94179	0.94295	0.94408

Contd.

Table A.2: The standard normal distribution values ($x \geq 0$) *(Contd.)*

x	0.00	0.01	0.02	0.03	0.04	0.05	0.06	0.07	0.08	0.09
1.6	0.94520	0.94630	0.94738	0.94845	0.94950	0.95053	0.95154	0.95254	0.95352	0.95449
1.7	0.95543	0.95637	0.95728	0.95818	0.95907	0.95994	0.96080	0.96164	0.96246	0.96327
1.8	0.96407	0.96485	0.96562	0.96638	0.96712	0.96784	0.96856	0.96926	0.96995	0.97062
1.9	0.97128	0.97193	0.97257	0.97320	0.97381	0.97441	0.97500	0.97558	0.97615	0.97670
2.0	0.97725	0.97778	0.97831	0.97882	0.97932	0.97982	0.98030	0.98077	0.98124	0.98169
2.1	0.98214	0.98257	0.98300	0.98341	0.98382	0.98422	0.98461	0.98500	0.98537	0.98574
2.2	0.98610	0.98645	0.98679	0.98713	0.98745	0.98778	0.98809	0.98840	0.98870	0.98899
2.3	0.98928	0.98956	0.98983	0.99010	0.99036	0.99061	0.99086	0.99111	0.99134	0.99158
2.4	0.99180	0.99202	0.99224	0.99245	0.99266	0.99286	0.99305	0.99324	0.99343	0.99361
2.5	0.99379	0.99396	0.99413	0.99430	0.99446	0.99461	0.99477	0.99492	0.99506	0.99520
2.6	0.99534	0.99547	0.99560	0.99573	0.99585	0.99598	0.99609	0.99621	0.99632	0.99643
2.7	0.99653	0.99664	0.99674	0.99683	0.99693	0.99702	0.99711	0.99720	0.99728	0.99736
2.8	0.99744	0.99752	0.99760	0.99767	0.99774	0.99781	0.99788	0.99795	0.99801	0.99807
2.9	0.99813	0.99819	0.99825	0.99831	0.99836	0.99841	0.99846	0.99851	0.99856	0.99861
3.0	0.99865	0.99869	0.99874	0.99878	0.99882	0.99886	0.99889	0.99893	0.99896	0.99900
3.1	0.99903	0.99906	0.99910	0.99913	0.99916	0.99918	0.99921	0.99924	0.99926	0.99929
3.2	0.99931	0.99934	0.99936	0.99938	0.99940	0.99942	0.99944	0.99946	0.99948	0.99950
3.3	0.99952	0.99953	0.99955	0.99957	0.99958	0.99960	0.99961	0.99962	0.99964	0.99965
3.4	0.99966	0.99968	0.99969	0.99970	0.99971	0.99972	0.99973	0.99974	0.99975	0.99976
3.5	0.99977	0.99978	0.99978	0.99979	0.99980	0.99981	0.99981	0.99982	0.99983	0.99983

Table A.3: Error function and complementary error function

x	*erf*(x)	*erfc*(x)	x	*erf*(x)	*erfc*(x)
0.00	0	1	1.3	0.9340079	0.0659921
0.05	0.056372	0.943628	1.4	0.9522851	0.0477149
0.10	0.1124629	0.8875371	1.5	0.9661051	0.0338949
0.15	0.167996	0.832004	1.6	0.9763484	0.0236516
0.20	0.2227026	0.7772974	1.7	0.9837905	0.0162095
0.25	0.2763264	0.7236736	1.8	0.9890905	0.0109095
0.30	0.3286268	0.6713732	1.9	0.9927904	0.0072096
0.35	0.3793821	0.6206179	2.0	0.9953223	0.0046777
0.40	0.4283924	0.5716076	2.1	0.9970205	0.0029795
0.45	0.4754817	0.5245183	2.2	0.9981372	0.0018628
0.50	0.5204999	0.4795001	2.3	0.9988568	0.0011432
0.55	0.5633234	0.4366766	2.4	0.9993115	0.0006885
0.60	0.6038561	0.3961439	2.5	0.999593	0.000407
0.65	0.6420293	0.3579707	2.6	0.999764	0.000236
0.70	0.6778012	0.3221988	2.7	0.9998657	0.0001343
0.75	0.7111556	0.2888444	2.8	0.999925	0.000075
0.80	0.742101	0.257899	2.9	0.9999589	0.0000411
0.85	0.7706681	0.2293319	3.0	0.9999779	0.0000221
0.90	0.7969082	0.2030918	3.1	0.9999884	0.0000116
0.95	0.8208908	0.1791092	3.2	0.999994	0.000006
1.00	0.8427008	0.1572992	3.3	0.9999969	0.0000031
1.10	0.8802051	0.1197949	3.4	0.9999985	0.0000015
1.20	0.910314	0.089686	3.5	0.9999993	0

Appendix B

Some Important Distribution and Density Functions

B.1 Discrete Random Variables

Bernoulli (p): For $0 \le p \le 1$,

$$f_X(x) = \begin{cases} (1-p) & \text{for } x = 0 \\ p & \text{for } x = 1 \\ 0 & \text{otherwise} \end{cases} \quad \text{...(B.1)}$$

$$E[X] = p \quad \text{...(B.2)}$$

$$\text{var}[X] = p(1-p) \quad \text{...(B.3)}$$

$$\phi_X(\omega) = 1 - p + pe^{j\omega} \quad \text{...(B.4)}$$

Binomial (n, p): For a positive integer n and $0 \le p \le 1$,

$$f_X(x) = \sum_{k=0}^{n} {}^nC_k p^k q^{(n-k)} \delta(x-k) \quad \text{...(B.5)}$$

$$F_X(x) = \sum_{k=0}^{n} {}^nC_k p^k q^{(n-k)} u(x-k) \quad \text{...(B.6)}$$

$$E[X] = np \quad \text{...(B.7)}$$

$$\text{var}[X] = np(1-p) \quad \text{...(B.8)}$$

$$\phi_X(\omega) = (1 - p + pe^{j\omega})^n \quad \text{...(B.9)}$$

Discrete uniform (m, n): For integers m and n such that $m < n$,

$$f_X(x) = \begin{cases} 1/(n-m+1) & \text{for } x = m, m+1, \ldots, n \\ 0 & \text{otherwise} \end{cases} \quad \text{...(B.10)}$$

$$E[X] = \frac{m+n}{2} \quad \text{...(B.11)}$$

$$\text{var}[X] = \frac{(n-m)(n-m+2)}{12} \quad \text{...(B.12)}$$

$$\phi_X(\omega) = \frac{e^{j\omega m} - e^{j\omega(n+1)}}{(n-m+1)(1-e^{j\omega})} \quad \text{...(B.13)}$$

Pascal: For positive integer k, $0 < p < 1$ and $N = 1, 2...$

$$f_X(x) = \sum_{k=N}^{\infty} {}^{k-1}C_{N-1} p^N q^{(k-N)} \delta(x-k) \quad ...(B.14)$$

$$F_X(x) = \sum_{k=N}^{\infty} {}^{k-1}C_{N-1} p^N q^{(k-N)} u(x-k) \quad ...(B.15)$$

$$E[X] = \frac{N}{p} \quad ...(B.16)$$

$$\text{var}[X] = \frac{N(1-p)}{p^2} \quad ...(B.17)$$

$$\phi_X(\omega) = p^N e^{j\omega N}[1-(1-p)e^{j\omega}]^{-N} \quad ...(B.18)$$

Poisson: For $b > 0$,

$$f_X(x) = e^{-\mu} \sum_{k=0}^{\infty} \frac{\mu^k}{k!} \delta(x-k) \quad ...(B.19)$$

$$F_X(x) = e^{-\mu} \sum_{k=0}^{\infty} \frac{\mu^k}{k!} u(x-k) \quad ...(B.20)$$

$$E[X] = \mu \quad ...(B.21)$$

$$\text{var}[X] = \mu \quad ...(B.22)$$

$$\phi_X(\omega) = \exp[\mu(e^{j\omega} - 1)] \quad ...(B.23)$$

B.2 Continuous Random Variables

Beta(i, j): For positive integers i and j, the beta function is defined as

$$\beta(i, j) = \frac{(i+j-1)!}{(i-1)!(j-1)!} \quad ...(B.24)$$

For a $\beta(i, j)$ random variable X,

$$f_X(x) = \begin{cases} \beta(i,j)x^{i-1}(1-x)^{j-1} & 0 < x < 1 \\ 0 & \text{otherwise} \end{cases} \quad ...(B.25)$$

$$E[X] = \frac{i}{i+j} \quad ...(B.26)$$

$$\text{var}[X] = \frac{ij}{(i+j)^2(i+j+1)} \quad ...(B.27)$$

Cauchy: For $b > 0$ and $-\infty < a -\infty$

$$f_X(x) = \frac{(b/a)}{b^2 + (x-a)^2} \quad ...(B.28)$$

$$F_X(x) = \frac{1}{2} + \frac{1}{\pi}\tan^{-1}\left(\frac{x-a}{b}\right) \quad ...(B.29)$$

$$E[X] = \text{undefined} \quad \text{...(B.30)}$$
$$\text{var}[X] = \text{undefined} \quad \text{...(B.31)}$$
$$\phi_X(\omega) = e^{ja\omega - b|\omega|} \quad \text{...(B.32)}$$

***Erlang*(*n*, λ):** For $\lambda > 0$ and a positive integer n,

$$f_X(x) = \begin{cases} \dfrac{\lambda^n x^{n-1} e^{-\lambda x}}{(n-1)!} & x \geq 0 \\ 0 & \text{otherwise} \end{cases} \quad \text{...(B.33)}$$

$$E[X] = n/\lambda \quad \text{...(B.34)}$$
$$\text{var}[X] = n/\lambda^2 \quad \text{...(B.35)}$$
$$\phi_X(\omega) = \left(\frac{\lambda}{\lambda - j\omega}\right)^n \quad \text{...(B.36)}$$

***Exponential* (λ):** For $\lambda > 0$,

$$f_X(x) = \begin{cases} 0 & x \leq 0 \\ \lambda e^{-\lambda x} & x > 0 \end{cases} \quad \text{...(B.37)}$$

$$F_X(x) = \begin{cases} 0 & x \leq 0 \\ 1 - e^{-\lambda x} & x > 0 \end{cases} \quad \text{...(B.38)}$$

$$E[X] = 1/\lambda \quad \text{...(B.39)}$$
$$\text{var}[X] = 1/\lambda^2 \quad \text{...(B.40)}$$
$$\phi_X(\omega) = \left(\frac{\lambda}{\lambda - j\omega}\right) \quad \text{...(B.41)}$$

***Gamma* (*a*, *b*):** For $a > -1$ and $b > 0$,

$$f_X(x) = \begin{cases} \dfrac{x^a e^{-x/b}}{a!\, b^{a+1}} & x > 0 \\ 0 & \text{otherwise} \end{cases} \quad \text{...(B.42)}$$

$$E[X] = (a+1)/b \quad \text{...(B.43)}$$
$$\text{Var}[X] = (a+1)/b^2 \quad \text{...(B.44)}$$
$$\phi_X(\omega) = \frac{1}{(1 - bj\omega)^{a+1}} \quad \text{...(B.45)}$$

***Gaussian* (μ, σ):** For constants $\sigma > 0$, $-\infty < \mu < \infty$

$$f_X(x) = \frac{1}{\sqrt{2\pi\sigma^2}} e^{-(x-\mu)^2/2\sigma^2} \quad \text{...(B.46)}$$

$$E[X] = \mu \quad \text{...(B.47)}$$
$$\text{var}[X] = \sigma^2 \quad \text{...(B.48)}$$
$$\phi_X(\omega) = e^{j\omega\mu - \omega^2\sigma^2/2} \quad \text{...(B.49)}$$

Laplace (a, b): For constants $a > 0, -\infty < b < \infty$

$$f_X(x) = \frac{a}{2}e^{-a|x-b|} \quad \text{...(B.50)}$$

$$E[X] = b \quad \text{...(B.51)}$$

$$\text{var}[X] = 2/a^2 \quad \text{...(B.52)}$$

$$\phi_X(\omega) = \frac{a^2 e^{j\omega b}}{a^2 + \omega^2} \quad \text{...(B.53)}$$

Rayleigh (a): For $a > 0$,

$$f_X(x) = \begin{cases} \frac{2}{b}(x-a)e^{-(x-a)^2/b} & x \geq a \\ 0 & x < a \end{cases} \quad \text{...(B.54)}$$

$$F_X(x) = \begin{cases} 1 - e^{-(x-a)^2/b} & x \geq a \\ 0 & x < a \end{cases} \quad \text{...(B.55)}$$

$$E[X] = a + \sqrt{\frac{\pi b}{4}} \quad \text{...(B.56)}$$

$$\text{var}[X] = \frac{b(4-\pi)}{4} \quad \text{...(B.57)}$$

Uniform (a, b): For constants $a < b$,

$$f_X(x) = \begin{cases} \frac{1}{b-a} & a \leq x \leq b, \\ 0 & \text{elsewhere} \end{cases} \quad \text{...(B.58)}$$

$$F_X(x) = \begin{cases} 0 & x < a \\ \frac{x-a}{b-a} & a \leq x < b \\ 1 & x \geq b \end{cases} \quad \text{...(B.59)}$$

$$E[X] = \frac{a+b}{2} \quad \text{...(B.60)}$$

$$\text{var}[X] = \frac{(b-a)^2}{12} \quad \text{...(B.61)}$$

$$\phi_X(\omega) = \frac{e^{j\omega b} - e^{j\omega a}}{j\omega(b-a)} \quad \text{...(B.62)}$$

Appendix C

Unit Impulse and its Properties

Impulse or Dirac delta function is represented by δ. It is an idealization of a signal that:

- Is very large near $t = 0$
- Is very small away from $t = 0$
- Has integral 1.

Case 1: Continuous time

Unit impulse is defined as

$$\delta(t) = \lim_{\tau \to 0} \delta_p(t) \qquad \text{...(C.1)}$$

$$\delta(t) = \int_{-\tau/2}^{\tau/2} \frac{1}{\tau} dt = 1 \qquad \text{...(C.2)}$$

Case 2: Discrete time

$$\delta(n) = \begin{cases} 1 & n = 0 \\ 0 & \text{otherwise} \end{cases} \qquad \text{...(C.3)}$$

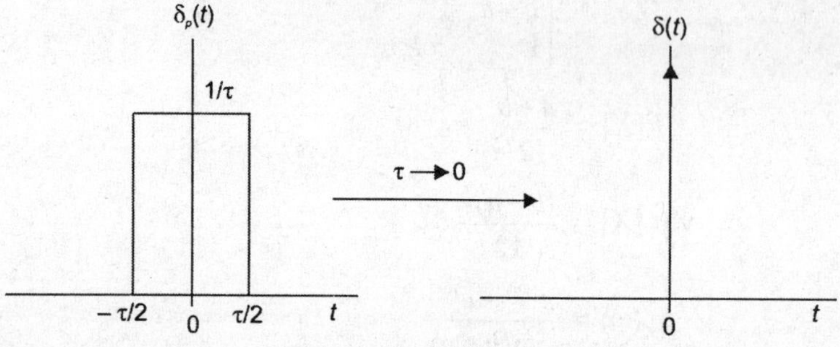

Fig. C.1: Unit impulse signal

Properties

1. δ(0) is not really defined.

2. $\int_{-\infty}^{\infty} f(t)\delta(t-T)dt = f(T)$

3. $\delta(t) = \dfrac{du(t)}{dt}$

where $u(t)$ is the unit step signal. It is defined as

$$u(t) = \begin{cases} 1 & \text{for } t \geq 0 \\ 0 & \text{otherwise} \end{cases}$$

4. $\int_a^b \delta(t)dt = 1$ if $a < 0$ and $b > 0$.

5. $\int_a^b \delta(t)dt = 1$ if $a > 0$ and $b < 0$.

6. $\int_a^b \delta(t)dt =$ ambiguous if $a = 0$ and $b = 0$.

Appendix D

Fourier Transform

The Fourier transform $X(\omega)$, of a signal $x(t)$ is given by

$$X(\omega) = \int_{-\infty}^{\infty} x(t)e^{-j\omega t}dt \qquad ...(D.1)$$

The inverse Fourier transform is given by

$$x(t) = \frac{1}{2\pi}\int_{-\infty}^{\infty} X(\omega)e^{j\omega t}\,d\omega \qquad ...(D.2)$$

Some Important Fourier Transform Pairs

Table D.1: Important pairs of continuous time Fourier transform

S.No.	$x(t)$	$X(\ )$
1.	Unit impulse $\delta(t)$	1
2.	Unit step $u(t)$	$\frac{1}{j\omega} + \pi\delta(\omega)$
3.	1	$2\pi\delta(\omega)$
4.	$\cos \omega_0 t$	$\pi[\delta(\omega - \omega_0) + \delta(\omega + \omega_0)$
5.	$\sin \omega_0 t$	$-j\pi[\delta(\omega - \omega_0) - \delta(\omega + \omega_0)$
6.	$e^{-at}u(t), a > 0$	$\frac{1}{a + j\omega}$
7.	$e^{-a\lvert t\rvert}u(t), a > 0$	$\frac{2a}{a^2 + \omega^2}$
8.	$\frac{1}{a^2 + t^2}$	$e^{-a\lvert\omega\rvert}$
9.	$\operatorname{sgn} t = \begin{cases} +1 & t > 0 \\ -1 & t < 0 \end{cases}$	$\frac{2}{j\omega}$
10.	$\sum_{k=-\infty}^{\infty} \delta(t - kT)$	$\omega_0 \sum_{k=-\infty}^{\infty} \delta(\omega - k\omega_0), \omega_0 = \frac{2\pi}{T}$

Table D.2: Important properties of Fourier transform

Property	*Signal*	*Fourier transform*
	$x(t)$	$X(\omega)$
	$x_1(t)$	$X_1(\omega)$
	$x_2(t)$	$X_2(\omega)$
Linearity	$ax_1(t) + bx_2(t)$	$aX_1(\omega) + bX_2(\omega)$
Time shifting	$x(t - t_0)$	$e^{-j\omega t_0}X(\omega)$
Frequency shifting	$e^{j\omega_0 t}x(t)$	$X(\omega - \omega_0)$
Time scaling	$x(at)$	$\frac{1}{\|a\|}X\left(\frac{\omega}{a}\right)$
Time reversal	$x(-t)$	$X(-\omega)$
Duality	$X(t)$	$2\pi x(-\omega)$
Time differentiation	$\frac{dx(t)}{dt}$	$j\omega X(\omega)$
Frequency differentiation	$-jtx(t)$	$\frac{dX(\omega)}{d\omega}$
Integration	$\int_{-\infty}^{t} x(\tau)d\tau$	$\pi X(0)\delta(\omega) + \frac{1}{j\omega}X(\omega)$
Convolution	$x_1(t) * x_2(t) = \int_{-\infty}^{\infty} x_1(\tau)x_2(t-\tau)d\tau$	$X_1(\omega)\, X_2(\omega)$
Multiplication	$x_1(t)\, x_2(t)$	$\frac{1}{2\pi}X_1(\omega) * X_2(\omega)$
Parseval's theorem	$\int_{-\infty}^{\infty} \|x(t)\|^2 dt$	$\frac{1}{2\pi}\int_{-\infty}^{\infty} \|X(\omega)\|^2 d\omega$
Real signal	$x(t) = x_e(t) + x_0(t)$	$X(\omega) = A(\omega) + jB(\omega)$ $X(-\omega) = X^*(\omega)$
Even component	$x_e(t)$	$\text{Re}\{X(\omega)\} = A(\omega)$
Odd component	$x_0(t)$	$j\,\text{Img}\{X(\omega)\} = jB(\omega)$

Discrete Time Fourier Transform

The discrete time Fourier transform $X(\Omega)$, of a signal $x(t)$ is given by

$$X(\Omega) = \sum_{n=-\infty}^{\infty} x(n)e^{-jn\Omega} \qquad \text{...(D.3)}$$

The inverse Fourier transform is given by

$$x(n) = \frac{1}{2\pi}\int_{-\pi}^{\pi} X(\Omega)e^{jn\Omega}\, d\omega \qquad \text{...(D.4)}$$

Table D.3: Important pairs of discrete time Fourier transform

S.No.	$x(n)$	$X(\Omega)$
1.	Unit impulse $\delta(n)$	1
2.	Unit step $u(n)$	$\pi\delta(\Omega)+\frac{1}{1-e^{-j\Omega}}$
3.	1	$2\pi\delta(\Omega)$
4.	$\cos\omega_0 n$	$\pi[\delta(\Omega-\Omega_0)+\delta(\Omega+\Omega_0)$
5.	$\sin\omega_0 n$	$-j\pi[\delta(\Omega-\Omega_0)-\delta(\Omega+\Omega_0)$
6.	$a^n u(n), \lvert a\rvert<1$	$\frac{1}{1-ae^{-j\Omega}}$
7.	$\lvert a^n\rvert, \lvert a\rvert<1$	$\frac{1-a^2}{1-2a\cos\Omega+a^2}$
8.	$(n+1)a^n u(n), \lvert a\rvert<1$	$\frac{1}{\left(1-ae^{-j\Omega}\right)^2}$
9.	$\sum_{k=-\infty}^{\infty}\delta(n-kN_0)$	$\Omega_0\sum_{k=-\infty}^{\infty}\delta(\Omega-k\Omega_0), \Omega_0=\frac{2\pi}{N_0}$

Appendix E

Useful Mathematical Formulae

Trigonometric Identities

$$\cos(A \pm B) = \cos A \cos B \mp \sin A \sin B \quad \text{...(E.1)}$$

$$\sin(A \pm B) = \sin A \cos B \pm \cos A \sin B \quad \text{...(E.2)}$$

$$\cos(2A) = \cos^2 A - \sin^2 A \quad \text{...(E.3)}$$

$$\sin(2A) = 2 \sin A \cos A \quad \text{...(E.4)}$$

$$\sin A \cos B = \frac{1}{2}[\sin(A+B) + \sin(A-B)] \quad \text{...(E.5)}$$

$$\cos A \cos B = \frac{1}{2}[\cos(A-B) + \cos(A+B)] \quad \text{...(E.6)}$$

$$\sin A \sin B = \frac{1}{2}[\cos(A-B) - \cos(A+B)] \quad \text{...(E.7)}$$

$$2\cos^2\theta = 1 + \cos 2\theta \quad \text{...(E.8)}$$

$$2\sin^2\theta = 1 - \cos 2\theta \quad \text{...(E.9)}$$

$$4\cos^2\theta = 3\cos\theta + \cos 3\theta \quad \text{...(E.10)}$$

$$4\sin^3\theta = 3\sin\theta - \sin 3\theta \quad \text{...(E.11)}$$

The Euler Formula

$$e^{j\theta} = \cos\theta + j\sin\theta \quad \text{...(E.12)}$$

$$\cos\theta = \frac{e^{j\theta} + e^{-j\theta}}{2} \quad \text{...(E.13)}$$

$$\sin\theta = \frac{e^{j\theta} - e^{-j\theta}}{2} \quad \text{...(E.14)}$$

Sequences and Series

$$\sum_{i=0}^{n} a^i = 1 + a^1 + a^2 + \cdots + a^n = \frac{1-a^{n+1}}{1-a} \quad \text{...(E.15)}$$

When $|a| < 1$,

$$\sum_{i=0}^{\infty} a^i = \lim_{n\to\infty} \sum_{i=0}^{n} a^i = \lim_{n\to\infty} \frac{1-a^{n+1}}{1-a} = \frac{1}{1-a} \quad \text{...(E.16)}$$

$$\sum_{i=1}^{N} n = \frac{N(N+1)}{2} \quad \text{...(E.17)}$$

$$\sum_{i=1}^{N} n^2 = \frac{N(N+1)(2N+1)}{6} \quad \text{...(E.18)}$$

$$\sum_{i=1}^{N} n^3 = \frac{N^2(N+1)^2}{4} \quad \text{...(E.19)}$$

Integration

$$\int \cos x\, dx = \sin x \quad \text{...(E.20)}$$

$$\int \sin x\, dx = -\cos x \quad \text{...(E.21)}$$

$$\int x \cos x\, dx = \cos x + x \sin x \quad \text{...(E.22)}$$

$$\int u dv = u\int dv - \int u dv \quad \text{...(E.23)}$$

$$\int e^{ax} dx = \frac{e^{ax}}{a} \quad \text{...(E.24)}$$

$$\int_0^{\infty} \frac{\sin x}{x} dx = \frac{\pi}{2} \quad \text{...(E.25)}$$

The Gamma function is defined as

$$\Gamma(z) = \int_0^{\infty} t^{z-1} e^{-t} dt \quad \text{...(E.26)}$$

If $z = n$, a positive integer, $\Gamma(n) = (n-1)!$ and $\Gamma(1/2) = \sqrt{\pi}$, $\Gamma(3/2) = \sqrt{\pi}/2$ and $\Gamma(5/2) = 3\sqrt{\pi}/4$.

Leibniz's Rule

Let $\quad G(u) = \int_{\alpha(u)}^{\beta(u)} H(x,u)dx$

then $$\frac{dG(u)}{du} = H[\beta(u),u]\frac{d\beta(u)}{du} - H[\alpha(u),u]\frac{d\alpha(u)}{du} + \int_{\alpha(u)}^{\beta(u)} \frac{\partial H(x,u)}{\partial u} dx \quad \text{...(E.27)}$$

Derivatives

$$(uv)' = u'v + uv' \quad \text{...(E.28)}$$

$$\left(\frac{u}{v}\right)' = \frac{u'v - uv'}{v^2} \quad \text{...(E.29)}$$

$$\frac{d(e^{g(x)})}{dx} = e^{g(x)}g'(x) \quad ...(E.30)$$

$$\frac{d\cos x}{dx} = -\sin x \quad ...(E.31)$$

$$\frac{d\sin x}{dx} = \cos x \quad ...(E.32)$$

$$\frac{de^{ax}}{dx} = ae^{ax} \quad ...(E.33)$$

$$\frac{dx^n}{dx} = nx^{n-1} \quad ...(E.34)$$

References

1. Peebles PZ, *Probability Random Variables and Random Signal Processing*, McGraw Hill, 4th edn, 2000.
2. Haykin Simon, *Communication Systems*, John Wiley, 4th edn, 2001.
3. Papoulis A and Pillai SU, *Probability, Random Variables and Stochastic Processes*, McGraw Hill, 4th edn, 2001.
4. Woods JW and Stark H, *Probability and Random Processes with Applications to Signal Processing*, Prentice Hall, 3rd edn, 2001.
5. Yates Roy D and Goodman David J, *Probability and Stochastic Process: A Friendly Introduction for Electrical and Computer Engineers*, John Wiley and Sons, 2nd edn, 2005.
6. Hsu Hwei, *Probability, Random Variables and Random Processes*, Schaums Outlines.
7. Krishnan Venkatarama, *Probability and Random Processes*, John Wiley and Sons.
8. Hayes MH, *Statistical Digital Signal Processing and Modeling*, 1996.
9. Drake AW, *Fundamentals of Applied Probability Theory*, McGraw Hill, New York, 1967.
10. Ross SM, *Introduction to Probability Models*, Academic Press, 8th edn, 2003.

Index